T0133394

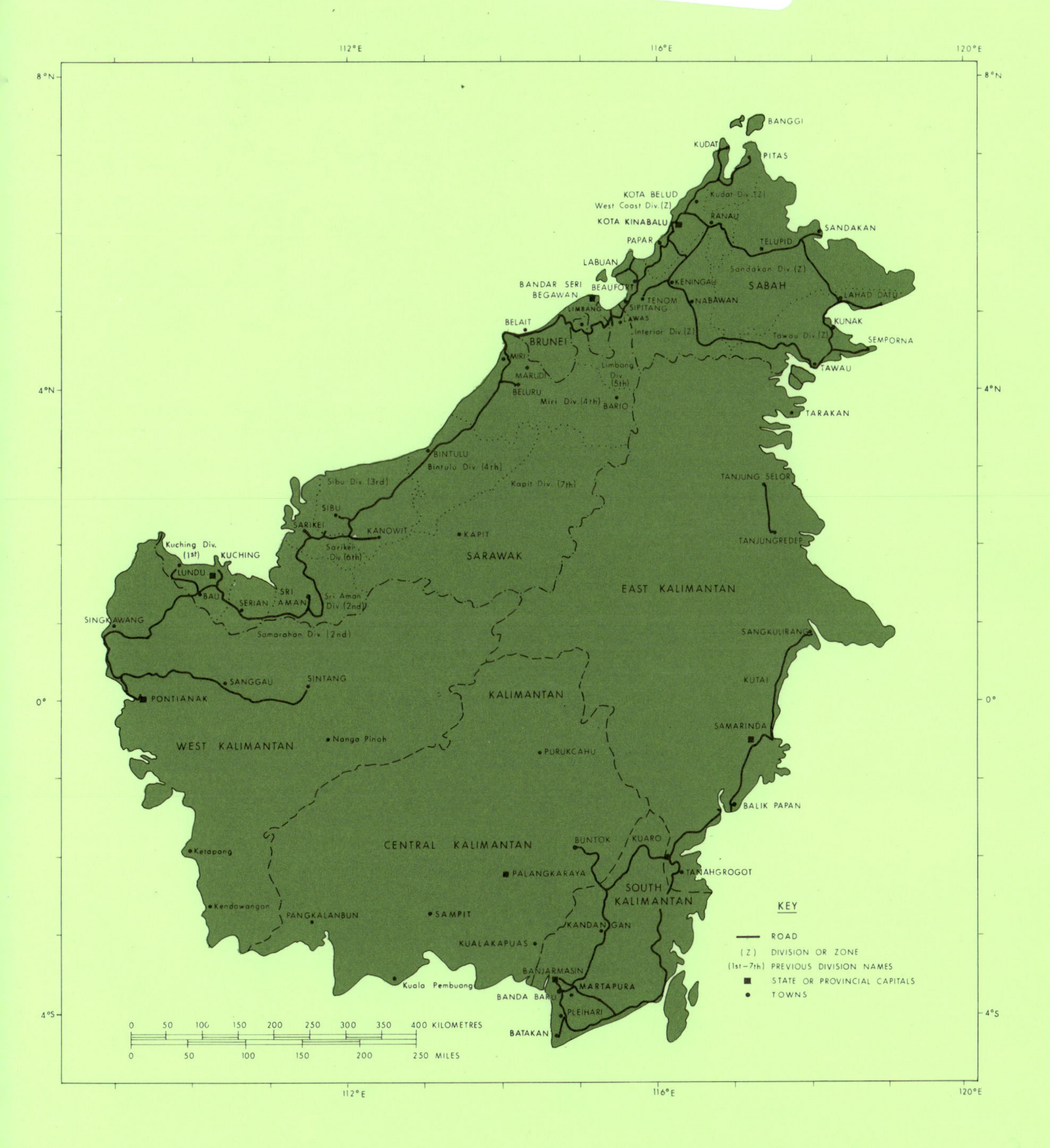
112°E
116°E
120°E
8°N
4°N
0°
4°S
BANGGI
KUDAT
PITAS
KOTA BELUD
West Coast Div.(Z)
Kudat Div.(Z)
KOTA KINABALU
RANAU
SANDAKAN
PAPAR
TELUPID
LABUAN
Sandakan Div.(Z)
BANDAR SERI
BEGAWAN
BEAUFORT
KENINGAU
SABAH
LAHAD DATU
LIMBANG
SIPITANG
TENOM
NABAWAN
BELAIT
LAWAS
Interior Div.(Z)
KUNAK
Tawau Div.(Z)
SEMPORNA
BRUNEI
MIRI
TAWAU
MARUDI
Limbang
Div.
(5th)
BELURU
Miri Div.(4th)
BARIO
TARAKAN
BINTULU
Bintulu Div.(4th)
Sibu Div.(3rd)
Kapit Div.(7th)
TANJUNG SELOR
SIBU
SARIKEI
KANOWIT
KAPIT
Kuching Div.
(1st)
KUCHING
Sarikei
Div.(6th)
SARAWAK
TANJUNGREDEP
LUNDU
BAU
SERIAN
SRI
AMAN
Sri Aman
Div.(2nd)
EAST KALIMANTAN
SINGKAWANG
Samarahan Div.(2nd)
SANGKULIRANG
SANGGAU
SINTANG
KUTAI
PONTIANAK
KALIMANTAN
SAMARINDA
WEST KALIMANTAN
Nanga Pinoh
PURUKCAHU
BALIK PAPAN
BUNTOK
KUARO
CENTRAL KALIMANTAN
Ketapang
PALANGKARAYA
TANAHGROGOT
SOUTH
KALIMANTAN
Kendawangan
PANGKALANBUN
SAMPIT
KANDANGAN
KUALAKAPUAS
BANJARMASIN
MARTAPURA
Kuala Pembuang
BANDA BARU
PLEIHARI
BATAKAN
KEY
ROAD
(Z) DIVISION OR ZONE
(1st–7th) PREVIOUS DIVISION NAMES
STATE OR PROVINCIAL CAPITALS
TOWNS
0 50 100 150 200 250 300 350 400 KILOMETRES
0 50 100 150 200 250 MILES

Orchids Of Borneo

VOL. 1 INTRODUCTION AND A SELECTION OF SPECIES

Cleisocentron merrillianum

Orchids Of Borneo

VOL. 1 INTRODUCTION AND A SELECTION OF SPECIES

By C.L. CHAN, A. LAMB, P.S. SHIM and J.J. WOOD

Series Editor
P.J. Cribb

Series Co-ordinators
A. Lamb & C.L. Chan

Published by

The Sabah Society
Kota Kinabalu

in association with

Royal Botanic Gardens
Kew, England

28th March 1994

Published by

The Sabah Society
P. O. Box 10547
88806 Kota Kinabalu
Sabah, Malaysia

in association with

The Bentham-Moxon Trust
The Royal Botanic Gardens
Kew, Richmond, Surrey TW9 3AB
England

First published 1994

ORCHIDS OF BORNEO Vol. 1. Introduction and a selection of species.

ISBN 967-99947-3-2

Printed and bound in Malaysia
by Print & Co. Sdn. Bhd., Kuala Lumpur.

CONTENTS

LIST OF COLOUR PLATES

(taxon number in this account given in brackets)

Dendrobium lamellatum

FOREWORD

It is a privilege for the Sabah Society to co-publish with the Royal Botanic Gardens, Kew, Volume 1 of the Orchids of Borneo which documents a unique natural endowment of Borneo with special reference to Sabah. This will be another important contribution to the knowledge and appreciation of the vast array of wild orchids that the island is fortunate to possess. This effort also contributes to a major objective of the Sabah Society which is to increase the awareness of the natural heritage of Sabah.

The Sabah Society sees the proper documentation of Borneo's natural assets as an effective way to stimulate a wider interest in their importance. Apart from its scientific value, this publication will also be useful in enhancing the growing realisation of conserving both species and habitats where natural resources are concerned. As such, this publication will be of value to all natural historians: serious researchers, special interest groups and orchid enthusiasts.

The authors of this volume, namely, Messrs C.L. Chan, A. Lamb, P.S. Shim and J.J. Wood are acknowledged experts on Bornean orchids and their authoritative contributions are major strengths of this publication. This special work has drawn on the generous interest and support of many: the series editor Dr P.J. Cribb, himself an acknowledged authority on the orchids of this region, has given much of his time and knowledge. The Sabah Society is also greatly indebted to Mr Tan Jiew Hoe who has magnanimously provided the financial support for the production of this volume.

Dr Ti Teow Chuan
President,
The Sabah Society

Ceratochilus jiewhoei

PROLOGUE

Orchids represent a culmination of evolution in the Plant Kingdom. In Borneo we are in the midst of one of the richest floristic regions on earth. Here, the diversity of the orchids has attracted much attention but until now efforts at documenting the orchid flora have been restricted in scope. The collaboration among Malaysian orchidologists and their counterparts at the great herbaria at Kew and Leiden have made this series a very special one. For the first time, this taxonomically difficult group of plants is receiving coordinated attention from specialists in different areas of the orchid world.

The series was conceptualised in 1982, some ten years ago. It will be a long-term documentation of the interesting orchids of Borneo, spurred on by progress in research in different fields. This first volume includes a selection of a hundred species of orchids, principally of the horticulturally important genera, including *Arachnis, Dimorphorchis, Paphiopedilum, Paraphalaenopsis, Phalaenopsis* and *Vanda*. Many of the species are illustrated here for the first time, and every species and most varieties included are accompanied by a line illustration and a colour photograph. The second volume in the series is devoted to the genus *Bulbophyllum*, the largest in Borneo. Further volumes are in preparation.

I wish to pay special tribute to the early pioneers in orchid hybridisation: to Richard Eric Holttum, Tan Hoon Siang, John Laycock, Emile Galistan and many others who have inspired me to admire orchids since childhood. More importantly, I view with equal admiration the efforts of those who so tirelessly document their diversity and biology. By publishing this immensely valuable knowledge and documentation, the conservation of orchids and indeed of plant life and habitats in general will have an even stronger basis for support. This volume is a very special one for me and I am happy to have supported its production.

Tan Jiew Hoe
Singapore

Vanda lamellata

INTRODUCTION TO THE SERIES

From the time that Sir Hugh Low first climbed to the summit in 1851, Mount Kinabalu in Sabah became known for its extremely rich flora, which attracted visits from collectors sent out by horticultural firms who were looking for new plants for cultivation in tropical greenhouses in Europe. Low's most important introductions were pitcher plants *(Nepenthes)*. The mountain's most spectacular orchids were discovered by later travellers, and they represent only a tiny percentage of the orchid flora of Borneo.

A majority of tropical orchid plants are adapted to grow on the branches of trees and are called epiphytes. Most epiphytes can only flourish in a climate of high rainfall with no prolonged dry season such as that of Borneo. Other orchids grow on the ground, often in the shade of forest. Both epiphytes and ground orchids depend on the existence of the continuous canopy produced by the branches of forest trees. Few other countries in the world have climatic conditions and natural forest more suitable for the growth of orchids, and the result, over a few million years, has been the development of a very abundant, very diverse and unique orchid flora.

Tropical orchids were first scientifically described from plants grown in Europe where there were botanists to study them. Later, botanical explorers preserved dried specimens from which many more new species were described. But the structure of orchid flowers is so intricate, and often so delicate, and their colours so varied and distinctive, that description from dried specimens alone is not enough. The plants need to be studied alive in their country of origin, where botanists can examine, describe and illustrate them in a living state, either in the wild or as cultivated plants. In view of the rapid destruction of forests in Borneo, many local orchid species are in danger of being exterminated before an adequate record of them has been made. Such a record is thus now urgently needed. The task is so large and so complex that a single person could not achieve it. The island is fortunate that a local team of dedicated field workers and experienced describers and illustrators are collaborating to undertake this work.

The records of all previously described orchids are very copious and new descriptions need to be checked with old ones to avoid duplication of names and consequent confusion. Such records have been compiled over many years in the great herbaria at Kew and Leiden. The cooperation of botanists in these herbaria has been an important part of the whole process of recording the orchids of Borneo, and I am sure that the local team would wish to express their appreciation of this help.

In the present volume one hundred varied species are described, with an illustration for each showing significant detail of the kind that photographs cannot achieve, the result of prolonged study, thought and artistic skill. I express my admiration for this achievement and the hope of all concerned that many more similar volumes will follow. The undertaking of which this is a beginning is a vast and courageous one.

R.E. Holttum
Kew
21st May 1986

ACKNOWLEDGEMENTS

When first mooted, the idea for this series was enthusiastically received at the Royal Botanic Gardens, Kew, in England, and the Singapore Botanic Gardens. The late Professor R.E. Holttum, author of "Orchids of Malaya" (1964), gave much support to the project. He had emphasised the need for a team effort to enable such a large task to be undertaken.

For Volume One, the generosity of Mr Tan Jiew Hoe of Singapore in providing the printing costs needs to be specially acknowledged. As a gesture of appreciation, a new orchid species, *Ceratochilus jiewhoei*, is dedicated to him in this volume.

We would like to acknowledge the help of the former Ministers of Environmental Development and Manpower, YB Datuk Yap Pak Leong and YB Datuk Joseph Pairin Kitingan (now the Chief Minister of Sabah) and the former Permanent Secretary, William Baxter, for the support and help given in the initial setting up of the Tenom Orchid Centre and its facilities, which provided living material for almost all the orchid species dealt with in this volume. We are grateful to the Director of the Agriculture Department, Datuk Aripen Bin Hj. Ampong and the former Assitant Director for Research, Dr Tay Eong Beok and the present Assistant Director for Research, Mahinder Singh Kalsi for support and encouragement. We have benefitted greatly, both in the Tenom Orchid Centre and in the field, from the help provided by the Supervisor, Herbert Lim Keng Leong, and former supervisors Awang Osman, Harry Lohok and their staff. These include Jutili Kiki, Aninguh Surat, and King Kong Raipan, a superb tree climber.

We are especially grateful to, and have benefitted from, the enthusiasm and cooperation given by the Director of Sabah Parks, Datuk Lamri Ali, Deputy Director Francis Liew and former Park Ecologist Anthea Phillipps and the present Park Ecologist Jamili Nais. The help of Eric Wong, the Park Warden at the Kinabalu Park, Senior Park Ranger Gabriel Sinit, Justin Jukian, Park Naturalists Ansow Gunsalam and Thomas Yusof and former Park Naturalist Tan Fui Lian are also acknowledged.

We also received assistance from the Forestry Department's Forest Research Centre, Sepilok, Sandakan, for which we are grateful to past Forest Botanists Tiong King Keong and Lee Ying Fah, and the present Senior Forest Botanist Dr Wong Khoon Meng, who has generously given time to improve the text. We wish also to thank the Director of Forestry, Datuk Miller Munang and the Head of the Centre, Robert C. Ong for permission to continue using the herbarium there.

At the Singapore Botanic Gardens, Dr Chang Kiaw Lan, the former curator of the Herbarium, George Alphonso, the former acting Director, and the present Director, Dr Kiat W. Tan, provided assistance and access to the Herbarium, which houses the important Kinabalu orchid collections and notes of C.E. Carr. At Kew, Peter Taylor and the editor for this series and Curator of the Orchid Herbarium, Dr Phillip Cribb, also provided assistance in identifying specimens at the herbarium.

The Bentham-Moxon Trust at Kew agreed to assist not only with distribution, but also with jointly publishing later volumes.

We should also like to make special mention of Dr Phillip Cribb for not only fulfilling the demanding responsibility of being the series editor for this work, but also contributing significantly to the present volume. Professor Grenville Lucas, the Keeper of the Herbarium at Kew, has kindly allowed us free access to the Kew Orchid Herbarium and Library.

Lucy Liew Fui Ling, Chin Wan Wai, Dr Jaap J. Vermeulen and Dr. Phillip Cribb helped with the line illustrations. William W.W. Wong and Mark Fothergill contributed one plate each to the present volume. Chan Kwong Choi, Dr Andrew J. Bacon, Professor John H. Beaman, Dr Teofila E. Beaman, Reed S. Beaman, Kamarudin Mat Salleh, Martin Sands, Datuk Lee Shong Mai, the late John Hepburn, and Au Yang Nang Yip from Kuching, Sarawak freely allowed us to utilise their living plants or specimens. In the final selection of photographs, the expertise of Poon Wai Ming is much appreciated. Dr Gunnar Seidenfaden, Dr Ed de Vogel, Dr Peter van Welzen, Dr John Dransfield, Christopher Bailes, Sarah Thomas, Dr George Argent, Jim Comber, Dr Eric Christenson, Yong Lee Ming, Lee Su Win, Chee Fui Nee, Awang Jamal Bin Hassim, Gregory Silak and many friends and colleagues involved in the preparation of this work are gratefully acknowledged.

The Sabah Society has given much impetus to this project, in particular its former President, Tengku D.Z. Adlin, whose enthusiasm together with the interest of its members has given great support.

CHAPTER 1

INTRODUCTION

The decision to compile information on the native orchids of Borneo was stimulated by the great variety of species found, as well as the dearth of readily available literature and reference specimens in Borneo, which made the identification of orchids difficult. In particular, rapid forest clearance and logging over many areas on the island made the organisation of this work urgent.

The last checklist of Borneo orchids was completed in 1942 by the Japanese botanist G. Masamune, who enumerated 1082 species. Prior to that, the main reference works on Borneo orchids had been those of Henry Ridley (1896) and of Oakes Ames and Charles Schweinfurth (1921) who prepared checklists, and other papers on the orchids. They and Robert Allen Rolfe and Cedric E. Carr worked mainly on the large collections of orchids from Mt. Kinabalu, particularly those of Joseph and Mary Strong Clemens and Lillian Gibbs. Carr (1935) also studied his own Kinabalu collection and those of the 1932 Oxford University expedition to the Dulit Range in Sarawak, whilst the great Dutch orchidologist J.J. Smith (1931) studied collections from Kalimantan as well as described some of those made by the Clemenses in Sabah and Sarawak.

Since then, orchidologists living in Borneo have been able to refer to the much more comprehensive accounts of the orchids of surrounding countries, such as those of R.E. Holttum (1964) for Peninsular Malaysia, recently revised by G. Seidenfaden & J.J. Wood (1992), J.J. Smith (1905) for Java, recently revised and updated by J.B. Comber (1990), and an enumeration of the orchids of Sumatra by J.J. Smith (1933). The orchids of the Philippines were covered by Ames (1905-1925), Quisumbing (1981) and Helen Valmayor (1984). Even New Guinea orchids have been relatively well covered by J.J. Smith (1908-1934), Schlechter (1914, translation published 1982) and Millar (1978). The series of thorough monographs by G. Seidenfaden covering the orchids of Thailand and published from 1975 to 1988, are also a useful reference for Bornean orchids.

Some monographs have also been key references, such as those on *Phalaenopsis* by H. Sweet (1980), and *Apostasia* by E. F. de Vogel (1969). Since the project began, further useful monographs on genera with species found in Borneo have appeared, and several have benefitted from this project. Among these are "*Corybas* West of Wallace's Line" by J. Dransfield, J. B. Comber and G. Smith (1986); "The genus *Paphiopedilum*" by P.J. Cribb (1987), and "The genus *Cymbidium*" by D. DuPuy and P.J. Cribb (1988). Shortly after the project began, the series Orchid Monographs was started in 1986 and published by the Rijksherbarium, Leiden. The recently revised genera *Chelonistele, Entomophobia, Geesinkorchis, Nabaluia* (de Vogel, 1986), *Pholidota* (de Vogel, 1988) and *Coelogyne* section Tomentosae (de Vogel, 1992), *Acriopsis* (Minderhoud & de Vogel, 1986), *Ania, Mischobulbum* and *Tainia* (Turner, 1992), are all represented by species in Borneo.

In the course of the present work, we have had to consult the valuable orchid collections made by many parties in the past. Many of these are housed at the herbaria at Kew, Singapore, Bogor and Leiden, the last being the centre of activity for the Flora Malesiana project which includes Borneo. Of particular importance are the recent collections made by E.F. de Vogel and J.J. Vermeulen (Leiden) and those by J.J. Wood, P.J. Cribb, J. Dransfield, M. Sands and C. Bailes (Kew). J.H. Beaman and W. Meijer from the USA have also added many important collections in their field work in Sabah. J.J. Wood, J.H. Beaman and R.S. Beaman (1993) have recently completed an enumeration of the orchids of Mt. Kinabalu based on the study of over 4800 specimens by 93 collectors in 14 herbaria. The checklist includes 686 species in 121 genera with many collections still to be named for just this mountain alone, and this would lend support to predictions that Borneo may have as many as 2000 species in up to 150 genera.

Many important collections have also been made by scientists and herbarium staff based in Brunei, Sabah and Sarawak, and in Bogor. The majority are collections made by the staff of the various Forest Departments who have collected orchids during their field trips. The collections of W.H. Lim and H. Saharan (Peninsular Malaysia), J. Fowlie, E. Ross, J. Asher, E.A. Christenson (USA), Purseglove, S. Collenette, J.B. Comber and E.J.H. Corner (UK), G. Schoser (Germany) and Meta Held (Switzerland) are also noteworthy.

The living collections established in Borneo have helped our studies and will play a significant role in continuing studies. The Sabah Government has taken firm steps to conserve its orchid flora and the Department of Agriculture set up the Tenom Orchid Centre (TOC) in 1980 in the Interior Zone. It now contains nearly 500 lowland species. The Sabah Parks Department has established a Mountain Garden at 1500 m in the Kinabalu Park, to include the higher mountain orchid species, and another Orchid Centre at Poring (POC) at 600 m near Ranau, for orchids from all over Sabah. The latter now has over 400 species in its collection. Another highland mountain garden, on the Sinsuran Road at 1600 m and managed by the District Council in Tambunan, has just been started. It already has over 200 species including some from other mountains in Sabah.

These centres contain nearly all the known species of horticultural interest and also many of botanical interest. Some of these are rare or endangered species such as those in the genera *Paphiopedilum, Phalaenopsis, Renanthera* and *Vanda.* At the Tenom and Poring Centres, micro-propagation laboratories are being set up to propagate many of the rarer species. These Centres all have facilities for botanists to study the orchids in the collections.

In Sarawak, an orchid centre has been set up at Semenggok, near Kuching, by the Forestry Department. The oldest centre is, however, at the Bogor Botanic Garden in Java where living collections of orchids from each of the four provinces of Kalimantan can be seen.

In this first volume 100 species are covered. A brief explanation of the layout of the species descriptions, and how to use the book precede the species accounts. One of the species belongs to a new genus described here for the first time, 11 others are new, and a further three are new varieties recorded for the first time. The taxa are treated in alphabetical order.

Ornithochilus difformis var. kinabaluensis

CHAPTER 2

BORNEO, THE ORCHID ISLAND

Borneo is the third largest island in the world and lies on the Sunda shelf in the eastern half of the Malay Archipelago in South East Asia. Straddling the equator between 108° 50' E to 119° 20' E, and from 7° 4' N to 4° 10' S, the island is over 1300 km long and 950 km wide with an area of nearly 740,000 square kilometres. Politically Borneo is shared by three countries, Malaysia, Indonesia and Brunei Darussalam.

Borneo is a centre of diversity of many edible fruit genera and is truly an"island of fruits", but could equally be referred to as the "Orchid Island". Whitmore (1984) points out that Malesia, made up of the eastern and western Malay archipelagos on two continental shelves, supports an exceedingly rich flora estimated by van Steenis (1971) at 25,000 species of flowering plants, or 10% of the world's flora. The largest family in the region is the Orchidaceae with 3000-4000 species, comprising 12-16% of the flora. Lamb (1991) has estimated that 2500-3000 orchid species are found in Borneo, equivalent to 10% of the world's orchids, 10-12% of the Malesian flora and 75% of the Malesian orchid flora. Of these, 30-40% are thought to be endemic to the island. Wood and Cribb (in press) list over 1400 species in 147 genera. Undoubtedly many more await discovery and this figure will probably rise to between 1500 and 1600 species in around 150 or more genera. Some localities in Borneo can be exceedingly rich. Wood, Beaman and Beaman (1993) documented nearly 700 species of orchids for Mt. Kinabalu, in 121 genera. This represents nearly a fifth of all the vascular plants, and perhaps nearly a half of the Bornean orchids on this mountain alone. However, other habitats in Borneo such as the mangroves can be much poorer in orchid species.

The rich orchid flora, reflecting the overall highly diverse flora of Borneo, may have come about because of several major factors. The Bornean flora, like that of the region in which it lies, is enriched by the intermingling of the floristic elements of more northerly (Laurasian) and southerly (Gondwanic) land masses across an early dissected land bridge joining Australia to Asia. Thus the Sunda shelf (Borneo) species *Calanthe triplicata, Phaius tankervilleae* and *Phalaenopsis amabilis* reach as far east as the Cape York Peninsula in Australia and the Southwest Pacific islands, whereas New Guinea taxa such as *Corybas, Cryptostylis, Diplocaulobium* and *Porphyrodesme papuana* have reached Borneo. Superposed onto this is a complex series of tectonic and geological events that have isolated Borneo as an island with a wide variety of rock and soil formations, and a diverse physiography ranging from rugged mountains to alluvial plains. These factors have engendered differentiation of plant form and speciation. The historical changes in the equatorial forest environment during the last Ice Ages, with periods of aridity, inundations, retreating and expanding rain forest and limits of tree distribution on mountains (Flenley, 1979), have produced a very dynamic forest vegetation and flora. This may explain why many orchid species can be found over a great altitudinal range and can also survive in a surprising range of different forest types.

The Ecological Distribution of Orchids

Tropical Evergreen Lowland Forest. The diversity of orchid species generally increases with altitude in dipterocarp forest but no detailed analysis of this has been undertaken. Epiphytic orchids are quite common in the emergent layer of the rain forest, often in the forks and along the main branches of trees. The huge leaf-litter-gathering orchid *Grammatophyllum speciosum* is often conspicuous in the forks of large emergents such as *Koompassia.* Along the large branches of dipterocarps with thick fissured bark, several species are found which are adapted to the drier, sunnier conditions, especially forms with large pseudobulbs to store moisture or having thick leathery leaves. Particularly common are *Bulbophyllum vaginatum, Coelogyne foerstermannii* and *Epigeneium treacherianum* that can form large masses along branches in hill dipterocarp forest. Other species found up in the emergents are thick-leaved *Phalaenopsis*, such as *P. cornucervi* and *P. pantherina*, several *Eria* and *Dendrobium* species, such as *D. secundum*, and the endemic *Vanda hastifera* and *V. scandens.* In the mid-canopy with more shade are found the broader leaved epiphytes, as well as many more *Cymbidium, Dendrobium, Eria, Trichoglottis* and Borneo's endemic *Dimorphorchis* species. In the hill forests *Phalaenopsis amabilis* and several *Coelogyne* species can be found in the more shaded lower strata. The rare *Phalaenopsis gigantea* with its huge leaves is another species that has evolved to cope with the poor light in the lower canopy. In hill dipterocarp forest the large climbing *Vanilla kinabaluensis* is one of the few orchids adapted to climb as high as 30 m up the large boles of the trees.

Terrestrial orchids are not common on the shaded forest floor, but increase in number as light improves near to streams or on ridges. *Calanthe triplicata, Claderia viridiflora* and *Nephelaphyllum pulchrum* are frequently found but these are more common in the hill dipterocarp forest. *Phaius, Tainia* and *Tropidia* species are less frequent. Often saprophytic orchids, such as the large *Erythrorchis altissima* and *Galeola nudiflora,* are found on large rotting stumps. *Stereosandra javanica, Gastrodia* and *Lecanorchis* species are other saprophytic orchids less frequently found.

Secondary forests and open grassy areas provide a habitat for several orchids that seem to favour the improved light. Terrestrial orchids that do particularly well in more open grassy areas are *Arundina graminifolia, Bromheadia finlaysoniana* and *Phaius tankervilleae,* while *Corymborkis veratrifolia* and *Eulophia spectabilis* do well under shade such as in old rubber plantations and secondary hill forest. Common epiphytes are *Aerides odorata, Cymbidium bicolor* and *C. finlaysonianum, Dendrobium crumenatum* and several species of *Bulbophyllum* and *Thrixspermum.*

Riverine and Riparian Forests. Along the river banks are found a variety of tree species, including *Dipterocarpus oblongifolius, Eugenia, Pometia pinnata, Tristaniopsis* and *Saraca declinata.* These trees are particular favourites of epiphytic orchids that prefer river-bank microclimates. Species of *Bulbophyllum, Coelogyne, Dendrobium, Eria, Flickingeria* and *Vanda* are frequently found along the upper reaches of rivers. Of particular note for Borneo are the riverine taxa *Arachnis breviscapa, Bulbophyllum mandibulare, Coelogyne pandurata, C. septemcostata, Dimorphorchis lowii* with its long pendulous inflorescences, *Flickingeria fimbriata, Phalaenopsis maculata, P. modesta, P. sumatrana, Pholidota imbricata,* and *Vanda dearei.*

Along the river banks on mossy boulders and rock outcrops can be found a whole range of terrestrial orchids, particularly species of *Calanthe*, *Peristylus*, *Phaius*, *Plocoglottis*, *Spathoglottis* and *Tainia* while *Eulophia graminea* is often found in the sandy deposits along more open sunny riverbanks. Along the lower reaches the microclimate changes since the forest cannot shade the broad slow-moving rivers, and the orchid diversity is less than along the cooler upper reaches.

Mangrove Forests. In the coastal mangrove, orchids are rare, one record being *Vanda hastifera* on a very large old mangrove tree. Little effort has been made to survey the orchid flora of the mangrove forests due to the difficulty of walking through them.

Nipa-Dungun vegetation near river mouths. Here *Nypa* (nipa) palms forming dense narrow bands along the river edge have an emerging canopy of dungun trees (*Heritiera globosa*), often shading the *Nypa*, and an understorey of *Ceriops*. This is a much more suitable habitat for epiphytic orchids and here *Eria javanica*, a trunk epiphyte, *Coelogyne rochussenii*, various *Bulbophyllum* and *Thrixspermum* species are encountered.

Freshwater Swamp Forest. As the ground is constantly water-logged, terrestrial orchids are not present. Even epiphytic orchids are rare but they have been little studied. In the main mixed lowland freshwater swamp forests the epiphytic orchids are common along the branches of the larger trees in the upper canopy and the species are similar to those of the lowland dipterocarp forest. In the upper canopy, *Coelogyne* species such as *C. foerstermannii*, are common. *Bulbophyllum*, *Dendrobium*, *Eria*, *Thrixspermum*, *Trichoglottis* and *Vanda* species are also frequently seen. In some more open swamp areas now largely stripped, the showy orchid *Papilionanthe hookeriana* (syn. *Vanda hookeriana*), the so-called "Kinta Valley Weed" of Peninsular Malaysia, is found. *Thrixspermum amplexicaule*, another monopodial scrambling orchid, can be found in the hummocks of scrub vegetation. No detailed studies of the orchid flora of the freshwater swamp forest have been undertaken but, in general, the diversity is lower than that of the lowland dryland forest and this is true for the forest composition as a whole.

Peatswamp Forests. Only a few ground orchids are found, such as *Claderia viridiflora*, *Habenaria singapurensis* and species of *Bromheadia*, *Malaxis* and *Plocoglottis*. Epiphytic orchids are often the same species as found in freshwater swamp forest and lowland dipterocarp forest with probably between 50 to 100 species found in mixed peatswamp forest throughout Borneo, but with probably less than 50 species in any one forest area. Other orchids from this forest type recorded by Anderson (1963) include *Adenoncos sumatrana*, *Appendicula pendula*, *Cystorchis variegata* var. *purpurea*, *Dendrobium cumulatum*, *Dipodium pictum*, *Eria pannea*, *Eulophia spectabilis* (syn. *Eulophia squalida*), *Grammatophyllum speciosum* and *Liparis lacerata*. Many epiphytic plants, including orchids in both heath and peatswamp forest, are associated with ants from whence they derive faecal inorganic ions. Other orchids and epiphytic plants, including *Acriopsis*, *Cymbidium* and *Grammatophyllum* species and *Bulbophyllum beccarii* have adapted leaf, flower and twig fall "catching" techniques.

Leaf-litter-gathering orchids such as *Bulbophyllum beccarii* can be found on the large *Shorea* trunks in some "alan" forest (peatswamp forest dominated by *Shorea albida*) in west Borneo (Wong, 1990). *Cystorchis variegata* var. *purpurea*, *Eulophia spectabilis* (syn. *E. squalida*) and *Zeuxine violascens* are also recorded (Anderson, 1963).

In the stunted pole forest of some peatswamp areas, many epiphytic orchids are found near to the ground. These include several *Bulbophyllum* species such as *B. apodum, B. botryophorum, B. macranthum, B. vaginatum* and several species in the section *Cirrhopetalum*. The genus *Eria* includes several species found both in the heath forests and the peatswamp padang forest such as *Eria neglecta, E. nutans, E. pannea* and *E. pudica*. *Robiquetia spathulata* is also recorded, together with species of *Appendicula, Dendrobium, Pomatocalpa, Thelasis* and *Thrixspermum*. Generally less than 40 species of orchid have been recorded for the more infertile phases of the peatswamp forest. *Bromheadia finlaysoniana*, an orchid typical of open secondary heath forest, is restricted to "padang keruntum" (pole forest dominated by *Combretocarpus* or keruntum trees) in peatswamps.

Limestone. Generally limestone areas are very rich in orchid species and endemics. Throughout Borneo there are outcrops of limestone which are sometimes extensive, such as at Mulu in Sarawak, and are surrounded by lowland forests. Most of these outcrops are karst formations with steep sheer cliffs and spectacular pinnacles or dome-like formations. However, they provide a very different habitat for plants and are usually rich in species with a large number of endemics. The mossy rocks and forest floor are often rich in terrestrial orchids such as *Acanthephippium, Anoectochilus* and *Goodyera,* whereas the epiphytes are similar to the lowland forest.

On cliffs and steep slopes some trees and shrubs have adapted by clinging with long roots to the steeper slopes, crevices, sinkholes or ledges, where pockets of alluvial rendzina soils have formed. Where moister shade pockets occur, several rare species of *Paphiopedilum* that are restricted to limestone are found. These include *P. sanderianum* that grows on cliffs; *P. kolopakingii, P. stonei* and *P. supardii* that grow in pockets of leaf litter on ledges; and *P. philippinense* on more exposed cliffs by the sea. Other terrestrial orchids in this unique habitat growing in leaf litter on boulders and in crevices are the jewel orchids such as the Borneo endemic *Dossinia marmorata*, with golden net-veined leaves, and *Goodyera ustulata, Hetaeria* species, a lithophytic *Taeniophyllum* and another semi-lithophytic endemic orchid, *Ania ponggolensis*.

The trees and roots of the cliffs are home to several newly discovered species of *Trichoglottis* and the new genus *Spongiola*. *Paphiopedilum lowii* and *Vanda dearei* also favour the base of trees and the forks of trees on limestone. *Phalaenopsis maculata* is often found on mossy roots on the cliffs. *Vanda lamellata* and *Rhynchostylis gigantea*, recorded growing on trees on limestone cliffs by the sea, are subject to salt spray.

As limestone hills can be isolated outcrops in lowland forest the flora is never uniform, each outcrop having a different diversity of plants often with its own endemics. This is true for the trees, shrubs and herbs, including the orchids, as mentioned by Kiew (1991) who has indicated that for these reasons no limestone hill harbours more than a fraction of the total limestone flora.

On the higher limestone mountains of Gunung Benarat and Gunung Api in Sarawak that rise respectively to over 1200 m and 1700 m, the cloud layer produces a thick acidic peat to 0.6 m that separates the plant roots from the limestone parent rock. Here the orchids are similar to those of the lower montane zone with many species of *Coelogyne, Dendrochilum* and *Pholidota.*

Strand and Coastal Forests. On the sandy beach areas under the mixed strand forest, occasional terrestrial orchids such as *Geodorum densiflorum* can be found. The cliffs and rocky outcrops along shorelines are exposed to hot sunny conditions, salt spray and strong winds and, therefore, are subject to periods of intense drying out. A few terrestrial and some epiphytic orchids are found in these conditions. The leafless *Taeniophyllum obtusum* is found on tree trunks within contact of occasional sea spray which is usually washed off during rain storms. *Aerides odorata, Cymbidium finlaysonianum, Dendrobium crumenatum* and *D. secundum* are usually epiphytic but *C. finlaysonianum* and *D. crumenatum* are sometimes found on rocky ledges. *Arachnis hookeriana,* a scrambling orchid on rocky promontories, has largely disappeared from its habitat due to indiscriminate collectors. *A. flosaeris* is also found along cliff tops.

Dendrobium parthenium, Pteroceras unguiculatum and *Vanda lamellata* are rare orchids on ultramafic shorelines. Forms of *Phalaenopsis amabilis* and *Pteroceras teres* also occur on islands on high cliffs above the sea. Off the east coast of Sabah, elements of the Philippine flora, such as *Trichoglottis wenzelii,* have been recorded on trees on rocky coasts and shorelines.

Heath Forests. Heath forests are very extensive in Borneo, particularly in Kalimantan, and are a particular feature of the island. Two types are recognised, generally called "kerangas" and "kerapah", Iban terms denoting infertile land and swampy land which will not grow hill rice. They consist of stunted forests on poor podsolic white sandy soils.

The poor nutritional aspects of the habitat are also reflected in a large number of ant and insectivorous plants. Some ants utilize the root masses of the orchid *Acriopsis* as a home, and all five Bornean species of this genus have been recorded in "kerangas" forest. These species, like some species of *Cymbidium* and *Grammatophyllum,* have mats of sharp, stiff, erect roots that point upwards, and spear, trap and collect leaf-litter falling from the canopy above. This method is also used by those species in the sunny upper canopy of the emergents in dipterocarp forest with a similar microclimate. The litter gathered decomposes to a humus which acts as a sponge for storing moisture as well as providing nutrients. A rare species that does this in "kerangas" forest, where it is endemic, is *Porphyroglottis maxwelliae.* Another "kerangas" endemic that has this habit is *Cymbidium rectum. C. finlaysonianum* and *Grammatophyllum speciosum* are also commonly encountered in "kerangas" but at only a few metres above the forest floor.

Another system of leaf-litter-gathering occurs in the large-leaved orchid *Bulbophyllum beccarii* in shadier "kerapah" forest. Its huge leaves are cupped against tree trunks up which it spirals and into which leaves and flowers fall from the canopy to form a compost that is invaded by its roots for moisture and nutrients. This is often supplemented from stem-flow rain water as well.

"Kerangas" and "kerapah" forests, though they may be poor in species diversity overall compared to mixed lowland forest, are proportionally often much richer in orchid species, especially in heath forest at higher altitudes where the open sunny forest floor with a stunted canopy provides a similar climate to the upper canopy of emergents. Species found in this canopy such as *Coelogyne foerstermannii* can be found growing at ground level in "kerangas". Many *Bulbophyllum, Coelogyne, Dendrobium, Dendrochilum, Epigeneium, Eria* and *Pholidota* species are also found at ground level in humps of mosses and leaf litter that accumulate at the base of the trees.

Terrestrials are common, particularly *Bromheadia* in "kerangas" and *Calanthe* in "kerapah". Nearly 200 species of orchid have been recorded in Sabah for this forest formation in one area of mixed heath forest at 450 m (Lamb, unpublished data).

Recently a number of new species of orchids have been recorded from these forests, many adapted to the drier sunnier conditions of "kerangas", others to the shadier peat conditions of "kerapah".

Ultramafic (ultrabasic) Forests. These forests grow on soils low in available calcium, potassium and phosphorus but with high concentrations of iron and nickel. They are considered too toxic for many plants. However, where there is a deeply weathered profile, such as on lower slopes with deep colluvial soils, forest similar except for its species to evergreen lowland dipterocarp forest is found. On steeper slopes, where the soils are shallow, the forest can be distinct and very close to a typical heath forest. Like other heath forests it is rich in orchids, several being new species and records made in recent years. Several species, including *Bulbophyllum botryophorum, B. planibulbe, Coelogyne zurowetzii, Dendrobium microglaphys,* and *D. pachyphyllum,* usually only found in heath forest, have now been recorded for this formation. It is rich in endemics and contains many rare and endangered orchids such as *Arachnis longisepala, Bulbophyllum vinaceum, Corybas serpentinus, Dendrobium parthenium, D. sculptum, Paphiopedilum dayanum, P. hookerae, P. rothschildianum, Paraphalaenopsis labukensis, Phaius reflexipetalus* and *Renanthera bella.* Other species, absent from heath forest, are also found here including several species of *Bulbophyllum* and *Dendrobium,* and ground orchids such as the primitive *Apostasia* and *Neuwiedia* as well as the attractive jewel orchid *Macodes lowii.*

Tropical Lower Montane Rain Forest. Wood, Beaman and Beaman (1993) have provided considerable information on the distribution of orchid species and genera for the different zones and habitats on Mt Kinabalu. In general they confirm similar findings by Comber (1990) for Java, that the greatest diversity of orchid species occurs in the lower montane forest from 900 m to 1800 m, and from our observations this is also true for other montane areas in Borneo. Wood, Beaman and Beaman (1993) noted over 130 taxa between 1500 and 2000 m on the ultramafic soils. Among the terrestrials are many species of *Anoectochilus, Calanthe, Chrysoglossum, Collabium, Corybas, Cryptostylis, Cystorchis, Goodyera, Kuhlhasseltia, Malaxis, Nephelaphyllum, Phaius,* and some species of *Paphiopedilum.* Several saprophytic orchid genera are often found in the peaty top layer and these include species of *Aphyllorchis, Cyrtosia, Gastrodia* and *Lecanorchis,* with other species from partly saprophytic genera such as *Cystorchis* and *Tropidia.* The epiphytic orchids are in great abundance, notably the necklace orchids

(Coelogyne) and allied genera including *Chelonistele, Dendrochilum, Entomophobia, Nabaluia,* and *Pholidota.* Also in great abundance are species of the large genera *Bulbophyllum, Dendrobium* and *Eria.* The large showy flowers of species of *Bulbophyllum* section *Sestochilus* are also common in this forest zone, often as epiphytes on tree trunks and branches. Of the *Dendrobium* species, *D. spectatissimum* with large white flowers with an orange or yellow blotch on the lip, is the most striking. This species is often found on branches in the upper canopy together with *D. uniflorum.* Species of *Dilochia, Eria, Liparis* and *Trichotosia* are also commonly epiphytic on trunks or large branches often in mosses with other orchids. The lavender-blue-flowered *Cleisocentron merrillianum* is another branch epiphyte of note. Twig epiphytes are represented by species of the leafless *Taeniophyllum* and the leafy genera *Adenoncos, Bulbophyllum, Dendrochilum, Dendrobium, Microtatorchis, Oberonia, Thrixspermum* and the newly described *Ceratochilus jiewhoei.*

Of the better known horticultural orchids, fewer species are found compared with hill-dipterocarp, riverine and other lower forest formations. No *Phalaenopsis, Renanthera* or *Vanda* occur, and of the monopodial species, only *Arachnis calcarata* is truly a lower montane forest species. Several *Cymbidium* species are found in upland dipterocarp forest and lower montane forest. These include *Cymbidium chloranthum, C. dayanum, C. ensifolium* subsp. *haematodes* and *C. lancifolium.*

On poor sandy soils over sandstone in Sabah and Sarawak are lower montane forests in which the canopy is more open and there is more light. These forests are not rich in orchids but contain unusual and rare species, a high proportion of which are likely to be endemic, including *Dendrobium* species such as *D. sanguineum, D. sculptum,* species of *Coelogyne, Dilochia* and *Arachnis calcarata.* We still have a rather poor knowledge of the orchid flora of these forests.

Tropical Upper Montane Forest. In general vascular epiphytes, mainly orchids, are frequent but not nearly so abundant, particularly in the more windy exposed canopy. Many epiphytic orchids are found just above ground level or growing in the clumps of moss at ground level where they are more protected. For orchids lower light levels and temperatures are limiting factors and hence they are less abundant than in the lower montane forest. *Dendrochilum* and, to a lesser extent, *Bulbophyllum, Chelonistele, Coelogyne,* and *Eria* are the commonest genera here, with a few species of *Dendrobium, Dilochia, Epigeneium, Liparis* and *Thrixspermum* also present. Terrestrial genera such as *Calanthe, Corybas, Cryptostylis, Nephelaphyllum, Pristiglottis, Spathoglottis* and *Tainia* are rare and are often only found by streams. However, on ultramafic soils on Mt. Kinabalu, the orchids are much more distinct and many endemics are found such as some *Coelogyne* species, *Cymbidium elongatum, Dendrobium maraiparense, D. piranha, D. spectatissimum, Platanthera stapfii* and *Spathoglottis gracilis.*

Tropical Sub-Alpine Vegetation. This vegetation is restricted in Borneo to Mt. Kinabalu only, from about 2900 m to about 3400 m. The shallow soils have a thin humus layer and consequently, there is a much thinner herb layer. The forest is drier and, although mosses and liverworts are found on the tree-trunks and ground, it does not compare to the moss forests lower down. The orchid diversity is very much lower. In the lower reaches, a few *Bulbophyllum* species are common, such as

B. coriaceum and species of section *Monilibulbus*, especially *B. montense*. *Eria* species such as *E. grandis, Coelogyne, Dendrochilum* and *Epigeneium* species are some of the few orchids recorded from this forest. The epiphyte *Thrixspermum triangulare* extends from the mossy forest into this sub-alpine zone. *Eria grandis* can form a thick ground herb layer in the forest and *Bulbophyllum coriaceum* is very common as an epiphyte as well as on rocks or on the ground. Dendrochilums are also common, including *D. pterogyne*, with its reddish pseudobulbs, and *D. alatum* which is commonly epiphytic on *Leptospermum*. *D. grandiflorum* and *D. haslamii* are also often seen in this zone. *Coelogyne plicatissima* still occurs but is replaced by *C. papillosa* at higher altitudes. Although orchid diversity is low, orchids constitute a significant proportion of the herbaceous vegetation.

The tropical upper sub-alpine evergreen vegetation from 3400 m to 3700 m is an open forest. In open grassy areas the truly terrestrial *Platanthera* species are found, the commoner ones being *P. gibbsiae* and *P. kinabaluensis*, but *P. crassinervia* has also been recorded. *Coelogyne papillosa* and *Eria grandis* are terrestrial under the shade of the trees and the same *Dendrochilum* species are the most common epiphytes. However, *D. stachyodes,* another endemic, forms large clumps in the open grassy areas, and grows with thick clumps of moss along cracks in more open granite rocky plateau areas. *Bulbophyllum* species and a leafless *Taeniophyllum* can also be observed at this altitude.

Tropical Alpine Vegetation. The summit area of Mt. Kinabalu is a mosaic of pockets of alpine scrub, where any depth of soil has formed, surrounded by an alpine rock-desert. The species diversity is very low, and largely comprises some sub-alpine constituents that have become dwarfed by the climatic conditions. Only two species of orchid have been found at this lower limit of the alpine zone, in the alpine scrub. On shaded rocks and branches of the larger shrubs, *Dendrochilum alpinum* has been found, whereas *Eria grandis* occurs in rock crevices. But both are rarely seen in this zone and are considered to be sub-alpine species.

CHAPTER 3

THE LIFE HISTORY OF ORCHIDS

It is perhaps commonplace to say that all life on earth is interdependent but, with orchids, this aphorism is particularly well illustrated. The life cycle of an orchid depends upon its relationship with a particular fungus and also in most cases with a particular pollinating animal, usually an insect.

Our knowledge of the role of the fungus in the life cycle of an orchid was discovered in the early years of this century by the French scientist Noel Bernard and elucidated by the German Hans Burgeff. Only in recent years has their pioneering work been continued, most notably by Australian botanists Warcup (1985) and Clements (1982, 1987, 1988). Orchids produce seed that is dust-like and is amongst the smallest of all flowering plants. The seed comprises a seed coat or testa which encloses the embryo of 100 to 200 cells. Orchid seeds contain no endosperm, the food store that comprises the bulk of most other seeds. Therefore, the orchid seed has no integral food store to enable the seed to germinate and grow on its own when given the right environmental conditions. This potentially disastrous state of affairs has been overcome by the development of an intimate symbiosis between the orchid and a fungus which is called a *mycorrhiza*.

In appropriate conditions a mycorrhizal thread or *hypha* will penetrate one end of the orchid embryo through its large suspensor cell. As the fungus enters the inner cells of the embryo, the orchid begins to digest it, thereby releasing nutrients to fuel the growth of the orchid. The germinating orchid develops into a small elongate, top-shaped or spindle-shaped body called a *protocorm*. The protocorm continues to grow until the orchid has produced green chlorophyll-bearing leaves when the dependency upon the fungal partner declines and may even cease. In most tropical epiphytic species the role of the mycorrhizal fungus appears to be confined to the early stages of the development of the orchid. However, there is strong empirical evidence to suggest that, in terrestrial species, the symbiotic relationships may be important throughout the life of the orchid. Other orchids such as *Cystorchis aphylla* and *Epipogium roseum*, for example, never produce leaves and are consequently unable to manufacture their own food and only appear above ground to flower. Their entire development depends upon the ability of their mycorrhizae to provide them with nutrients, a testimony to the potential of this bizarre relationship. Such orchids are termed *saprophytes*.

We still have a great deal to discover about the relationships of orchids and their mycorrhizal fungi. The specificity of the relationship is still debated, although recent work has indicated at least some degree of specificity in the European and Australian terrestrial species. One of the main problems in this respect is our poor knowledge of the taxonomy of the fungi involved. Fungi that look alike do not necessarily produce equivalent germination in the same batch of orchid seed. A second area that needs further study is that illustrated by Clements (1987) when he suggested that the fungi may be symbionts of other plants at the same time as their relationship with an orchid. The observation that the two species of Australian underground orchid, *Rhizanthella*, are always associated with *Melaleuca* bushes is strongly suggestive that this may sometimes be the case.

At the other end of the life cycle, orchids are also dependent upon rather specific relationships for their successful reproduction. Orchids have unusual flowers. There are species whose flowers fancifully resemble bees, wasps, flies, spiders, monkeys and dancing ladies. However, an important function lurks behind this charming menagerie. The orchid flowers have evolved to attract pollinators that will ensure fertilisation and seed production for the continuation of the species. The pollination of *Paphiopedilum rothschildianum* has been studied in the wild by Atwood (1985) who found that the orchid was pollinated by a hover fly. The female fly is deceived into thinking that the hairs on the staminode are aphid eggs which are the normal prey of the hover fly larvae. It lays its own eggs among them and will occasionally fall into the pouch-like lip of the orchid. The only escape from this slippery sided trap is along a ladder of hairs within the lip that pass successively beneath the stamens and stigma. Cross-pollination is ensured when the hover fly visits another flower and repeats its abortive egg-laying attempt there.

Most other orchids are cross-pollinated by attracting insects by a variety of means, some of which border on the bizarre. Our knowledge of the pollination of Bornean orchids is still very limited and very little work has been undertaken on this fascinating subject. However, the extraordinary diversity found in the flowers of Bornean orchids often allows us to guess at the probable pollinator. The putrid smell of the many small flowers of *Bulbophyllum beccarii* suggests that the flowers are pollinated by carrion flies. Flies are also the most likely pollinators of other small-flowered bulbophyllums. In *Cymbidium atropurpureum* and *C. finlaysonianum,* bumble bees have been seen visiting the flowers and it seems likely that they are the pollinators. Many questions, however, remain unanswered to tease botanists in the future. What, for example, is the function of the floral dimorphism found in *Dimorphorchis,* and why does *Paphiopedilum sanderianum* have such long petals?

Pollination mechanisms are usually relatively species-specific although hybridisation is not infrequent in the wild in some groups of orchids. Natural hybrids between *Paphiopedilum dayanum* and *P. rothschildianum* and *P. javanicum* var. *virens* have been recorded from the slopes of Mt. Kinabalu.

Some orchids have eliminated the uncertainties in achieving cross-pollination for the higher chance of reproductive success through self-pollination. Several widespread species are very successful with this mode of reproduction, for example, the saprophytic *Epipogium roseum* is self-pollinating throughout most of its range.

Whatever the mode of reproduction, successful pollination and fertilisation leads to the production of large numbers of seeds. A single seed pod of most Bornean orchids will produce hundreds or even thousands of dust-like seeds. The Leopard Orchid, *Grammatophyllum speciosum*, for example, produces over a million seeds in a single pod. Many orchid seeds lack an embryo, so that perhaps a high seed production potentially offsets the many seeds which will not germinate. Fortunately, most of the mycorrhizal fungi seem to be widespread species and the production of large numbers of seed ensures that at least some will meet an appropriate fungus.

CHAPTER 4
THE STRUCTURE OF ORCHIDS

The diversity of form found in the orchid family in Borneo is amazing. The smallest orchids to be found there are possibly some of the tiny *Bulbophyllum* species which are no more than a few millimetres tall while the liana-like *Vanilla* species that climb up into the tallest forest trees may reach 30 m or more long. The vanillas are by no means the bulkiest of Borneo's orchids. Some of the giant specimens of *Grammatophyllum speciosum* may weigh many tens of kilogrammes with inflorescences two or more metres long and flowers ten centimetres across. Others, such as *Dendrochilum microscopicum,* have flower parts under one millimetre long. What then unites these diverse species into the family Orchidaceae? The distinctive features of orchids which separate them from other flowering plants lie primarily in their flowers.

The Flower

Orchid flowers are simple in structure and yet highly modified from the more typical monocotyledon flower as exemplified by a *Lilium* or *Trillium* , to which orchids are very distantly allied. These characteristically have their floral parts arranged in threes or multiples of three. Orchids are no exception and this can most easily be seen in the two outer whorls of the flower. Let us take as an example the common and widely grown Bornean orchid *Cymbidium finlaysonianum*, which is similar in general floral structure to the majority of Bornean orchids. Its floral parts are situated at the apex of the ovary which itself can be seen to be tripartite. The lowest whorl of the flower is the calyx which consists of three sepals which are petal-like and coloured yellow with a red stripe in the middle. The two lateral sepals differ slightly from the third, called the dorsal or median sepal. In some orchids such as dendrobiums and bulbophyllums, the lateral sepals form at the base a more or less conical chin called a *mentum*. In some, such as the slipper orchids (paphiopedilums), for example, the lateral sepals are united to the apex forming a *synsepal.*

The corolla of *C. finlaysonianum* comprises three petals which are brightly coloured. The two lateral petals, resembling the dorsal sepal in colouration and shape, are uppermost in the flower and differ markedly from the third petal which lies at the bottom of the flower. The third petal, called the *lip* or *labellum*, is highly modified, 3-lobed and with a two-ridged callus on its upper surface. In other orchids the upper surface of the lip may be adorned with a callus of raised ridges, lamellae or tufts of hairs or glands. The lip is an important adaptation of the orchid to facilitate cross-pollination. It can be imagined as a brightly coloured flag to attract potential and specific pollinators which are then guided towards the pollen and stigmatic surface by the form of the callus. The lip, therefore, can be supposed to act as a landing platform and the callus structure as a guidance system for the pollinator. In many species, but not in *C. finlaysonianum*, the lip is extended at the base to form a saccate to filiform *spur* which may or may not contain nectar. This again is

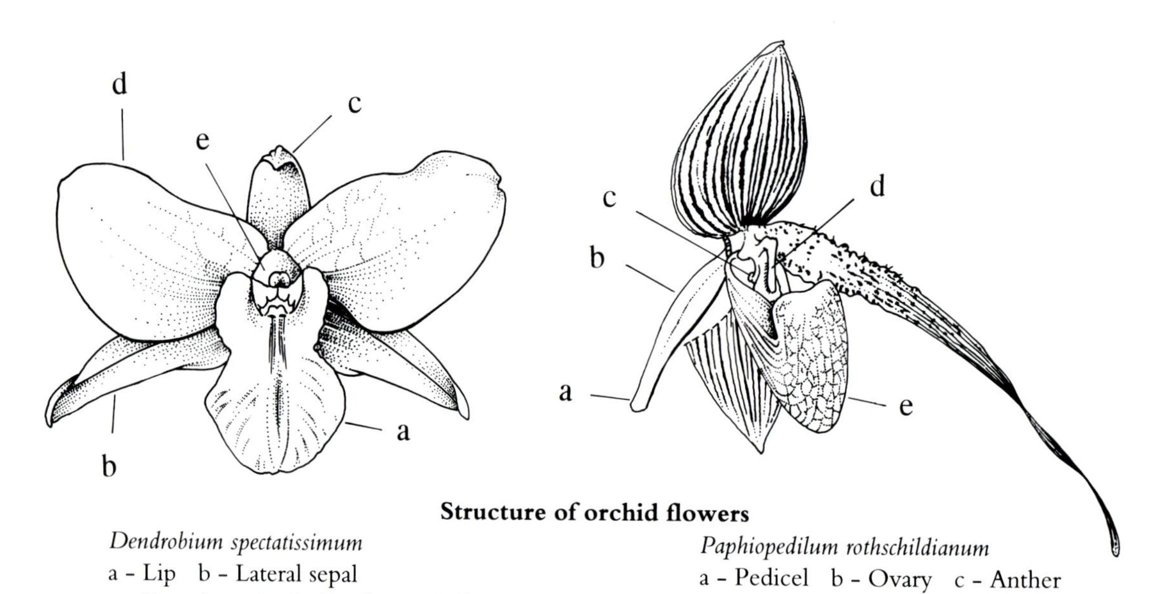

Structure of orchid flowers

Dendrobium spectatissimum
a - Lip b - Lateral sepal
c - Dorsal sepal d - Petal e - Column

Paphiopedilum rothschildianum
a - Pedicel b - Ovary c - Anther
d - Staminode e - Lip

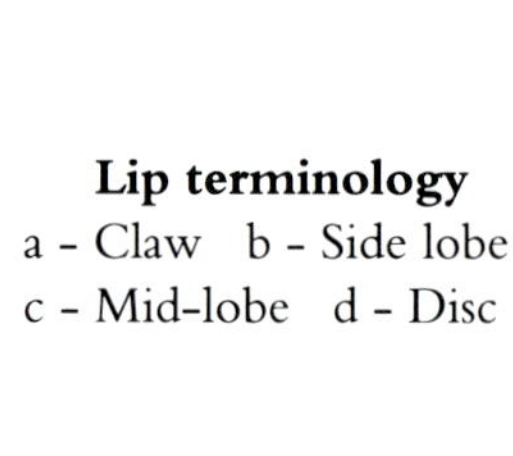

Lip terminology
a - Claw b - Side lobe
c - Mid-lobe d - Disc

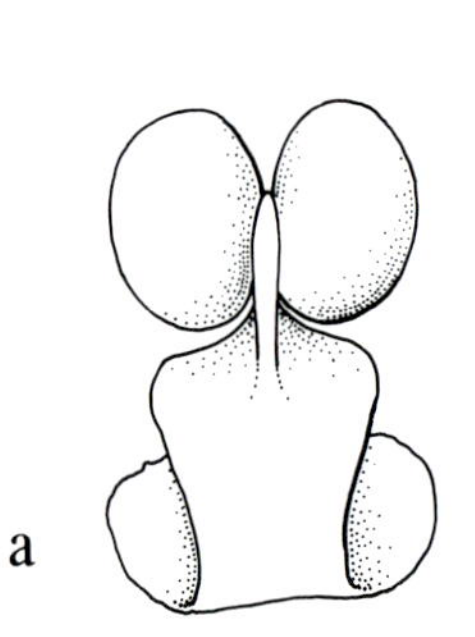

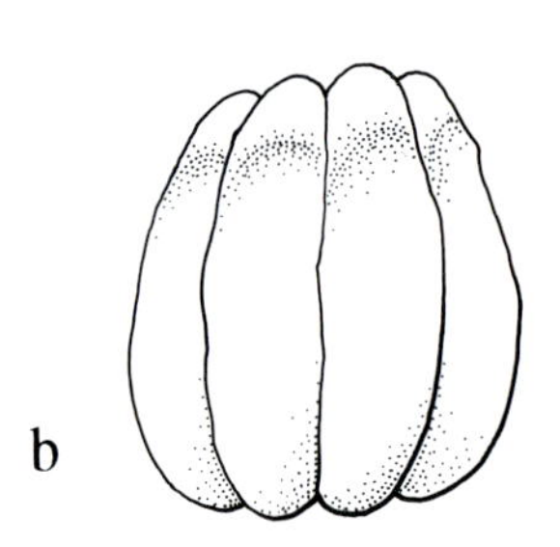

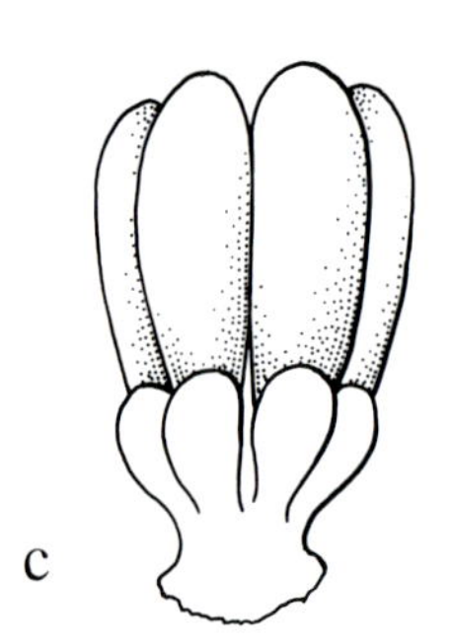

Variation in number of pollinia
a - Two, *Vanda* b - Four, *Dendrobium* c - Eight, *Acanthephippium*

important in attracting a potential pollinator. In some of the vandoid orchids the spur may contain fleshy callosities within and these are important for identification purposes.

The central part of the orchid flower shows the greatest modifications to the basic monocotyledon pattern. Reduction in the number of floral parts and fusion of the male and female organs into a single structure have been the major evolutionary trends in the orchid family. The fused organ in the centre of the flower is called the *column*. In *Cymbidium finlaysonianum,* and in most Bornean orchids, a single anther lies at the apex of the column. The anther does not contain powdery pollen as in most plants but bears the pollen in four discrete masses, called *pollinia*. The pollinia are attached to a sticky disc called the *viscidium*. In other species the pollinial number may be two, four, six or rarely eight and these are attached to the viscidium either directly or by a stalk called a *stipes* in most epiphytic orchids and a *caudicle* in most terrestrial ones.

In the slipper orchids (paphiopedilums), which are considered to be more primitive in floral structure, there are two anthers on the column, placed on each side behind a large shield-shaped sterile anther, the *staminode*. Another group also found in Borneo are the Apostasioideae, comprising the two genera *Apostasia* and *Neuwiedia* with flowers that look more like some lilies than other orchids. The former also has two anthers while the latter has three. In both of these genera, the fusion between the anthers and female organ is only partial and most authorities consider them to be the most primitive of all orchids. The stigma in *Cymbidium finlaysonianum* is positioned on the ventral surface of the column in the centre of the flower. The stigma is a sticky lobed depression situated below and behind the anther but in some terrestrial genera, such as *Habenaria*, the stigma is bilobed with the receptive surfaces at the apex of each lobe. In many species the pollen masses are transferred to the stigmatic surface by a modified beak-like lobe of the stigma called the *rostellum*. This is developed in *C. finlaysonianum* as a projecting flap that catches the pollen masses as the pollinator passes beneath on its way out of the flower.

An interesting feature of the development of most orchid flowers is the phenomenon of *resupination*. In bud, the lip lies uppermost in the flower while the column lies lowermost. In species with a pendent inflorescence, such as *C. finlaysonianum*, the lip will, therefore, naturally lie lowermost in the flower when it opens. However, this would not be the case in the many species with erect inflorescences, such as *Phaius tankervilleae*. Here the opening of the flower would naturally lead to the lip assuming a place at the top of the flower above the column. This is the situation in the terrestrial orchid, *Nephelaphyllum pulchrum*. In most species this is not the case and the lip is lowermost in the flower. This position is achieved by means of a twisting of the flower stalk or ovary through 180 degrees as the bud develops. This twisting is termed *resupination*.

The Inflorescence

Orchids carry their flowers in a variety of ways. Even within the same genus different species have different ways of presenting the flowers. Most Bornean orchids have inflorescences bearing two or more flowers, usually borne on a more or less elongate floral axis comprising a stalk called the *peduncle* and a portion bearing the flowers, the *rachis*. In *Cymbidium finlaysonianum* the flowers are borne in an elongate pendent raceme which is unbranched with the flowers arranged in a lax spiral around the rachis. In *C. ensifolium* and *C. rectum* the raceme is still many-flowered but erect, while in *C. lancifolium* the inflorescence is reduced to two or three flowers in an erect raceme. In a raceme the individual flowers are attached to the floral axis by a stalk called the *pedicel*. In some species, such as *Neuwiedia veratrifolia* and *Peristylus hallieri*, pedicels are absent and the flowers are sessile on the axis; such inflorescences are termed *spicate*.

In the genus *Bulbophyllum* we find some interesting variations on the multi-flowered inflorescence. In several species the flowers are borne all facing to the same side of the rachis, this being called a *secund* inflorescence. The most spectacular group, however, are those in which the rachis is so contracted that the flowers all appear to come from one point at the top of the flower stalk in an umbel, with the inflorescence rather resembling the head of a daisy. Formerly these bulbophyllums, such as *Bulbophyllum mastersianum*, were considered for this reason to belong to a separate genus *Cirrhopetalum*.

Compound inflorescences with many flowers are also not uncommon in Bornean orchids, particularly among the vandoid genera. In genera such as *Pomatocalpa* and *Renanthera* we find branching inflorescences which are termed *panicles*. However, paniculate inflorescences are also found in many other epiphytic genera but are rare among terrestrials.

In many species the flowers are borne one at a time either sessile or on a shorter or longer peduncle. Solitary flowers can be found in many genera such as *Bulbophyllum*, the well-known *Bulbophyllum lobbii* being a good example, and among terrestrials such as *Corybas* and *Paphiopedilum hookerae* and *P. lawrenceanum*.

Vegetative Morphology

The vegetative features of orchids are, if anything, more variable than their floral ones. This is scarcely surprising when the variety of habitats in which orchids are found is considered. Orchids grow in almost every situation: on the permanently moist floor of the lowland tropical rainforest; in the uppermost branches of tall forest trees where heavy rainfall is followed by scorching sun for hours on end; on rocks near the summit of Mt. Kinabalu; and in the grassy areas found on landslips and roadsides. The major adaptations seen in orchid vegetative morphology have evolved to combat adverse environmental conditions, in particular, the problems of water conservation on a daily and seasonal basis.

That tropical orchids might suffer from periodic water deficits is not immediately obvious. Rainfall is not continuous even in the wettest habitats and in many places, even in the tropics, the rainfall patterns are markedly seasonal. Furthermore, most tropical orchids are epiphytic or lithophytic, growing on trunks, branches and twigs of trees or on rocks. In these situations water run-off is rapid and the orchids will

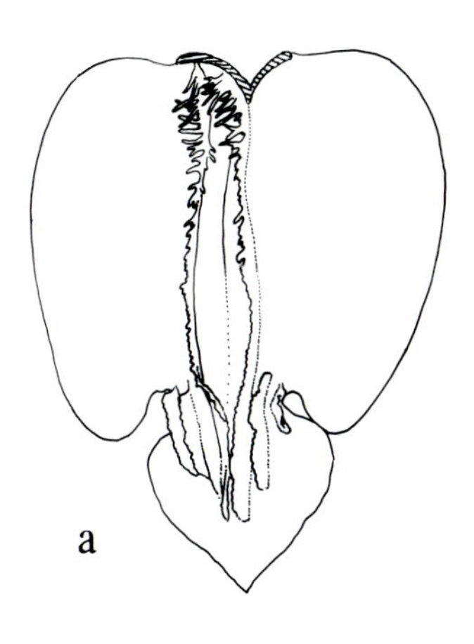

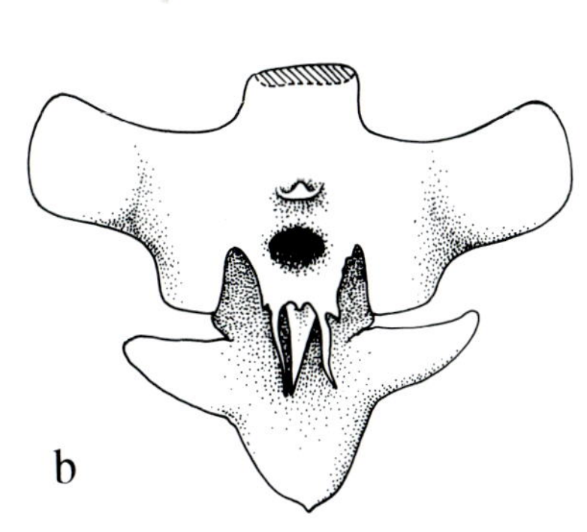

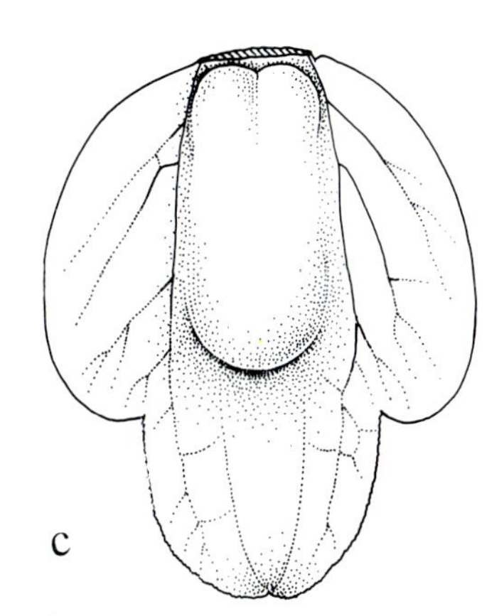

Callus types
a - Lamellate, *Coelogyne odoardi*
b - Complex, *Phalaenopsis cornucervi*
c - Simple, *Epigeneium longirepens*

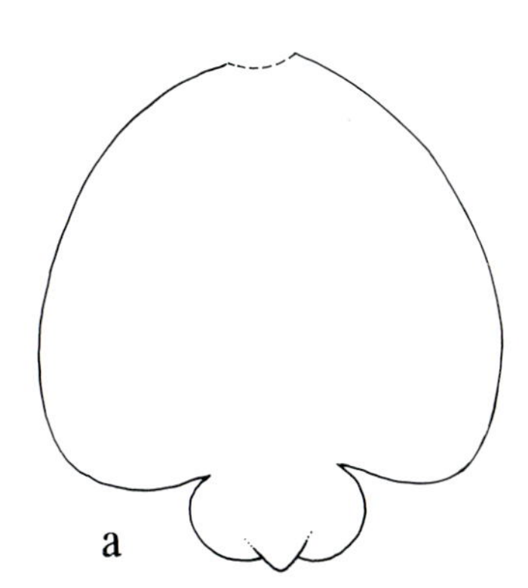

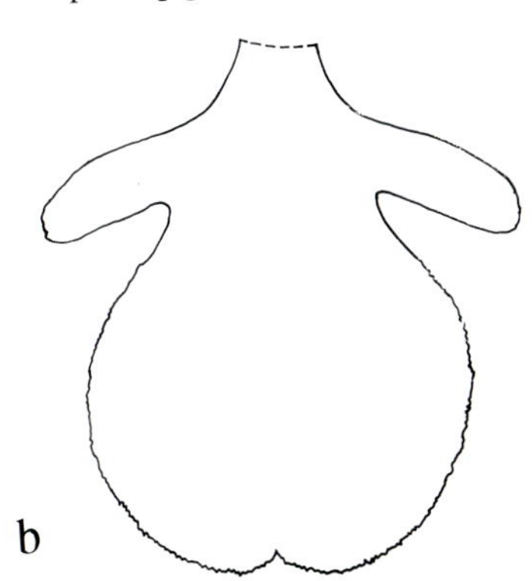

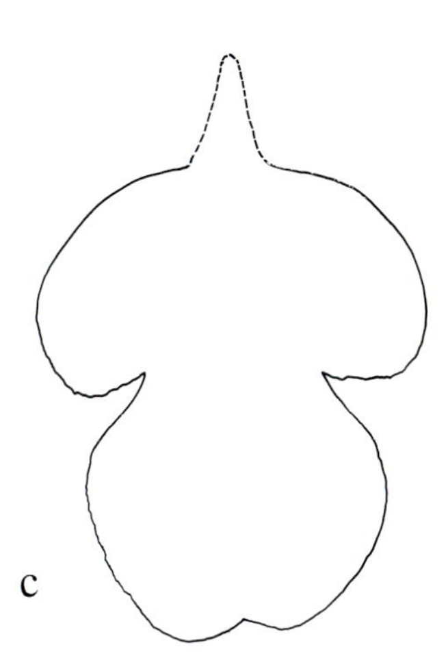

Lip lobing
a - Lobed at apex, *Coelogyne latiloba*
b - Lobed at base, *Collabium simplex*
c - Lobed in middle, *Dendrobium spectatissimum*

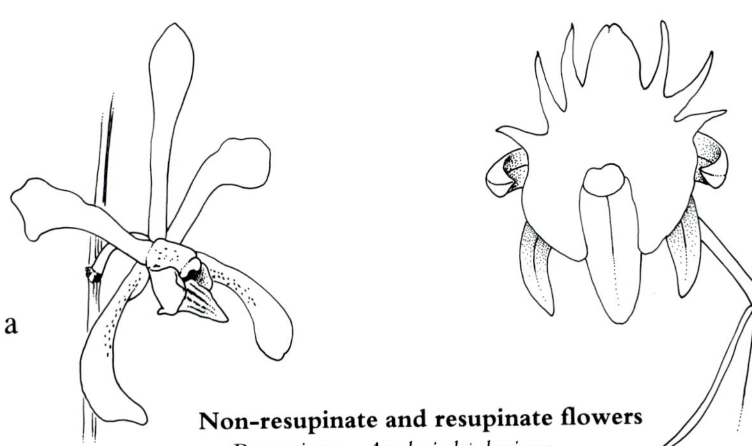

Non-resupinate and resupinate flowers
a - Resupinate, *Arachnis hookeriana*
b - Non-resupinate, *Malaxis lowii*

dry out quickly in the sunshine that follows the rain. Many orchids have therefore developed marked adaptions of one or more organs to allow them to survive these periodic droughts. Some of these adaptations are as dramatic as those encountered in the *Cactaceae*. The stem can develop into a water storage organ and this is so common in tropical orchids that the resulting structure has been given a technical name, a *pseudobulb*. *Cymbidium finlaysonianum*, which is often found growing on rocks, has pseudobulbous stems like large eggs. In *Cymbidium*, *Dendrobium* and *Eria* the pseudobulbs comprise several nodes while in *Bulbophyllum* they are of one node only. Pseudobulbs are also found in many terrestrial orchids and can grow either above the ground as in *Calanthe* or underground as in many *Eulophia* species.

In a few terrestrial orchids a different adaptation to surviving adverse conditions has been adopted. *Habenaria* and *Peristylus*, for example, lack pseudobulbs but have underground tuberoids with which they survive drought. The new growth grows from one end of the tuberoid in suitable conditions. In others such as *Goodyera* and *Zeuxine* the stems are succulent but not swollen. The horizontal stem or rhizome creeps along the ground in the leaf litter and erect shoots bearing the leaves are sent up periodically.

The leaf is another organ that has undergone dramatic modification in the orchids. Fleshy or leathery leaves with restricted stomata, such as those of *Cymbidium finlaysonianum*, are common. The leaves of species growing in drier places are often terete as in *Luisia* and *Papilionanthe*, while in genera such as *Taeniophyllum* and *Chiloschista*, the leaves have been reduced to scales and their photosynthetic function taken over by the often flattened green roots.

A few terrestrial orchids are also leafless but in their case they lack chlorophyll altogether and are saprophytic. Saprophytes can be found in several genera as diverse as *Cystorchis*, *Eulophia*, *Galeola* and *Gastrodia*. The most impressive of these is undoubtedly *Galeola* which can form liana-like plants many metres in length. Lacking chlorophyll the plants cannot photosynthesise and must obtain all of their nutrition from the mycorrhizal fungus with which they are associated. In the case of *Galeola* the fungal associate is said to be the wood-rotting fungus *Armillaria*. Although less spectacular and much smaller, the other saprophytes are just as interesting, none more so than *Cystorchis salmoneus*. This rare orchid is the closest approach in Borneo to an underground orchid and closely resembles the famous Australian species of the genus *Rhizanthella*. Its rhizome is subterranean and the inflorescence much reduced in length so that the flowers are clustered in a short pyramidal head that just emerges above the surface of the leaf litter.

The autotrophic terrestrial species usually have much thinner textured leaves than their epiphytic cousins. In lowland forest, the perpetually moist atmosphere and lack of direct sunlight means that such leaves are not vulnerable to drought. Some of the terrestrial species of the forest floor have beautifully marked leaves. In *Anoectochilus*, *Dossinia*, *Goodyera*, *Ludisia* and their relatives, the leaves can range from green to deep purple or black and may be reticulately veined with silver, gold or red. They are, not surprisingly, popularly known as "jewel orchids".

The roots themselves are much modified in most epiphytic orchids. They have been adapted for two functions, attachment to the substrate and water and nutrient uptake in a periodically dry environment. The tip is actively growing but the remainder of the root is covered by an envelope of dead empty cells called a *velamen*. The velamen protects the inner conductive tissue of the roots and may also aid the uptake of moisture from the atmosphere acting almost as blotting paper for the orchid.

As can be seen life in the tropics can be inhospitable even for orchids. In those regions with a more marked seasonality conditions may be positively hostile for orchids at certain times of the year. Even tropical forests experience periods of relative drought where the orchids have to survive days or even weeks without rainfall. In these conditions, tropical orchids without water-storage capabilities in their stems or leaves can drop their leaves and survive on the moisture stored in their roots which are protected by their cover of velamen. Without such adaptations most orchids would never have survived the three-month long drought of 1983.

An appreciation of the vegetative structure of orchids can provide the grower with the clues needed to provide optimal conditions for growth. If the seasonal nature of the growth that can be found in many orchids is ignored then they will perish rapidly. A knowledge of the floral morphology is just as critical for naming orchids because orchids are classified into genera and species on the finer details of the structure of their sepals, petals, lip and column. The floral dissections provided with each illustration in this volume provides the essential information for identification. For most species the shape of the sepals, petals and especially the lip will provide all of the information the reader needs. However, for the more critical taxa, details of the column, anther, pollinia and rostellum may be needed before accurate identification is possible.

Cymbidium borneense

CHAPTER 5
CLASSIFICATION

The orchid family is probably the largest in the Flowering Plant Kingdom containing upward of 20,000 species. In such a large family, it is useful to understand a little of how they have been classified because a knowledge of the relationships of orchid genera can be useful for identification and have useful predictive value for the botanist and orchid grower.

The twenty thousand or so species of orchid fall into about nine hundred genera which in turn have been grouped into subtribes, tribes and subfamilies in ascending order. The most recent fully published classification of the orchids is that of Dressler (1981, 1983, 1990 a-d) and it will be followed here. Dressler and his predecessors have attempted to group similar orchids together and to show how these groups are related to each other. The theory of evolution has provided a rationale for most authors who have attempted to construct classifications of related groups and most recent classifications have attempted to suggest evolutionary relationships.

Most classifications of orchids have concentrated upon features of the flowers, especially of the sexual parts, to provide clues as to relationships. Dressler has also added much additional information from vegetative morphology, anatomy, cytology and micromorphology to substantiate his classification. He divides the orchids into five subfamilies: Apostasioideae, Cypripedioideae, Spiranthoideae, Orchidoideae and Epidendroideae.

The Apostasioideae, comprising the genera *Apostasia* and *Neuwiedia*, are usually considered to be the most primitive orchids, or, by some, a distinct family the Apostasiaceae. Both genera are found in Borneo. They are terrestrials with plicate leaves, an erect spicate many-flowered inflorescence and flowers that are in many ways more like those of lilies than orchids. The flowers have three similar sepals and three more or less similar petals, one of which, the lip, may be slightly larger. The column bears a style to which two or three stamens are only partially fused at the base. They do, however, have root-stem tuberoids, an endomycorrhizal association and dust-like seeds. On the balance we prefer to include them within the Orchidaceae.

The Cypripedioideae have also, at times, been treated as a separate family. They are mainly terrestrial but a few are epiphytic. The stems may be short or long and the leaves either plicate or conduplicate. The flowers are quite distinctive with a large dorsal sepal and usually fused lateral sepals. The lateral petals are spreading or pendulous and range from subcircular to elongate and ribbon-like. The lip is always calceolate or slipper-shaped, hence their popular name of "slipper orchids". The column is short with two lateral ventral fertile anthers and a large sterile, usually shield-shaped, anther or staminode. The stigma is stalked and ventral. The Cypripedioideae do, however, have dust-like seeds, a mycorrhizal association, and a column with fused stamens and style. On balance, therefore, we include them in the orchid family. The sole Bornean genus is *Paphiopedilum*.

All of the other orchid subfamilies have a column with a single anther and pollen massed into discrete pollinia. The Spiranthoideae are well represented in Borneo and include the pretty jewel orchids that are such a feature of the flora of the forest floor. Genera such as *Anoectochilus*, *Cryptostylis*, *Cystorchis*, *Goodyera*, *Ludisia* and *Zeuxine* are all included in this subfamily which is characterised by the dorsal erect anther which is subequal to the rostellum, the mealy pollinia being attached to a viscidium at the apex of the rostellum, and usually by the creeping fleshy rhizome.

The Orchidoideae is poorly represented in Borneo but includes *Aphyllorchis*, *Corybas*, *Habenaria*, and *Peristylus*. They possess root-stem tuberoids (often referred to as tubers), sectile pollinia and an anther firmly attached by its base to the column.

The remaining and majority of the Bornean orchids are placed in the fifth subfamily, the Epidendroideae, characterised by hard discrete pollinia in an apically attached anther. It is not only the largest subfamily but also the most diverse and has been divided into two subfamilies by some authors. Those genera with lateral inflorescences, and an anther with reduced partitions, superposed pollinia, viscidia and stipes have been included in the Vandoideae. This distinction is, however, far from clear-cut and Dressler (1983) has concluded that the Vandoideae should therefore be included in the Epidendroideae.

Genera in all of the subfamilies are found in Borneo and are included in these volumes. We list below these genera in their appropriate subfamilies, tribes and subtribes and encourage readers to study their orchids closely to give them some insight into the relationships suggested here.

Subfamily **APOSTASIOIDEAE**

Apostasia, Neuwiedia

Subfamily **CYPRIPEDIOIDEAE**

Paphiopedilum

Subfamily **SPIRANTHOIDEAE**

Tribe Tropidieae — ***Corymborkis, Tropidia***

Tribe Cranichideae

Subtribe Goodyerinae — ***Anoectochilus, Cheirostylis Cystorchis, Dossinia, Erythrodes, Goodyera, Hetaeria, Hylophila, Kuhlhasseltia, Lepidogyne, Macodes, Myrmechis, Pristiglottis, Vrydagzynea, Zeuxine***

Subtribe Spiranthinae — ***Spiranthes***

Subtribe Cryptostylidinae — ***Cryptostylis***

Subfamily **ORCHIDOIDEAE**	
Tribe Diurideae	
Subtribe Acianthinae	***Corybas, Pantlingia***
Tribe Orchideae	
Subtribe Orchidinae	***Platanthera***
Subtribe Habenariinae	***Habenaria, Peristylus***
Subfamily **EPIDENDROIDEAE**	
Tribe Gastrodieae	
Subtribe Gastrodiinae	***Didymoplexiella, Didymoplexis, Gastrodia, Neoclemensia***
Subtribe Epipogiinae	***Epipogium, Stereosandra***
Tribe Neottieae	
Subtribe Limodorinae	***Aphyllorchis***
Tribe Nervilieae	***Nervilia***
Tribe Vanilleae	
Subtribe Galeolinae	***Cyrtosia, Erythrorchis, Galeola***
Subtribe Vanillinae	***Vanilla***
Subtribe Lecanorchidinae	***Lecanorchis***
Tribe Malaxideae	***Hippeophyllum, Liparis, Malaxis, Oberonia***
Tribe Cymbidieae	
Subtribe Bromheadiinae	***Bromheadia***
Subtribe Eulophiinae	***Dipodium, Eulophia, Geodorum, Oeceoclades***
Subtribe Thecostelinae	***Thecopus, Thecostele***
Subtribe Cyrtopodiinae	***Chrysoglossum, Claderia, Collabium, Cymbidium, Grammatophyllum, Pilophyllum, Porphyroglottis***
Subtribe Acriopsidinae	***Acriopsis***
Tribe Arethuseae	
Subtribe Bletiinae	***Acanthephippium, Ania, Calanthe, Mischobulbum, Nephelaphyllum, Pachystoma, Phaius, Plocoglottis, Spathoglottis, Tainia***
Subtribe Arundinae	***Arundina, Dilochia***

Tribe Glomereae	
Subtribe Glomerinae	***Agrostophyllum***
Subtribe Polystachyinae	***Polystachya***
Tribe Coelogyneae	
Subtribe Coelogyninae	***Chelonistele, Coelogyne, Dendrochilum, Entomophobia, Geesinkorchis, Nabaluia, Pholidota***
Tribe Podochileae	
Subtribe Eriinae	***Ascidieria, Ceratostylis, Eria, Porpax, Sarcostoma, Trichotosia***
Subtribe Podochilinae	***Appendicula, Poaephyllum, Podochilus***
Subtribe Thelasiinae	***Octarrhena, Phreatia, Thelasis***
Tribe Dendrobieae	
Subtribe Dendrobiinae	***Dendrobium, Diplocaulobium, Epigeneium, Flickingeria***
Subtribe Bulbophyllinae	***Bulbophyllum, Trias***
Tribe Vandeae	
Subtribe Aeridinae	***Abdominea, Adenoncos, Ascocentrum, Aerides, Arachnis, Ascochilopsis, Biermannia, Bogoria, Brachypeza, Chamaeanthus, Ceratochilus, Chroniochilus, Cleisocentron, Cleisomeria, Cleisostoma, Cordiglottis, Dimorphorchis, Doritis, Dyakia, Gastrochilus, Grosourdya, Kingidium, Luisia, Macropodanthus, Malleola, Micropera, Microsaccus, Microtatorchis, Ornithochilus, Papilionanthe, Paraphalaenopsis, Pennilabium, Phalaenopsis, Pomatocalpa, Porphyrodesme, Porrorhachis, Pteroceras, Renanthera, Renantherella, Rhynchostylis, Robiquetia, Sarcoglyphis, Schoenorchis, Smitinandia, Spongiola, Staurochilus,***

Taeniophyllum, Thrixspermum, Trichoglottis, Tuberolabium, Vanda

The Species Concept

One of the most difficult and contentious problems facing botanists is the judgement of where specific boundaries can be drawn. Plant species are sometimes very variable and this can lead to disagreement, even amongst experienced botanists, as to what to consider as a species and how to treat any infraspecific variability. Botanists generally look for marked discontinuities in morphological and micromorphological variation and some degree of breeding barrier to help them with the somewhat artificial preoccupation of tidily dividing the plants of the world into species. In temperate areas of the world where the plants are often well known, it is often possible to demonstrate discontinuities between closely allied species and even the breeding barriers that keep them distinct. Variable species can also be readily identified when the variability between populations can be seen to be continuous. In other species, it may be possible to show that there are minor disjunctions in one or two characters that allow infraspecific variants to be formally recognised. In such cases the botanist usually has copious material to study and often access to living populations, allowing detailed examination of not only the gross morphology but also of cytology, anatomy, biochemistry and breeding systems to help him arrive at a sound decision.

The position in the tropics is seldom as clear-cut because the botanist is usually working with little material ranging from a few specimens to even a single specimen. The study material, usually herbarium specimens, is often incomplete, lacking fruit, flowers or leaves. This applies to many Bornean orchids and can make the task of the orchid taxonomist difficult especially where the critical differences are small or obscure. He must then extrapolate from his experience of other better known groups of orchids and other plants to arrive at a satisfactory judgement. Often, therefore, such decisions may have to be revised when further material becomes available.

The classification of Borneo's orchids is still at an early stage and new species are being described every year. Few genera have yet been systematically revised and the variation of their constituent species assessed and analysed. Where this has happened as in *Cymbidium* and *Paphiopedilum*, it has been possible to reduce some names to synonymy. For example, we now know that *Cypripedium elliottianum* is the same as *Paphiopedilum rothschildianum* and, as a later name, is a synonym of the latter. It has also been possible to recognise some infraspecific taxa within variable species. The taxon known in much of the horticultural literature as *Paphiopedilum virens* has been shown to be a variant, with more horizontal petals, of the widespread species *Paphiopedilum javanicum* from Java, Sumatra, Bali and Flores. The Bornean plants from Mt. Kinabalu are now treated as *P. javanicum* var. *virens*.

As more of Borneo's hinterland is explored and plants are retrieved from logged areas, we are beginning to understand the orchids better and this will inevitably lead to reassessments of the species and consequent name changes that most of us find so irritating.

Name Changes

The discovery of new Bornean orchids has proceeded in a spasmodic fashion over the years since Borneo was visited by early collectors such as Low and Lobb. This has led to many orchids being described as new to science more than once, with a consequent increase in the number of available names for a particular species. The original descriptions are scattered through many dozens of journals and books while the original specimens are likewise to be found in many herbaria, often in Europe and North America. The application of an orchid name can only be confirmed by reference to the specimen upon which the botanist based his original description (the *type*). Access to these critical specimens is often difficult and some, indeed, have been destroyed. In broad outline, if two names are found to refer to the same species then the earlier name is considered the correct one, the other a synonym. This process has led to many name changes in orchids and a consequent frustration on the part of gardeners when a familiar name is changed.

Multiple names for a single species are most frequent in those species with a wide distribution that are not confined to Borneo. Some species such as *Calanthe triplicata* are found in Borneo but also as far west as Madagascar and as far east as Tahiti. It is scarcely surprising that it has been described many times as new from various parts of this extended range. It is only in recent times that a better understanding of this species has led to *C. triplicata* being recognised as the correct name, while the many other names are considered synonyms.

Names can also change for other reasons. As the orchids become better known, reclassification may be necessary to reflect our better understanding. A species, for example, may be transferred from the genus in which it was first described to another. Such was the case for most Bornean *Paphiopedilum* species which were originally placed in *Cypripedium*, now considered to be a North Temperate genus which differs in having plicate leaves and a different aestivation.

Another source of name change is that caused by the misidentification of a plant when it is introduced into cultivation. The plant may become well established under the incorrect name which will necessarily be changed when its correct identity becomes established. The Bornean *Pholidota imbricata* for example, has been widely grown as *P. pallida* which has recently been shown to be confined to the Himalayas and distinct from the Bornean species.

When using the accounts in the "Orchids of Borneo" it is worth checking the lists of synonymous and misapplied names because some well-known names have inevitably been reduced to synonymy.

CHAPTER 6

KEYS TO GENERA

KEY TO GENERA

(EXCLUDING SAPROPHYTES AND TRIBE VANDEAE, SUBTRIBE AERIDINAE)

1. Flowers with 2 or 3 fertile anthers **2**
1. Flowers with a single fertile anther **4**
2. Perianth segments similar, the lip never deeply saccate **3**
2. Perianth segments very unequal, the lip deeply saccate, slipper- or pouch-shaped. Anthers 2, lateral. Staminode median, large and shield-shaped .... **Paphiopedilum**
3. Anthers 2, with or without a staminode. Inflorescence usually branched, curved and spreading, never erect .... **Apostasia**
3. Anthers 3, staminode absent. Inflorescence simple, erect .... **Neuwiedia**
4. Anther erect or bending back, never short and operculate at apex of column. Leaves usually spirally arranged, convolute, not articulated at base **5**
4. Anther eventually bending downward over column apex to become operculate, or operculate at column apex but not bending downward. Leaves distichous, usually articulate at base **33**
5. Plants exclusively terrestrial **6**
5. Plants climbing **31**
6. Rostellum elongate, equalling the anther. Root-stem tuberoids (tubers) absent **7**
6. Rostellum usually shorter than the anther. Root-stem tuberoids (tubers) present or absent **25**
7. Stems tough and rigid. Leaves plicate **8**
7. Stems weak and fleshy, often brittle, never tough and rigid. Leaves convolute or conduplicate **9**
8. Lip widest at apex. Column long. Inflorescence often branched .... **Corymborkis**
8. Lip widest at base. Column short. Inflorescence simple .... **Tropidia**
9. Pollinia sectile. Roots scattered along rhizome **10**
9. Pollinia not sectile. Roots in a close fascicle **24**
10. Flowers resupinate **11**
10. Flowers non-resupinate **23**

11. Spur or saccate base of lip containing neither glands nor hairs (hairs may occur near mid-lobe only) **12**

11. Lip hairy within or having papillae or glands on either side near base or in spur or sac **13**

12. Lip with spur which projects between lateral sepals .... **Erythrodes**

12. Lip saccate, entirely enclosed by lateral sepals .... **Hylophila**

13. Lip hairy within .... **Goodyera**

13. Lip otherwise **14**

14. Apex of lip not abruptly widened into a distinct spathulate or transverse, bilobed blade **15**

14. Apex of lip abruptly widened into a distinct spathulate or transverse, bilobed blade **17**

15. Saccate base of lip with a transverse row of small calli. Plants robust, up to 100 cm tall .... **Lepidogyne**

15. Saccate base of lip or spur containing stalked or sessile glands. Plants much smaller **16**

16. Hypochile swollen at base into twin lateral sacs each containing a sessile gland. Epichile with fleshy involute margins, forming a tube .... **Cystorchis**

16. Hypochile otherwise, containing 2 stalked glands. Epichile otherwise .... **Vrydagzynea**

17. Claw of lip with a toothed or pectinate flange on either side .... **Anoectochilus**

17. Claw of lip otherwise **18**

18. Leaves dark green with greenish-yellow, golden or pink median and secondary nerves .... **Dossinia**

18. Leaves without such coloured secondary nerves, although median nerve sometimes coloured **19**

19. Sepals connate for half their length to form a swollen tube .... **Cheirostylis**

19. Dorsal sepal and petals connivent, forming a hood, or free **20**

20. Dorsal sepal and petals connivent, forming a hood **21**

20. Dorsal sepal and petals free **22**

21. Column without appendages .... **Kuhlhasseltia**

21. Column with 2 narrow wings which project into the base of the lip .... **Pristiglottis**

22. Lip with a long claw. Stigmas on short processes. Inflorescence 1-2-flowered .... **Myrmechis**

22. Lip with a short claw. Stigmas sessile. Inflorescence several-flowered .... **Zeuxine**

23. Lip and column twisted to one side .... **Macodes**

23. Lip and column straight .... **Hetaeria**

24. Flowers resupinate, small, arranged spirally in a dense inflorescence .... **Spiranthes**

24. Flowers non-resupinate, large, arranged in all directions in a lax inflorescence. .... **Cryptostylis**

25. Root-stem tuberoids (tubers) present 26

25. Root-stem tuberoids (tubers) absent 31

26. Lip without a spur 27

26. Lip spurred 28

27. Inflorescence produced with the leaf. Lip orbicular. Column with a tooth-like process below .... **Pantlingia**

27. Inflorescence produced before the leaf. Lip 3-lobed, the base embracing the column. Column without a tooth-like process .... **Nervilia**

28. Lip 2-spurred, tubular below. Flowers helmet-shaped .... **Corybas**

28. Lip with 1 spur, not tubular below. Flowers otherwise 29

29. Stigmas each on a stigmatophore extending from the column, free from hypochile .... **Habenaria**

29. Stigmas not freely extending in front of column, sometimes connate or adpressed to lip hypochile 30

30. Lip simple, strap-shaped. Spur cylindric, rather long, not swollen at apex. Stigmas joined to form a concave structure, free from hypochile .... **Platanthera**

30. Lip 3-lobed. Spur short, usually globular, saccate or fusiform. Stigmas convex, cushion-like, connate with or adpressed to hypochile .... **Peristylus**

31. Leaves fleshy, never plicate. Stems fleshy. Pollinia soft and mealy, as monads .... **Vanilla**

31. Leaves plicate, stem never fleshy, often rather tough, sometimes brittle. Pollinia 2, cleft 32

32. Habit monopodial. Stems not distant on a creeping rhizome. Leaves distichous, imbricate, ensiform. Flowers pale yellowish with pink to crimson blotches .... **Dipodium**

32. Habit sympodial. Stems placed distantly on a creeping rhizome. Leaves elliptic, neither distichous nor imbricate. Flowers green .... **Claderia**

33. Plants terrestrial 34

33. Plants epiphytic or lithophytic 61

34. Pollinia 2 or 4, naked, i.e. without caudicles, viscidia and stipes usually absent 35

34. Pollinia 2 to 8, with caudicles (sometimes reduced), or a stipes 38

35. Column-foot absent **36**

35. Column-foot prominent **37**

36. Column long. Flowers usually resupinate. Lip without 2 large basal auricles, apex rarely pectinate .... **Liparis**

(in part)

36. Column short. Flowers non-resupinate. Lip without 2 large basal auricles, apex often pectinate .... **Malaxis**

37. Rhizomatous part of shoot (sometimes also the non-rhizomatous part) carrying one-noded pseudobulbs .... **Epigeneium**

(in part, sometimes *E. kinabaluense*)

37. Non-rhizomatous part of shoot consisting of several internodes, wholly or partly fleshy, with or without pseudobulbs ....
.... **Dendrobium**

(in part, some species in sections *Conostalix* & *Distichophyllum*)

38. Inflorescences numerous or not, borne along a slender leafy stem, lateral or terminal **39**

38. Inflorescences never numerous, usually solitary, never borne along a slender, leafy stem, usually lateral, sometimes axillary or terminal **44**

39. Inflorescences lateral **40**

39. Inflorescences terminal **42**

40. Pollinia 2. Inflorescences not numerous. Flowers c. 4 cm across ....
.... **Cymbidium**

(in part, *C. elongatum*)

40. Pollinia 6 or 8. Inflorescences numerous. Flowers much smaller **41**

41. Pollinia 6. Leaf sheaths and flowers glabrous .... **Appendicula**

(in part)

41. Pollinia 8. Leaf sheaths and flowers usually covered in reddish brown hispid hairs .... **Trichotosia**

(in part)

42. Pollinia 2 .... **Bromheadia**

(in part, *B. borneensis, B. crassifolia, B. finlaysoniana*)

42. Pollinia 8. .... **43**

43. Flowers large, up to 8 cm across (sometimes peloric). Petals much broader than sepals. Inflorescence usually unbranched. Floral bracts small, acute, persistent
.... **Arundina**

43. Flowers much smaller. Petals similar to sepals. Inflorescence branching. Floral bracts conspicuous, concave, deciduous .... **Dilochia**

44. Pollinia 2 .... **45**

44. Pollinia 4 or 8 .... **47**

45. Plants densely covered in yellowish brown hairs. Flowers non-resupinate ....
.... **Pilophyllum**

45. Plants glabrous. Flowers resupinate **46**

46. Lip mobile. Column with 2 fleshy basal keels. Spur formed by column-foot
.... **Chrysoglossum**

46. Lip immobile. Column without basal keels. Mentum formed by column-foot, base of lateral sepals and base of lip .... **Collabium**

47. Pollinia 4 .... **48**

47. Pollinia 8 .... **53**

48. Inflorescence arcuate, strongly decurved .... **Geodorum**

48. Inflorescence otherwise **49**

49. Lip not spurred **50**

49. Lip spurred **52**

50. Lip divided into a distinct, somewhat saccate hypochile and a 2-lobed epichile. Pseudobulbs flattened, always 2-leaved .... **Geesinkorchis**

50. Lip otherwise. Pseudobulbs never flattened, sometimes elongated into a leafy stem, 1 to many-leaved **51**

51. Lip convex, adnate to sides and apex of column-foot to form a sac, usually with an elastic hinge that springs when touched. Pseudobulbs often elongated into a leafy stem .... **Plocoglottis**

51. Lip never convex, free or fused at base to base of column, without an elastic hinge. Pseudobulbs short, often enclosed in sheathing leaf-bases **Cymbidium**

(in part, *C. borneense, C. ensifolium* subsp. *haematodes, C. lancifolium*)

52. Lip entire, or 3-lobed (mid-lobe not bilobulate) .... **Eulophia**

(in part, *E. graminea, E. spectabilis*)

52. Lip '4-lobed', mid-lobe bilobulate .... **Oeceoclades**

53. Pseudobulbs absent, replaced by a fleshy subterranean rhizome, swelling into a horizontal fusiform, sometimes V-shaped, tuber. Leaves usually several, often withering before inflorescence appears. Lip with a small basal pouch ....
.... **Pachystoma**

53. Plants pseudobulbous **54**

54. Lip spurred, or gibbous and partially adnate to and embracing column to form a tube **55**

54. Lip not spurred **58**

55. Pseudobulbs always 1-leafed. Plants remaining green when damaged **56**

55. Pseudobulbs 2-to several-leaved. Plants turning bluish-black when damaged. Lip spurred or gibbous **57**

56. Inflorescence lateral .... **Ania**

56. Inflorescence terminal .... **Nephelaphyllum**

57. Column margins fused with the base of the lip over nearly their entire length. Lip spurred .... **Calanthe**

57. Column margins fused with lip only at or near the base. Lip shortly spurred or gibbous .... **Phaius**

58. Pseudobulbs 1-leafed **59**

58. Pseudobulbs 2-to several-leafed **60**

59. Leaf base cordate in mature plants, petiole absent .... **Mischobulbum**

59. Leaf base ± decurrent along a petiole .... **Tainia**

60. Flowers urn-shaped, sepals fleshy, fused to form a swollen tube, free at the apices. Lip movably hinged to a column-foot, not clawed or callose .... **Acanthephippium**

60. Flowers with free, usually spreading sepals. Lip not movably hinged, mid-lobe very narrowly clawed, with 2 ovoid, often pubescent basal calli. Column-foot absent .... **Spathoglottis**

61. Pollinia 2 or 4, naked, i.e. without caudicles **62**

61. Pollinia 2 to 8, with distinct, though sometimes reduced, caudicles **69**

62. Column-foot absent. Leaves equitant, distichous, bilaterally flattened **63**

62. Column-foot prominent. Leaves dorsiventral, or occasionally bilaterally flattened (in *Dendrobium* sections *Aporum, Oxystophyllum* and *Strongyle* only) **64**

63. Groups of leaves close together. Column short .... **Oberonia**

63. Groups of leaves 4 cm apart. Column long .... **Hippeophyllum**

64. Lip usually immobile, not hinged at base. Mentum often spur-like **65**

64. Lip movably hinged to column-foot. Mentum saccate **68**

65. Rhizomatous part of shoot (sometimes also the non-rhizomatous part) bearing one-noded, 1- or 2-, rarely 3-leaved pseudobulbs **66**

65. Non-rhizomatous part of shoot (when present) consisting of several internodes, with or without 1- to several-noded pseudobulbs. Flowers ephemeral or long-lasting **67**

66. Erect parts of shoot closely set, tufted, 15-25 cm high, consisting of a single internode tapering from a fleshy base into a slender neck, with 1 apical leaf and 1-to 2-flowered successive inflorescences. Flowers on very long pedicels, ephemeral. Sepals and petals narrowly caudate .... **Diplocaulobium**

66. Erect parts of shoot spreading or suberect, never tufted, consisting of several internodes bearing one-noded, 1-, 2-, or rarely 3-leaved pseudobulbs. Inflorescence 1- to several-flowered. Flowers on shorter pedicels, long-lived. Sepals and petals narrowly elliptic .... **Epigeneium**

67. Stems superposed, the non-rhizomatous part of the shoot consisting of several quite long thin internodes, the uppermost pseudobulbous and 1-leaved. Flowers always ephemeral .... **Flickingeria**

67. Stems not superposed; either 1) rhizomatous, 2) erect and many-noded, 3) erect and 1-noded or several-noded from a many-noded rhizome, or 4) rhizome absent, new stems of many nodes arising from base of old ones. Leaves 1 to many. Flowers long-lived or ephemeral .... **Dendrobium** (in part)

68. Anther-cap with a prolongation in front, of varying shape (deeply lacerate in *T. tothastes*). Column usually with insignificant stelidia .... **Trias**

68. Anther-cap otherwise. Column stelidia usually prominent ...... **Bulbophyllum**

69. Stems slender, leafy, without pseudobulbs **70**

69. Stems pseudobulbous, pseudobulbs sometimes small and entirely enclosed by imbricate leaf sheaths **83**

70. Pollinia 2 or 4 .... **71**

70. Pollinia 6 or 8 .... **73**

71. Pollinia 2. Leaves sometimes laterally flattened .... **Bromheadia** (in part)

71. Pollinia 4. Leaves never laterally flattened **72**

72. Stems slender, often branched, with many close, distichous leaves up to 1.2 cm long .... **Podochilus**

72. Stems very short, tufted, with 1-2 linear leaves 6-12 cm long..... **Sarcostoma**

73. Pollinia 6 .... **Appendicula**

73. Pollinia 8 .... **74**

74. Inflorescence terminal, usually globose, surrounded by bracts. Flowers white or yellow .... **Agrostophyllum**

74. Inflorescence lateral, terminal or subterminal, never of globose heads. Flowers variously coloured **75**

75. Column-foot absent **76**

75. Column with a short or long foot **78**

76. Leaves laterally compressed or terete, distichous. Flowers yellowish green ....
.... **Octarrhena**

76. Leaves dorsiventral, linear to linear-elliptic or strap-shaped **77**

77. Inflorescence and sepals white-tomentose. Flowers arranged in whorls, non-resupinate .... **Ascidieria**

77. Inflorescence and sepals glabrous. Flowers not arranged in whorls, resupinate.
.... **Thelasis**

(in part, e.g. *T. carinata, T. micrantha*)

78. Leaf sheaths covered with reddish brown, or rarely white, hispid hairs. Leaves never fleshy and subterete .... **Trichotosia**

78. Leaf sheaths glabrous. Leaves sometimes fleshy and subterete **79**

79. Stems one-leaved .... **Ceratostylis**

79. Stems few- to many-leaved **80**

80. Stems short, entirely enclosed by imbricate leaf sheaths. Inflorescence a densely flowered raceme with small bracts .... **Phreatia**

(in part, e.g. *P. amesii, P. densiflora, P. monticola, P. secunda*)

80. Stems elongate, leafy throughout entire length **81**

81. Inflorescence terminal or subterminal, usually densely many-flowered, densely hirsute. Floral bracts small .... **Eria**

(section *Mycaranthes*)

81. Inflorescence axillary, few-flowered, glabrous **82**

82. Floral bracts large and brightly coloured .... **Eria**

(section *Cylindrolobus*)

82. Floral bracts minute, green or brownish .... **Poaephyllum**

83. Pollinia 2 .... **84**

83. Pollinia 4 or 8 .... **87**

84. Lip joined at its base with an outgrowth from the column and with column-foot to form a tube at right angles to base of column .... **Thecostele**

84. Lip otherwise **85**

85. Flowers non-resupinate. Lip convex when viewed from above, scoop-shaped when viewed from below, hairy, bee-like. Column with large, curved stelidia. Habit similar to *Grammatophyllum* .... **Porphyroglottis**

85. Flowers resupinate. Lip otherwise. Column lacking stelidia **86**

86. Plants very large, with pseudobulbs up to 3 m or more long. Flowers up to 10 cm across. Sepals and petals up to 2.6 cm wide, with large irregular blotches. Stipes U-shaped .... **Grammatophyllum**

86. Plants much smaller. Flowers up to 5.7 cm across. Sepals and petals narrow, without blotching. Stipes absent .... **Cymbidium**

(all epiphytic species except *C. lancifolium*)

87. Pollinia 4 .... **88**

87. Pollinia 8 .... **98**

88. Inflorescence terminal **89**

88. Inflorescence lateral **96**

89. Flowers with a distinct column-foot, always non-resupinate .... **Polystachya**

89. Flowers without a column-foot, resupinate, or, more rarely, non-resupinate **90**

90. Pollinia attached to a stipes .... **Geesinkorchis**

(in part)

90. Stipes absent **91**

91. Basal half of the narrow, saccate lip adnate to basal half of column. Apical half of lip separated by a transverse, high, fleshy callus .... **Entomophobia**

91. Lip otherwise **92**

92. Lip hypochile with long, slender lateral front lobes .... **Nabaluia**

92. Lip hypochile without such lobes **93**

93. Column usually with lateral arms (stelidia) .... **Dendrochilum**

93.Column without lateral arms **94**

94. Lip hypochile saccate, distinctly separate from epichile. Lip rarely 3-lobed .... **Pholidota**

94. Lip hypochile, although often concave, not sharply distinct from epichile. Lip almost always 3-lobed **95**

95. Side-lobes of lip (when present) narrow, borne from front part of hypochile at right angles to the epichile. Hypochile narrow, saccate .... **Chelonistele**

95. Side-lobes of lip broad, widening gradually from base of lip. Hypochile ± concave, broader and rarely saccate .... **Coelogyne**

96. Lip joined at its base with an outgrowth from the column and with column-foot to form a tube at right angles to base of column .... **Thecopus**

96. Lip otherwise **97**

97. Lateral sepals united into a synsepalum. Stipes long, linear............. **Acriopsis**

97. Lateral sepals free. Stipes absent .... **Cymbidium**

(in part, *C. lancifolium* only)

98. Sepals connate to varying degrees, forming a tube. Pseudobulbs flattened .... **Porpax**

98. Sepals free **99**

99. Column with a prominent foot. Rachis usually hirsute or woolly. Pseudobulbs rarely flattened .... **Eria**

(in part)

99. Column absent or short. Rachis glabrous. Pseudobulbs sometimes flattened **100**

100.Column with a short foot. Anther-cap horizontal on top of column, not beaked .... **Phreatia**

(in part, e.g. *P. listrophora, P. sulcata*)

100.Column-foot absent. Anther-cap vertical behind column, beaked .... **Thelasis**

(in part, e.g. *T. capitata, T. carnosa, T. variabilis*)

KEY TO SAPROPHYTIC GENERA

(LEAFLESS TERRESTRIALS LACKING CHLOROPHYLL)

1. Flowers with sepals and petals fused (connate) to a varying degree, often appearing campanulate and always resupinate **2**

1. Flowers with free, spreading or connivent sepals and petals, not appearing campanulate, or lateral sepals connate; resupinate or non-resupinate **5**

2. Petals fimbriate, bright orange .... **Neoclemensia**

2. Petals otherwise **3**

3. Column with long decurved arms (stelidia), foot absent .... **Didymoplexiella**

3. Column without long decurved arms, with a short foot **4**

4. Pollinia 4. Dorsal sepal and petals adnate to form a single trifid segment forming a shallow cup or tube with the partially connate lateral sepals. Stigma near column apex .... **Didymoplexis**

4. Pollinia 2. Sepals and petals connate to form a 5-lobed tube which is sometimes gibbous at the base, and which may be split between the lateral sepals. Stigma at base of column. .... **Gastrodia**

5. Flowers always resupinate, lip lowermost **6**

5. Flowers non-resupinate or resupinate **11**

6. Stem branching, tough and wiry. Sepals and petals surrounded by a shallow denticulate calyculus (cup) .... **Lecanorchis**

6. Stem simple, slender or fleshy. Sepals and petals not encircled by a shallow denticulate calyculus (cup) **7**

7. Lip divided into a distinct hypochile and epichile. Hypochile with or without twin lateral sacs **8**

7. Lip not divided into a distinct hypochile and epichile **9**

8. Hypochile swollen at base into twin lateral sacs, each containing a globular sessile gland. Epichile with fleshy involute margins, tube-like. Flowers pink to reddish, tipped with white .... **Cystorchis**

(*C. aphylla, C. salmoneus, C. saprophytica*)

8. Hypochile without such sacs. Epichile 3-lobed. Flowers greenish white or creamy white and purple .... **Aphyllorchis**

9. Flowers large, reddish brown. Lip 3-lobed, saccate .... **Eulophia**

(*E. zollingeri* only)

9. Flowers small, white, or white flushed with purple at apex. Lip entire, with or without a spur **10**

10. Lip spurred, strap-shaped, margin not undulate .... **Platanthera**

(*P. saprophytica* only)

10. Lip not spurred, narrowly elliptic, margin undulate .... **Stereosandra**

11. Stem simple. Flowers non-resupinate. Lip with a short spur...... **Epipogium**

11. Stem branching. Flowers resupinate or non-resupinate. Spur absent.......... **12**

12. Stems long and climbing. Flowers resupinate. Fruits dry, dehiscent **13**

12. Stems short, never climbing. Flowers non-resupinate. Sepals brownish mealy or blackish ramentaceous on reverse. Fruits succulent and indehiscent or dry and dehiscent **14**

13. Rachis and flowers furfuraceous-pubescent. Stems stout. Column stout, arcuate, clavate .... **Galeola**

13. Rachis and flowers glabrous. Column slender, erect .... **Erythrorchis**

14. Plant robust, with several thick, fleshy stems borne from each rhizome. Sepals obtuse, concave, brownish mealy on reverse. Fruits succulent, indehiscent .... **Cyrtosia**

14. Plant slender, with a single narrow, wiry stem borne from each rhizome. Sepals acute, reflexed (*T. saprophytica*), or lateral sepals connate (*T. connata* ined.), blackish ramentaceous on reverse. Fruits dry, dehiscent .... **Tropidia**

(*T. connata* ined. and *T. saprophytica* only)

KEY TO GENERA OF THE SUBTRIBE AERIDINAE

1. Pollinia 4 .... **2**
1. Pollinia 2 .... **27**
2. Pollinia more or less equal, globular, free from each other (*Group 1*) **3**
2. Pollinia appearing as 2 pollen masses, each completely divided into either unequal, or more or less equal, semiglobular free halves (*Group 2*) **6**
3. Plants without leaves, or leaves reduced to minute brown scales. Stem minute. Roots terete or flattened, containing chlorophyll .... **Taeniophyllum**
3. Plants with normal leaves. Roots lacking chlorophyll **4**
4. Large terrestrial, *Phalaenopsis*-like. Leaves radical. Inflorescences long, erect, many-flowered .... **Doritis**
4. Small epiphytes. Leaves borne along a distinct stem. Inflorescence 1 to 4-flowered **5**
5. Leaves dorsiventral. Inflorescence 1 to 4-flowered. Flowers green .... **Adenoncos**
5. Leaves bilaterally flattened/compressed. Inflorescence 2-flowered. Flowers white .... **Microsaccus**
6. Flowers without a distinct column-foot **7**
6. Flowers with a distinct, though sometimes short column-foot **22**
7. Leaves bilaterally flattened, distichous, with sheathing bases, resembling those of *Microsaccus*. Mid-lobe of lip expanded into a broadly oblong-elliptic, emarginate blade .... **Ceratochilus**
7. Leaves otherwise **8**
8. Lip not adnate to column, movable. Sepals and petals narrow, usually rather spathulate. Spur short and conical .... **Arachnis**
8. Lip adnate to column, not movable **9**
9. Spur with a distinct longitudinal internal median septum **10**
9. Spur without a distinct longitudinal internal septum **13**
10. Rostellum projection short or long, turned obliquely sideward and upward, supporting a thin linear stipes, sometimes to 9 times as long as diameter of pollinia .... **Micropera**
10. Rostellum projection and stipes otherwise **11**
11. Column with a raised fleshy, laterally compressed rostellum which sits on top of the clinandrium and has a longitudinal furrow along its edge into which the stipes and dorsally placed pollinia recline .... **Sarcoglyphis**
11. Column without such a structure **12**

12. Floral bracts, ovary and flowers densely pubescent. Floral bracts large, longer than flowers .... **Cleisomeria**

12. Floral bracts, ovary and flowers glabrous. Floral bracts minute.... **Cleisostoma**

13. Back wall of spur without calli and/or outgrowths **14**

13. Back wall of spur ornamented with calli and/or outgrowths **20**

14. Hypochile of lip globose-saccate, the side-lobes reduced to low, often fleshy edges of the sac, mid-lobe fan-shaped .... **Gastrochilus**

(*G. patinatus* only, see also Group 4, couplet 43)

14. Hypochile otherwise **15**

15. Mid-lobe of lip distinctly pectinate-fringed. Stipes linear, about 4 times as long as diameter of pollinia .... **Ornithochilus**

15. Mid-lobe otherwise. Stipes about twice as long as diameter of pollinia...... **16**

16. Spur or sac separated from apical portion of lip by a fleshy transverse wall or ridge **17**

16. Spur or sac not separated from apical portion of lip by a fleshy transverse wall or ridge **18**

17. Flowers pale greenish yellow or cinnamon orange, spotted black. Rostellum projection large, narrow at base, rising in front of colum .... **Abdominea**

17. Flowers creamy white with lilac-pink patch on lip. Rostellum projection narrow, somewhat decurved... **Smitinandia**

18. Lip as long as or longer than dorsal sepal. Flowers small, white to pink, bluish or mauve. Leaves narrowly lanceolate or terete .... **Schoenorchis**

18. Lip much shorter than dorsal sepal. Flowers red or yellow, showy **19**

19. Leaves dorsiventral, bilobed. Column $^{1}/_{4}$ the length of dorsal sepal .... **Renanthera**

19. Leaves more or less semi-terete, acute. Column $^{2}/_{3}$ the length of dorsal sepal . ... **Renantherella**

20. Lip with a tongue or valvate callus, often forked at the tip, projecting diagonally from deep inside the spur .... **Pomatocalpa**

20. Lip with an often hairy ligulate tongue placed close to the spur entrance or at the base of the lip **21**

21. Inflorescence branched, scape long, several-flowered .... **Staurochilus**

21. Inflorescence unbranched, scape short, often several close together, one- to few-flowered .... **Trichoglottis**

22. Flowers large, showy, dimorphic, the basal two always strongly scented, differently coloured from those above .... **Dimorphorchis**

22. Flowers much smaller, not dimorphic **23**

23. Lip without a distinct spur or sac, but hypochile often somewhat concave. Flowers ephemeral **24**

23. Lip with a distinct spur or sac. Flowers long-lasting **25**

24. Leaves terete .... **Cordiglottis**

24. Leaves dorsiventral .... **Thrixspermum**

25. Spur or sac with a median longitudinal septum .... **Cleisostoma**

25. Spur or sac without a longitudinal septum **26**

26. Lip epichile with two forward pointing teeth emerging from base .... **Kingidium**

26. Lip epichile otherwise **27**

27. Stems very short. Inflorescence borne below the leaves. Flowers small, greenish yellow and white, marked with crimson on the lip. Lip deeply saccate .... **Bogoria**

27. Stems long, usually pendent. Inflorescences axillary. Flowers translucent lavender-blue or dark lilac-pink. Lip distinctly spurred .... **Cleisocentron**

28. Pollinia sulcate or porate **29**

28. Pollinia entire **44**

29. Pollinia sulcate, i.e. more or less, but not completely cleft or split (*Group 3*) **30**

29. Pollinia porate (*Group 4*) **41**

30. Column-foot absent or very indistinct **31**

30. Column-foot distinct, though sometimes short **35**

31. Flowers often large and showy, usually a few, well spaced on a raceme. Stipes short and broad, entire, shelf-like .... **Vanda**

31. Flowers small to medium-sized, crowded on to a usually densely many-flowered raceme or panicle. Stipes linear, spathulate, uncinnate, rarely hamate. Rostellum prominent, bifid or long and pointed **32**

32. Leaves linear, acute, fleshy. Inflorescence and flowers scarlet-red .... **Porphyrodesme**

32. Leaves broader, unequally bilobed. Inflorescence and flowers otherwise **33**

33. Plants small, stem and inflorescence less than 4 cm. Rachis very fleshy, clavate. Flowers borne in succession, minute .... **Ascochilopsis**

33. Plants larger. Rachis not clavate. Flowers not borne in succession, small to medium sized **34**

34. Stems short. Leaves borne close together, most often with many light-coloured nerves. Lip entire or obscurely 3-lobed, deeply saccate or with a short backward-pointing spur without interior ornaments .... **Rhynchostylis**

34. Stems rather long. Leaves distant, without pale nerves. Lip 3-lobed, spur often apically inflated and occasionally with callosities or scales within .. **Robiquetia**

35. Leaves terete, sometimes up to 165 cm long **36**

35. Leaves dorsiventral, much shorter **37**

36. Stems up to 2 m long. Lip spurred, ecallose. Column-foot entire .... **Papilionanthe**

36. Stems very short. Lip not spurred, with a conduplicate, plate-like callus situated at the junction of the mid- and side lobes. Column-foot 3-fingered.............. .... **Paraphalaenopsis**

37. Spur or sac, if present, developed from the hypochile. Epichile dorsiventral... .... **38**

37. Spur or sac borne centrally on lip. Epichile reduced, fleshy **39**

38. Spur absent, or rudimentary. Lip with at least one forward-pointing forked appendage. Flowers few, sometimes large and showy, distichous .... **Phalaenopsis**

38. Spur well developed. Forked appendages absent. Flowers many, facing in every direction, developing simultaneously .... **Aerides**

39. Rostellum projection long, slender. Lip bent upward so as to make a right angle with column-foot, distinctly unguiculate. Flowers long-lasting, developing simultaneously .... **Macropodanthus**

39. Rostellum projection inconspicuous. Lip continuing the line of and usually flush with column-foot. Flowers usually ephemeral, usually developing successively, a few open at a time **40**

40. Column-foot longer than the column proper................................. **Pteroceras**

40. Column-foot short, column proper elongate .... **Brachypeza**

41. Leaves terete. Lip neither saccate nor spurred .... **Luisia**

41. Leaves dorsiventral. Lip saccate or spurred **42**

42. Lip mobile on a short but distinct column-foot. Spur or sac absent .... **Biermannia**

42. Lip immobile. Column-foot absent. Spur or sac present.............................. **43**

43. Lip with a globose-saccate hypochile, the epichile separated from it by a transverse ridge connecting front edges of side-lobes .... **Gastrochilus**

(see also Group 2, couplet 14)

43. Lip with a rather long, cylindric or extinctoriform spur and a ligulate midlobe **44**

44. Spur without a prominent backwall callus. Rostellar projection broadly triangular. Stipes without apical appendages. Pollinator guides absent .... **Ascocentrum**

44. Spur with a prominent backwall callus. Rostellar projection elongate, attenuate, sigmoid. Stipes with apical appendages. A pair of crimson pollinator guides present at throat of spur .... **Dyakia**

45. Column with a distinct foot. Lip movable **46**

45. Column without a foot. Lip not movable **48**

46. Lip without a spur, side-lobes sometimes fimbriate .... **Chamaeanthus**

46. Lip spurred or saccate **47**

47. Spur-like conical portion of lip more or less solid. Peduncle short, glabrous .. .. **Chroniochilus**

47. Sac or spur thin-walled, without interior fleshiness. Peduncle longer, often prickly-hairy .... **Grosourdya**

48. Lip with a bristle or tooth inside near the apex. Floral bracts conspicuous, triangular, leafy .... **Microtatorchis**

48. Lip without an apical bristle or tooth. Floral bracts not leafy **49**

49. Side-lobes of lip very large, often fringed.................................... **Pennilabium**

49. Side-lobes of lip small, never fringed **50**

50. Midlobe of lip resembling a small spongy pouch, hollow above, solid towards apex .... **Spongiola**

50. Midlobe of lip otherwise **51**

51. Lip not truly spurred, but with a spur-like tubular cavity. Lateral sepals adpressed to lip .... **Porrorhachis**

51. Lip spurred. Lateral sepals not adpressed to lip **52**

52. Rachis slender, never thickened and sulcate, or clavate. Column hammer-shaped. Stipes linear-spathulate, much broadened at apex.............................. **Malleola**

52. Rachis fleshy, sulcate, or clavate. Column short and stout. Stipes linear, much reduced... **Tuberolabium**

CHAPTER 7
DESCRIPTIONS AND FIGURES

The information provided in the species' descriptions has been standardised into a simple layout that will be followed in all subsequent volumes.

Because species of several genera are described in Volume 1, they are arranged alphabetically according to genus, and then to the species. In subsequent volumes dealing with one genus, the species are arranged alphabetically or alphabetically within sections (eg. Vol. 2 Bulbophyllum).

The Species Name

For each species described the correct botanical name and authority is provided first, followed by the reference to the original description, and the latest published work or monographic treatment covering the species. Details of types are given, and the herbarium in which they are located when known. A list of synonyms follows. This is relevant to many Bornean species as many have already been described earlier under a different name from neighbouring countries. The earliest name is always the correct one under the International Code of Botanical Nomenclature and all subsequent names that are found to be referable to the same species become synonyms. However, synonyms must be validly published after study and comparison with taxa thought to be the same. Work of this kind is most important to clarify much of the past confusion that has arisen when one species has been described sometimes under several different names and from different countries. Access to the vast amount of literature and to herbarium specimens for comparison, such as are available at Kew and Leiden, is essential to complete this task. All published names can be found in Index Kewensis which is updated regularly.

Species' Descriptions

In the past many orchids were described as new to science without an illustration and many were only provided with a short Latin description. Some early illustrations were not entirely accurate and in some cases descriptions and illustrations were based on a single specimen. Great variation is often found in plant size, flower size and colour which may not be evident from the type collection. Often, when large differences occur, usually between plants from different localities or even countries, further minor distinguishing characters can be quantified, and varieties recognised. Wherever possible in this series the botanical descriptions provide a range of measurements that reflect plant size, flower size and colour variations made from living and preserved specimens.

Errors in habit, form and size can arise when specimens are illustrated from drawings of dried herbarium specimens, especially if only brief field notes are given with the specimen. Many dried specimens have little or no notes on flower colour whilst others may have a detailed painting or colour photograph. Hence, in this series illustrations have been made from living material where possible. Colour plates are provided in each volume, often depicting colour variations, varieties or forms.

Each line drawing where possible depicts the habit of the plant, with a scale for size, and details of the flowers, all to scale in centimetres (cm) or millimetres (mm).

A glossary of terms used in the descriptions is provided together with figures illustrating the botanical terms used to describe leaf shape, inflorescence type, flower parts and details of the column and pollinia (See Appendix 1, 2 & 3 or Figures A, B, C).

The glossary of terms is based on Daydon Jackson's "A glossary of botanic terms", (4th Edition, reprinted 1971), and "The Manual of Cultivated Orchid Species" by Bechtel, Cribb and Launert (3rd Edition, 1992). The diagnostic figures have also been taken from "The Manual of Cultivated Orchid Species" by Bechtel, Cribb & Launert (3rd Edition, 1992).

Detailed information on the shape, size and structure of the anther-cap, pollinia and fruit is not always provided in the descriptions, except when available, but generally they are nearly always illustrated in the line drawing, except the fruit. In many cases, the living plants on collection have missing pollinaria, while fruit are only fully formed three months or more after flowering. Most of the living material drawn for the series has been dried and flowers put into spirit, and passed to Kew, Leiden and other herbaria for future reference.

Following the botanical description other information on the species is grouped under five headings as follows:

Habitat & Ecology

Under this are brief notes on life form, eg. terrestrial, lithophytic, epiphytic on trees, shrubs or vines, or saprophytic. Where possible the altitudinal range is given and the forest type and soils or rocks on which the forest type occurs is provided. Notes on flowering are either given here or under notes and observations. Special note is made, for instance, if the species is restricted in habitat either to certain rock formations such as limestone, or to certain forest types such as tropical heath forest, or riverine habitats.

Distribution in Borneo

Information on the distribution in Borneo is based on information given on the herbarium material examined or seen by the authors in those herbaria visited. Where data is lacking, the record may only be for Borneo, but in most cases the country, and district, mountain range, etc. is provided. An identification list, based on a selection of specimens examined in herbaria, is provided after the glossary. Such a list may not be of interest to the general naturalist or orchidologist, but will help future monographers to locate material for reference and study. The herbaria in which the collections are located are indicated using standard abbreviations registered by the International Association of Plant Taxonomists. Other sources are given temporary codes for the present.

Herbaria cited in the series:

AAU : Herbarium Jutlandicum, Botanical Institute, University of Aarhus, Bygn. 137, Universitetsparken, DK-8000, Aarhus C, Denmark.

AMES : Orchid Herbarium of Oakes Ames, Botanical Museum, Harvard University, Cambridge, Massachusetts 02138, U.S.A.

B : Herbarium Botanischer Garten und Botanisches Museum Berlin-Dahlem, Königen-Luise-Strasse 6-8, D-1000 Berlin 33, Germany.

BM : Herbarium, Botany Department, The Natural History Museum, Cromwell Road, London SW7 5BD, England, U.K.

BO : Herbarium Bogoriense, Jalan Raya Juanda 22-24, Bogor, Java, Indonesia.

BR : Herbarium, Nationale Plantentuin van België, Jardin Botanique National de Belgique, Domein van Bouchout, B-1860 Meise, Belgium.

BRI : Herbarium, Plant Pathology Branch, Department of Primary Industries, Indooroopilly, Queensland 4068, Australia.

BRUN : Herbarium, Forestry Department, Bandar Seri Begawan, 2067, Brunei Darussalam.

C : Herbarium, Botanical Museum, University of Copenhagen, Gothersgade 130, DK-1123, Copenhagen, Denmark.

CAL : Central National Herbarium, P.O. Botanic Garden, Howrah, Calcutta 711 103, West Bengal, India.

CBG : Herbarium, Australian National Botanic Gardens, GPO Box 1777, Canberra, A.C.T. 2601, Australia.

E : Herbarium, Royal Botanic Garden, Edinburgh EH3 5LR, Scotland, U.K.

FI : Herbarium Universitatis Florentinae, Museo Botanico, Via G. La Pira 4, I-50121, Firenze, Italy.

JCB : Herbarium, Centre for Taxonomic Studies, St. Joseph's College, P.B. 5031, Bangalore 560 001, Karnataka, India.

K : Herbarium, Royal Botanic Gardens, Kew, Richmond, Surrey TW9 3AB, England, U.K. (**LINDLEY** refers to the John Lindley herbarium housed in the Orchid Herbarium; **WALLICH** refers to the Nathaniel Wallich herbarium housed at Kew).

L : Rijksherbarium, Postbus 9514, 2300 RA Leiden, The Netherlands.

LA : Herbarium, Biology Department, University of California, Los Angeles, California 90024-1606, U.S.A.

LAE : Papua New Guinea National Herbarium, Forest Research Institute, PO Box 314, Lae, Papua New Guinea.

LINN : Herbarium, Linnaean Society of London, Burlington House, Piccadilly, London W1V 0LQ, England, U.K.

M : Herbarium, Botanische Staatssammlung, Menzinger Strasse 67, D-8000, München 19, Germany.

MEL : National Herbarium of Victoria, Royal Botanic Gardens, Birdwood Avenue, South Yarra, Victoria 3141, Australia.

MO : Herbarium, Missouri Botanical Garden, PO Box 299, Saint Louis, Missouri 63166-0299, U.S.A.

NY : Herbarium, New York Botanical Garden, Bronx, New York 10458-5126, U.S.A.

P : Herbier, Laboratoire de Phanérogamie, Muséum National d'Histoire Naturelle, 16 rue Buffon, F-75005, Paris, France.

PNH : Philippine National Herbarium, National Museum, PO Box 2659, Manila, Philippines.

S : Herbarium, Botany Department, Swedish Museum of Natural History, PO Box 50007, S-104 05 Stockholm, Sweden.

SAN : Herbarium, Forest Research Centre, Forestry Department, PO Box 1407, 90008 Sandakan, Sabah, Malaysia.

SAR : Forest Herbarium, Department of Forestry, 93660 Kuching, Sarawak, Malaysia.

SEL : Herbarium, Marie Selby Botanical Gardens, 811 South Palm Avenue, Sarasota, Florida 34236, U.S.A.

SING : Herbarium, National Parks Board, Singapore, Botanic Gardens, Cluny Road, Singapore 1025, Singapore.

TAI : Herbarium, Botany Department, College of Science, National Taiwan University, Taipei 10760, Taiwan, Republic of China.

U : Herbarium, Institute of Systematic Botany, State University of Utrecht, Postbus 80.102, 3508 TC Utrecht, The Netherlands.

UKMS : Herbarium Jabatan Biologi, Universiti Kebangsaan Malaysia, Kampus Sabah, Locked Bag No. 62, 88996 Kota Kinabalu, Sabah, Malaysia.

UPS : Herbarium (Fytoteket), Uppsala University, PO Box 541, S-751 21, Uppsala, Sweden.

US : United States National Herbarium, Botany Department, NHB-166, Smithsonian Institution, Washington, D.C. 20560-0001, U.S.A.

W : Herbarium, Department of Botany, Naturhistorisches Museum Wien, Burgring 7, A-1014 Wien, Austria.

Z : Herbarium, Institut für Systematische Botanik, Universität Zürich, Zollikerstrasse 107, CH-8008 Zürich, Switzerland.

Other Sources and Collection Abbreviations:

RSNB : Royal Society Expedition North Borneo 1964 - specimens mostly at K.

SFN : Singapore Field Number - refers mainly to Carr collections at SING and BM

SNP : Herbarium of Sabah Parks at Kinabalu Park H.Q. (includes POC numbers = Poring Orchid Centre) Sabah, Malaysia.

TOC : Tenom Orchid Centre, Lagud Seberang, Tenom, Sabah, Malaysia.

General Distribution

Distribution outside Borneo is provided and endemic status indicated where applicable. Borneo is placed by botanists in the Malesian region and is included in the great floristic work Flora Malesiana. The region covers the tropical belt of South-east Asia between 10° N & S of the equator, but also includes those Philippine Islands above 10° N. Peninsular Malaysia is the only part of mainland Asia included, the remainder consisting of a great diversity of island groups with New Guinea to the east and Sumatra in the west. The region has been divided into 9 artificial Divisions or geographical units which are :

Div. I Sumatra and neighbouring islands.

Div. II Peninsular Malaysia and neighbouring islands.

Div. III Java and neighbouring islands.

Div. IV Lesser Sunda islands (Nusa Tenggara).

Div. V Borneo and neighbouring islands.

Div. VI Philippine Islands.

Div. VII Celebes (Sulawesi) and neighbouring islands.

Div. VIII Moluccas (Maluku)

Div. IX New Guinea and neighbouring islands excluding the Solomon Islands.

Most botanical collections in herbaria are organised or arranged following these divisions. Borneo as Region 5 is divided into 13 collection areas. North-west Borneo, Sarawak and Brunei, West Borneo, South and South-east Borneo, East and North-east Borneo (north of Balikpapan) Sabah (formerly British North Borneo), Balambangan Island, Banguey and Mangsi Islands, Labuan, Anambas and Natuna Islands, Tambelan Islands, Karimata Islands, Salemboe and Laoet Kecil Islands, Laoet and Seboeke Islands, Nanukan, Tarakan and Mandoel Islands.

Most of Borneo's orchid species are distributed within this region. However for this series use is made of the major political boundaries in Borneo, and not those used by Flora Malesiana.

Notes

Under this heading are discussed other particulars such as relationships with and differences from other species as well as natural variation. Details of known pollinators and pollination mechanisms are also included. In some cases the rarity and conservation status of the species is also discussed.

Derivation of Names

Where only one species in a genus is described in a volume both the derivation of the generic name and the species name is given. Where more than one species in a genus is covered the derivation of the generic name is provided under the first species. Derivation of the species' name only follows thereafter.

The derivation of generic names is taken from "Generic Names of Orchids" by Schultes and Pease (1963). For species' names that honour a collector or other distinguished person either in the orchid world or otherwise, the derivations are usually to be found in "Botanical Latin" by William T. Stearn (4th edition, 1992), or reference can be made to the original description in which the author generally explains the reason for the specific epithet.

Figure 1. Acanthephippium lilacinum J.J. Wood & C.L. Chan. - A: plant. - B: lip, spread out showing attachment to column. - C: lip mid-lobe, front view. - D: lip. - E: flower, lateral view with a petal and a lateral sepal removed. - F: petal. - G: dorsal sepal. - H: lateral sepal. - I: anther, front view. J: anther, lateral view. - K: pollinia. - L: ovary, transverse section. All drawn from cult. TOC by Chan Chew Lun.

1. ACANTHEPHIPPIUM LILACINUM J.J. Wood & C.L. Chan

Acanthephippium lilacinum *J.J. Wood et C.L. Chan* **sp. nov.** *A. mantiniano* L. Linden & Cogn. affinis sed sepalis petalisque intus roseis extra pallide ochreis, mento latiore, lobo medio labelli obtuso, et callo bipartito ad apicem eroso satis differt. Typus: Borneo, Sabah, Crocker Range, Sinsuron Rd., cult. Kew, *Lamb* K51 (holotypus K).

A large terrestrial herb. ***Pseudobulb*** 11-15 cm long, 1.5 cm in diameter, fleshy, cylindrical, green, 2- to 3-leaved at apex, covered when young by a tubular leaf sheath. ***Leaves*** up to 50 x 14 cm when mature, elliptic, acute to subacuminate; petiole up to 10 cm long. ***Inflorescence*** to 32 cm, erect from one of lower nodes on pseudobulb, up to 10-flowered; peduncle up to 11 cm long; floral bracts lanceolate, acute, up to 3 cm long, greenish white, flushed lilac-pink. ***Flowers*** fleshy, urceolate, unscented; sepals pale greenish fawn or pale ochre without, delicate lilac-pink within, fading to whitish below, spotted darker lilac-pink; petals lilac-pink, fading to whitish, spotted darker lilac-pink below; lip white with pale lilac-pink marks on mid-lobe; column white, its foot heavily flecked dark purple-lilac. ***Pedicel*** with ***ovary*** 3 cm long. ***Dorsal sepal*** 3 x 1.4 cm, oblong-lanceolate, rounded and recurved slightly at apex. ***Lateral sepals*** 3.2 x 1.8 cm, obliquely oblong-triangular, rounded and recurved slightly at apex, forming with the column-foot a broad, obscurely bilobed saccate mentum 1.5 cm long. ***Petals*** 2.8 x 1.4 cm, subrhombic, rounded at apex. ***Lip*** 1.4 x 1.6 cm, 3-lobed, very fleshy, articulate to apex of column-foot; side lobes erect, oblong, truncate; mid-lobe oblong-subspathulate, recurved; callus of two incurving ridges, erose at apex and enclosing a deep cavity. ***Column*** 1.5 cm long, column-foot geniculate in middle with apical half porrect, free and enclosed within lateral sepals; pollinia 8, pyriform, comprising four large and four small ones. Plate 1A.

HABITAT AND ECOLOGY: In deep leaf litter in montane primary forest; colluvial soils on steep ridges. Alt. 300 to 1300 m. Flowering recorded in February, March and November.

DISTRIBUTION IN BORNEO: SABAH: Crocker Range; Mt. Kinabalu.

GENERAL DISTRIBUTION: Endemic to Borneo (Sabah only).

NOTES: *A. lilacinum* is a most distinctive species readily recognised by its sepals which are subtle greenish fawn or pale ochre on the outside and, like the petals, lilac-coloured within. The lip is white spotted with lilac on the obtuse mid-lobe and the broad mentum apex. It is distinguished from *A. mantinianum* L. Linden & Cogn., from the Philippines, by the broader mentum, obtuse mid-lobe of the lip, a bipartite callus which is erose at the apex and the flower colour. It differs from *A. javanicum* Blume by the much taller inflorescence and lilac flowers.

DERIVATION OF NAME: The genus *Acanthephippium* derives its name from the Greek *akantha*, thorn, and *ephippion*, a saddle, referring to the lip which has toothed crests and resembles a saddle. The specific epithet is derived from the Latin *lilacinus*, lilac, and refers to the lilac-pink sepals and petals.

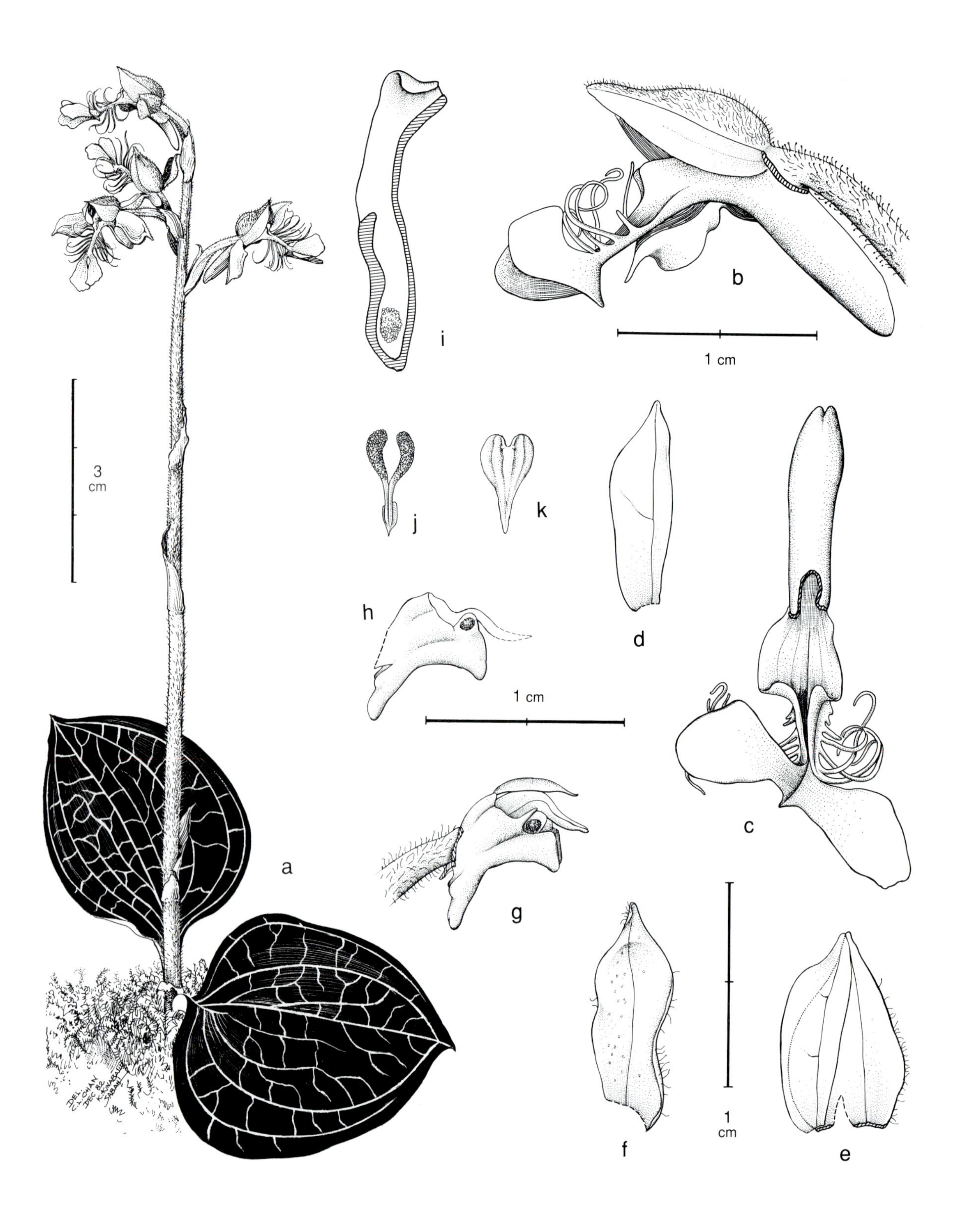

Figure 2. Anoectochilus longicalcaratus J.J. Sm. - A: plant. - B: flower, sideview with a lateral sepal removed. - C: lip. - D: petal. - E: dorsal sepal and petal. - F: lateral sepal. - G: column and ovary, lateral view. - H: column with anther and pollinia removed. - I: spur, longitudinal section showing gland at base. - J: pollinarium. - K: anther. All drawn from *Lamb* AL 9/82 by Chan Chew Lun.

2. ANOECTOCHILUS LONGICALCARATUS J.J.Sm.

Anoectochilus longicalcaratus *J.J.Sm.* in Bull. Jard. Bot. Buitzenzorg, ser. 3, 5: 18 (1922). Types: Sumatra, cult. E. Jacobsen sub/760; *Bünnemeijer* 765, 4036, 5350, 5445; *Groeneveldt* 1762; *Rothert* s.n. (syntypes BO).

A terrestrial ***herb*** up to 20 cm tall with a long, creeping, fleshy rhizome, often forming small colonies. ***Roots*** elongate, villose. ***Stem*** erect, short, up to 5 cm long, glabrous, fleshy, 3- to 5-leaved. ***Leaves*** 2-6 x 1-3.5 cm, oblong-elliptic to ovate, shortly apiculate, glabrous, very dark green, almost black above with gold, sometimes red or pink reticulate veins, pinkish purple beneath. ***Inflorescence*** erect, 10-15 cm tall, laxly 2- to 6-flowered; peduncle glandular-villose, terete, slender, bearing 2 ovate-acuminate bracts, which are sheathing at the base; floral bracts lanceolate, acuminate, 5-16 x 3-7 mm, pubescent, ciliate on lower margins, pink. ***Flowers*** showy, 2.2-2.5 cm wide, 2.5-3 cm long; sepals pink or white, suffused with pale pink; petals white, lip white with base of claw pale lemon yellow; spur base green; apex yellow-green, column yellow; anther-cap pink. ***Pedicel*** with ***ovary*** 1.3-1.5 cm long, glandular-pubescent. ***Dorsal sepal*** 0.9-1.1 x 0.3-0.6 cm, concave, lanceolate, acuminate, glandular pubescent on outside. ***Lateral sepals*** 1-1.3 x 0.3-0.6 cm, spreading, concave, oblong, acute, glandular pubescent on outside. ***Petals*** 0.9-1.2 x 0.3-0.4 cm, adnate to the dorsal sepal forming a hood over the column, obliquely oblong, acute, hyaline. ***Lip*** 1.2-1.4 x 1.2-1.4 cm, deflexed, obscurely tripartite; hypochile side lobes 4 x 1 mm, oblong, blunt; mesochile 3 mm long, narrow, concave, bearing several elongate, slender setae on each side and several short ones; epichile 6-7 x 3-6 mm, transversely pandurate, each side lobule obovate, the spur 1-1.4 cm long, cylindric, straight, bilobed at apex, bearing two rugulose glands internally at apex. ***Column*** 5 mm long, fleshy, with two blunt apical stelidia and two basal fleshy projections running into middle of spur; rostellum slender, porrect, acuminate; pollinia 2, mealy, clavate, attached to a cordate viscidium; anther-cap elongate-cordate, deeply concave, often red in colour. Plate 1B.

HABITAT AND ECOLOGY: In leaf litter in deep shade in montane forest of *Castanopsis*, *Lithocarpus*, etc. Alt. 600 to 1800 m. Flowering observed in March, April and September.

DISTRIBUTION IN BORNEO: SABAH: Mt. Alab; Mt. Kinabalu; Sipitang District.

GENERAL DISTRIBUTION: Sumatra.

NOTES: This species can be confused with *A. setaceus* but the venation of the leaves, the very long spur, and the broader lobes of the lip serve to distinguish it. Most Bornean plants are found on soils derived from sandstone, but some specimens have been found on ultramafic substrates.

DERIVATION OF NAME: The generic name is derived from the Greek *anoektos*, open, and *cheilos*, lip, as the apical part of the lip is bent down exposing it clearly. The specific epithet is derived from the Latin *longus*, long, and *calcaratus*, spurred.

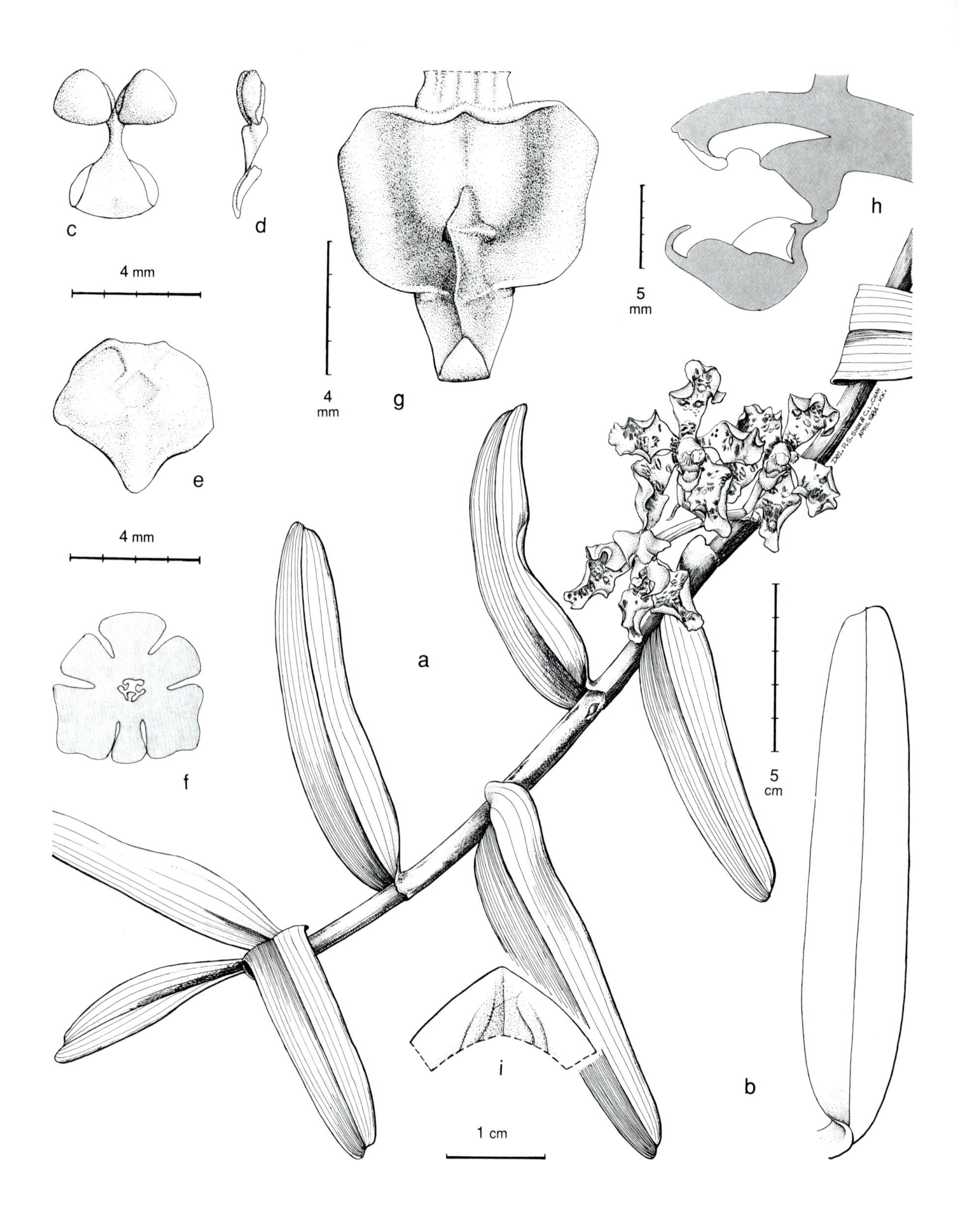

Figure 3. Arachnis breviscapa (J.J. Sm.) J.J. Sm. - A: plant. - B: leaf. - C: pollinarium. - D: pollinarium, lateral view. - E: anther. - F: ovary, transverse section. - G: lip. - H: column and lip, longitudinal section. - I: floral bract. All drawn from *Bacon* s.n. by Shim Phyau Soon and Chan Chew Lun.

3. ARACHNIS BREVISCAPA (J.J.Sm.) J.J.Sm.

Arachnis breviscapa *(J.J.Sm.) J.J.Sm.*, in Natuurk. Tijdschr. Ned.-Indië 72: 74 (1912). Type: Borneo, Sarawak, Quop, *Hewitt* s.n. (holotype BO, isotype K).

Arachnanthe breviscapa J.J.Sm. in Bull. Dép. Agric. Indes Néerl. 22: 48 (1909).

Vandopsis breviscapa (J.J.Sm.) Schltr. in Feddes Repert. 10: 196 (1911).

Large lithophytic or epiphytic ***herb*** with long pendulous stems up to 4.5 m long, sometimes branching, up to 0.9 cm in diameter, leafy along length; internodes 2-4 cm. ***Roots*** elongate, bootlace-like. ***Leaves*** 9-14 x 2-3.5 cm, coriaceous, semi-twisted at the base, linear, unequally roundly bilobed at the apex, articulated to a sheathing leaf-base 3-4 cm long. ***Inflorescences*** 1-6, lateral, unbranched, emerging through the base of the leaf sheath, laxly 2- to 3-flowered, up to 8.5 cm long; peduncle stout, 2-3 cm long; floral bracts 5-7.5 mm long, cucullate, almost tubular, broadly ovate, obtuse. ***Flowers*** fleshy, sweetly scented, about 5.5 cm across, flat; sepals and petals mustard-yellow with orange to rich red-brown markings; lip whitish with orange-brown stripes on the side lobes and a mauve-pink callus. ***Pedicel*** with ***ovary*** 1.5-2.4 cm long. ***Sepals*** 2.8-3.6 x 1.7-2 cm, subsimilar, clawed, obovate, rounded at the apex, with strongly undulate margins. ***Petals*** somewhat smaller than the sepals, falcate. ***Lip*** fleshy, small, lightly hinged to the column, saccate, 3-lobed, c. 1 cm long, 0.8 cm wide; side lobes erect, oblong-semicircular, 0.8 cm high; mid-lobe erect-incurved, linear-tapering, 0.8 cm long; callus a longitudinal bilobed ridge between the side lobes. ***Column*** 0.8-0.9 cm long, fleshy, subterete; pollinia 4, unequal, attached by a broadly triangular stipes to a large viscidium. Plate 1C.

HABITAT AND ECOLOGY: A lithophyte found growing on limestone outcrops from sea level to 600 m in Sarawak. In Sabah widespread and commonly epiphytic on trees by rivers.

DISTRIBUTION IN BORNEO: SABAH: Tambunan & Tenom Districts; Mt. Kinabalu. SARAWAK: Bidi Cave.

GENERAL DISTRIBUTION: Endemic to Borneo.

NOTES: Collections from Tenom and from the Lohan River area in the foothills of Mount Kinabalu, have flowers with yellow to orange markings on the sepals and petals and a sweet almond-like scent.

DERIVATION OF NAME: The generic name *Arachnis* is derived from the Greek *arachne*, spider, in allusion to the shape of the flowers. The specific epithet is derived from the Latin *brevis*, short, and *scapus*, scape.

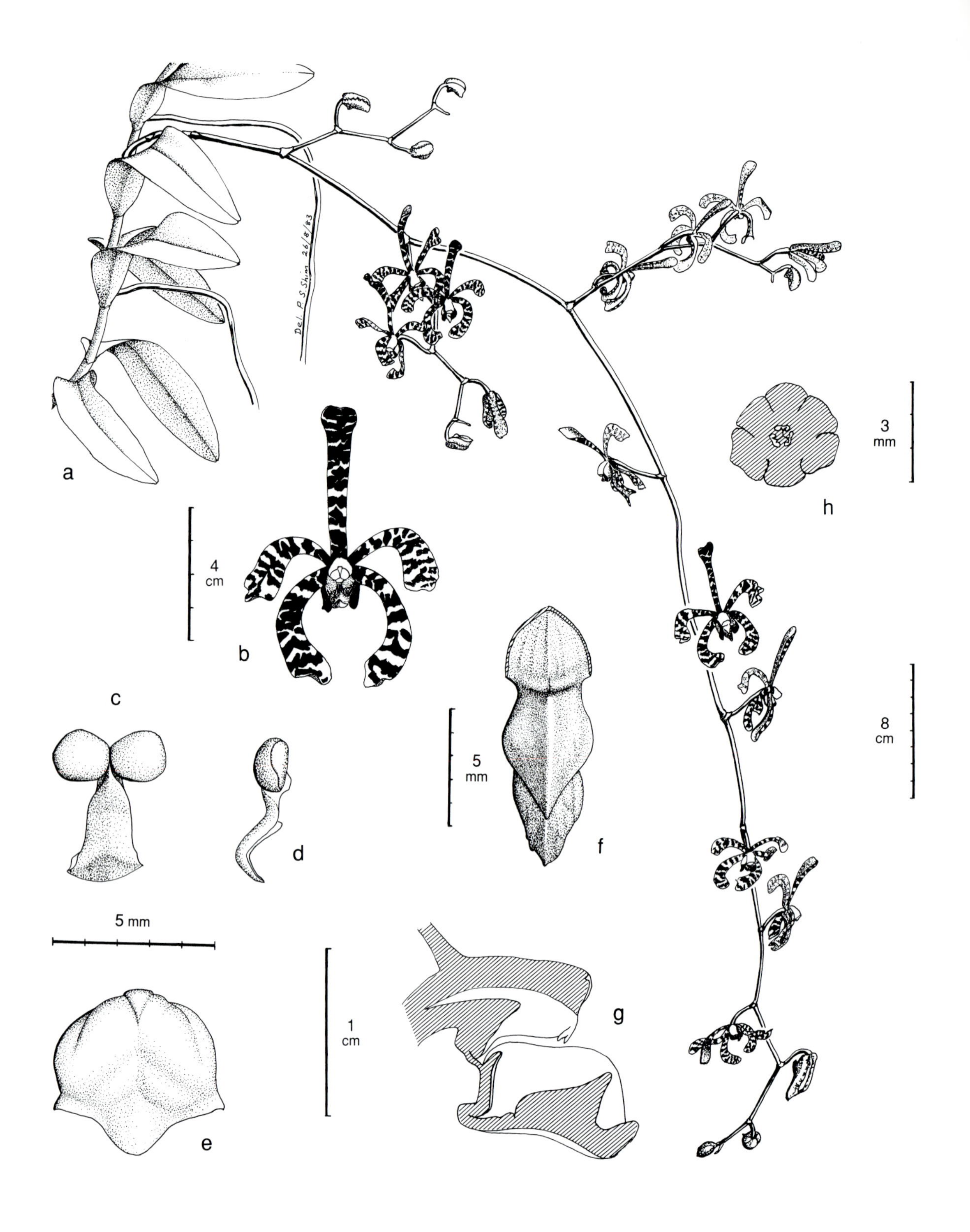

Figure 4. Arachnis flosaeris (L.) Rchb.f. - A: plant. - B: flower, front view. - C: pollinarium. - D: pollinarium, lateral view. - E: anther. - F: lip. - G: column and lip, longitudinal section. - H: ovary, transverse section. All drawn from cult. TOC by Shim Phyau Soon.

4. ARACHNIS FLOSAERIS (L.) Rchb.f.

Arachnis flosaeris *(L.) Rchb.f.* in Bot. Centralbl. 28: 343 (1886). Type: *Kaempfer* s.n.(holotype LINN).

Epidendrum flosaeris L., Sp. Plant.: 952 (1753).

Limodorum flosaeris (L.) Sw. in Nova Acta Regiae Soc. Sci. Upsal. 6: 80 (1799).

Aerides flosaeris (L.) Sw. in Schrader, J. Bot. 11: 233 (1799).

Arachnis moschifera Blume, Bijdr.: 365, t. 26 (1826). Type: Java, *Blume* s.n. (holotype L).

Renanthera arachnites Lindl., Gen. Sp. Orch. Pl.: 217 (1833), nom. nov. for *E. flosaeris*.

Arachnanthe moschifera (Blume) Blume, Rumphia 4: 55, t. 196, 199 (1848).

Renanthera moschifera (Blume) Hassk., Pl. Jav. Rar.: 130 (1848).

R. flosaeris (L.) Rchb.f., Xen. Orch. 1: 88 (1858).

Arachnanthe flosaeris (L.) J.J.Sm., Orch. Java 6: 584 (1905).

Arachnis flosaeris (L.) Schltr. in Feddes Repert. 10: 196 (1911).

A. flosaeris (L.) Rchb.f. var. *gracilis* Holttum in Malayan Orchid Rev. 2: 65 (1935).

Epiphytic ***herb***. ***Stems*** long, stout, scandent, the internodes 1.5-10 cm. ***Leaves*** to 18 x 5 cm, narrowing gradually towards apex, curved and slightly twisted, sheathing, aerial roots piercing sheaths at points above the base of the stem. ***Inflorescence*** simple or branched, ascending and drooping, to about 160 cm long, branches at right angles to the main rachis. ***Flowers*** 8-11 x 6-9 cm; sepals and petals varying from pale yellow-green with irregular maroon bars and spots to deep maroon. ***Dorsal sepal*** to 7 x 1.6 cm. ***Lateral sepals*** 2.8-4 x 0.8-1 cm, incurved, tips sometimes touching, wider than dorsal sepal when flattened. ***Petals*** 3-3.5 cm long, 1 cm wide at apex, ascending from base at 45° on either side of dorsal sepal, spathulate, falcate, strongly arching towards lateral sepals. ***Lip*** free, 3-lobed; side lobes 1 cm wide and as long as column, with several parallel smooth orange lamellae, ends of side lobes maroon, curved outwards; mid-lobe 1.5 cm long, at right angle to basal part of lip bearing side lobes, narrow, fleshy, raised down the middle into a low keel and extending back into an obscure spur, interrupted by a small transverse groove widening forward to 1 cm, tip broadly rounded with an abrupt narrow white point 4 mm long and a callus of nearly equal length below, tip and callus meeting at an acute angle when viewed from the side. ***Column*** 1.5 cm long, thick, white or cream; anther-cap terminal, white or yellow; pollinia 4, in 2 pairs, dorsally compressed. Plate 1D.

HABITAT AND ECOLOGY: Grows in a wide range of habitats from epiphytic on mangroves along the sea shore, scrambling up trees by rivers to epiphytic in lower montane forest. Alt. sea level to 1000 m. Flowering observed in August and September.

DISTRIBUTION IN BORNEO: KALIMANTAN SELATAN: Banjarmasin area. SABAH: Labuan; Mt. Kinabalu; Tenom District; Pun Batu; Tambunan.

GENERAL DISTRIBUTION: India(?), Thailand, Peninsular Malaysia, Sumatra, Java, Bali and the Philippines.

NOTES: The long sprays of maroon-blotched flowers usually have a 'musk'-like scent but varieties found on cliffs along the Crocker Range and the interior valleys in Sabah can vary from having flowers blotched with reddish brown to pinkish red, the latter having a sweet scent fitting the description of var. *gracilis* Holttum. The plants can form very large dense masses on sandstone cliffs. The natural hybrid *A.* x *maingayi* (Hook.f.) Schltr. (*A. hookeriana* x *A. flosaeris* var. *gracilis*) has also been recorded from Sabah.

DERIVATION OF NAME: The specific epithet is derived from the Latin *flos*, flower, and *aeris*, aerial, presumably referring to the flowers that hang down from tall trees.

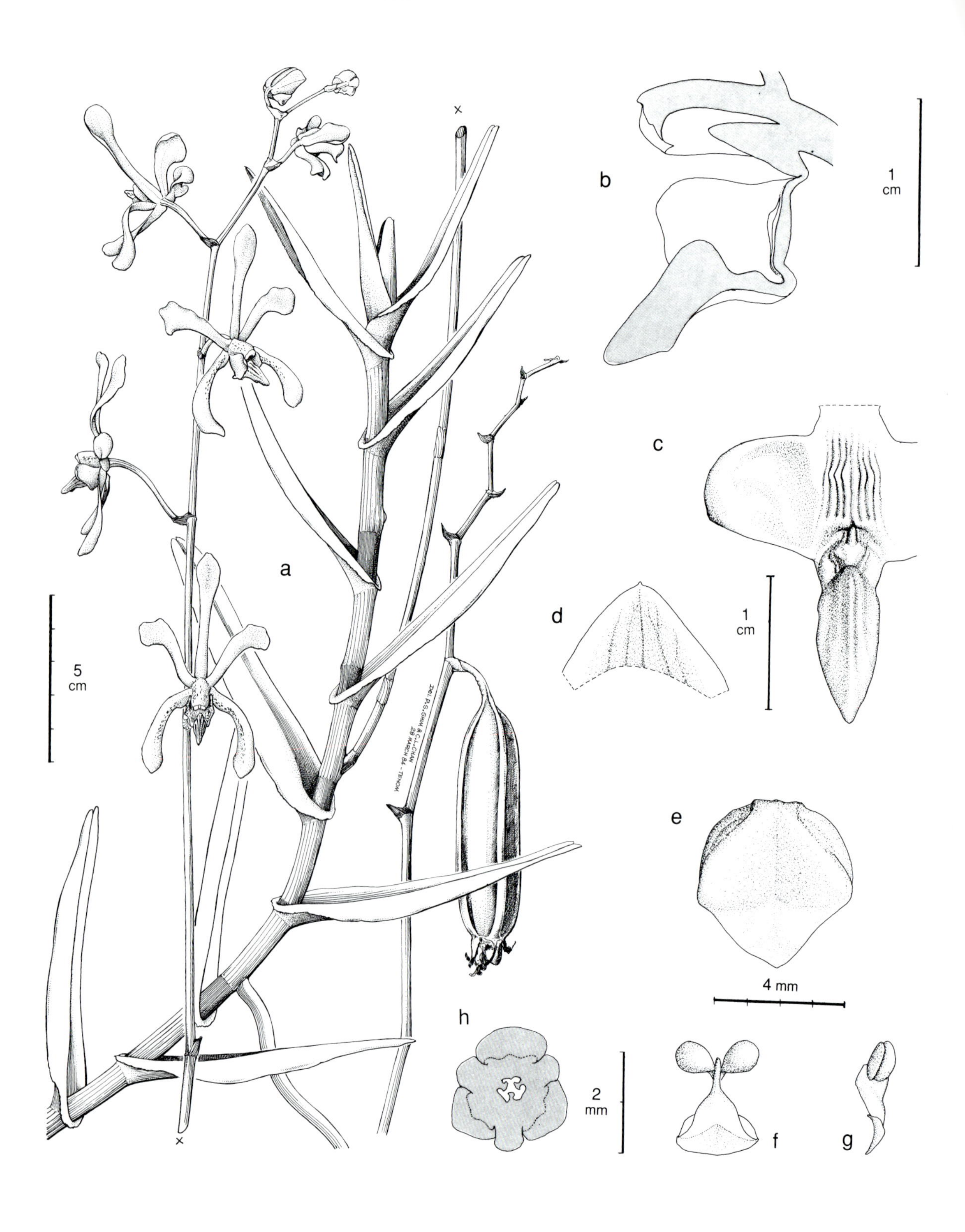

Figure 5. Arachnis hookeriana (Rchb. f.) Rchb. f. - A: plant. - B: column and lip, longitudinal section. - C: lip. - D: floral bract. - E: anther. - F: pollinarium. - G: pollinarium, lateral view. - H: ovary, transverse section. All drawn from cult. TOC by Shim Phyau Soon and Chan Chew Lun.

5. ARACHNIS HOOKERIANA (Rchb.f.) Rchb.f.

Arachnis hookeriana (Rchb.f.) Rchb.f., in Bot. Centralbl. 28: 343 (1886). Type: Borneo, Sabah, Labuan, *Motley* s.n. (holotype K).

Renanthera hookeriana Rchb.f., Xen. Orch. 2: 42, t. 113 (1862).

Arachnanthe alba Ridl. in Trans. Linn. Soc. London, Bot. 3: 369 (1893). Type: Peninsular Malaysia, *Ridley* s.n. (holotype BM).

Arachnis alba (Ridl.) Schltr. in Feddes Repert. 10: 197 (1911).

Epiphytic ***herb.*** ***Stems*** scandent; internodes 1-5 cm. ***Leaves*** to 10 x 2 cm, slightly narrowing towards apex, obliquely ascending, margins stiff, decurved, slightly toothed near base, continuing as a collar around the stem at the top of the sheath. ***Inflorescences*** erect, unbranched, to about 60 cm. ***Flowers*** 5-7 x 5-6 cm; sepals and petals creamy white to yellow, with or without fine purple mottling. ***Dorsal sepal*** to 4-5 cm long, 1 cm wide at apex when flattened, ligulate, obtuse. ***Lateral sepals*** and ***petals*** 2.5-3 cm long, 0.6-0.8 cm wide at apex, nearly equal, spreading, ligulate, obtuse, slightly decurved near tips. ***Lip*** free, 1.5 cm long, 3-lobed; side lobes 0.8 x 0.7-0.8 cm, slightly diverging, tips only slightly recurved, rounded, shorter than column; mid-lobe at right angles to basal part of lip bearing side lobes, 1 x 0.4- 0.5 cm, with a high narrow keel sloping downwards to the tip, and extending back into an obscure spur interrupted by a small transverse groove, entirely purple or with 6 purple stripes; callus very small, a little distance back from the tip, on the lower surface. ***Column*** 1 cm long, fleshy, white; anther-cap yellow; pollinia 4, in two pairs. Plate 1E.

HABITAT AND ECOLOGY: Growing over rocks amongst bushes and up trees on hills on sandy soils near the sea. Alt. sea level. Flowering observed in April and May.

DISTRIBUTION IN BORNEO: BRUNEI: Tutong District. SABAH: Labuan; Papar area. SARAWAK: locality unknown.

GENERAL DISTRIBUTION: Peninsular Malaysia, Singapore, Riau Archipelago and Borneo.

NOTES: Hewitt collected an 'alba' form of *A. hookeriana* in 1907 (*Hewitt* 1158, SAR). The Sabah plants have pure white sepals and petals, often with no purple spots but with yellowish cream tips, and with the normal pink and purple stripes on the lip.

DERIVATION OF NAME: Named after Sir William Jackson Hooker (1785-1865), the first Director of the Royal Botanic Gardens, Kew.

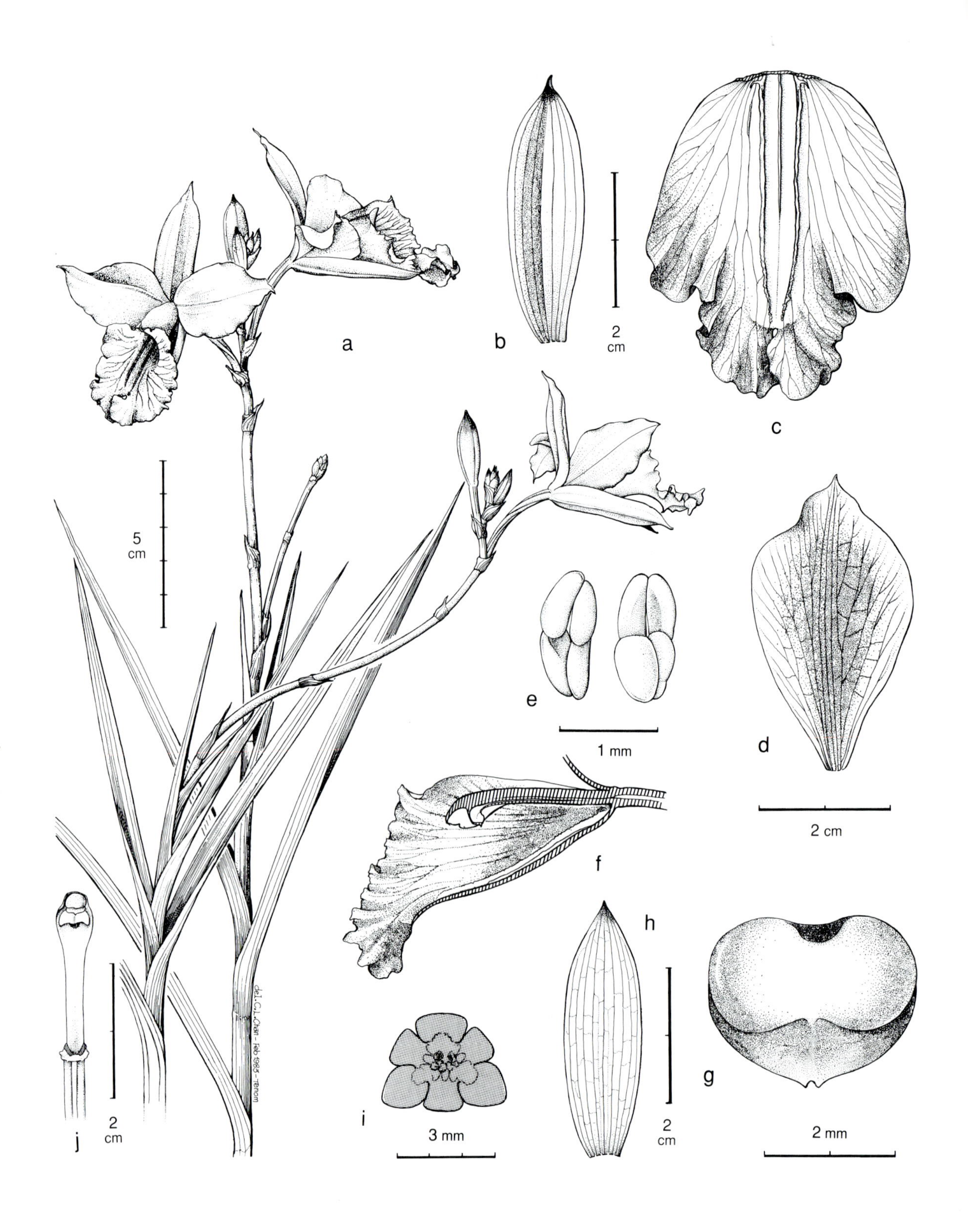

Figure 6. Arundina graminifolia (D. Don) Hochr. - A: plant. - B: lateral sepal. - C: lip, spread out. - D: petal. - E: pollinia. - F: column and lip, longitudinal section. - G: anther. - H: dorsal sepal. - I: ovary, transverse section. - J: column and ovary. All drawn from *Lamb* AL 77/83 by Chan Chew Lun.

6. ARUNDINA GRAMINIFOLIA (D.Don) Hochr.

Arundina graminifolia *(D.Don) Hochr.* in Bull. New York Bot. Gard. 6: 270 (1910). Type: Nepal, *Hamilton* s.n. (holotype BM).

Bletia graminifolia D.Don., Prod. Fl. Nepal.: 29 (1825).

Arundina speciosa Blume, Bijdr.: 401 (1825). Type: Java, *Blume* s.n. (holotype L).

A. chinensis Blume, Bijdr.: 402 (1825). Type: ex China, *Blume* s.n. (holotype L).

A. bambusifolia Lindl., Gen. Sp. Orch. Pl.: 125 (1831). Types: Nepal, Sylhet & Chittagong, *Wallich* s.n. (syntypes K).

Cymbidium bambusifolium (Lindl.) Roxb., Fl. India 3: 460 (1832).

Arundina densa Lindl. in Bot. Reg. 28: t. 38, misc. p. 25 (1832). Type: Singapore, *Loddiges* ex *Cuming* s.n. (holotype K).

Cymbidium meyenii Schauer in Nov. Act. Leop. Carol. Germ. Nat. Cur. 19: Suppl. 1.: 433 (1843). Type: China, Macao, *Meyen* s.n. (holotype not located).

Arundina affinis Griff., Notul. 3: 330 (1851). Type: India, Khasia Hills, *Griffith* s.n. (holotype CAL).

A. meyenii (Schauer) Rchb.f. in Linnaea 25: 227 (1852).

A. philippii Rchb.f. in Linnaea 25: 227 (1852). Types: China, *Philippi* s.n. & *Fortune* s.n. (syntypes W).

A. pulchella Teijsm. & Binn. in Ned. Kruidk. Arch. 3: 400 (1855). Type: China, hort. Bogor (holotype BO).

A. pulchra Miq. in Journ. Bot. Néerl. 1: 90 (1861). Type: China, *Krone* s.n. (holotype U).

A. densiflora Hook.f., Fl. Brit. India 5: 857 (1890) sphalm. for *A. densa* Lindl.

Donacopsis laotica Gagnep. in Bull. Mus. Hist. Nat. Paris, ser. 2, 4: 593 (1932), in part.

Large terrestrial ***herb.*** ***Stems*** erect, 2.5 m long, 1.5 cm. thick. ***Leaves*** 12-30 x 1.6-2.5 cm, distichous, narrowly oblong-lanceolate, grass-like. ***Inflorescence*** terminal, erect, simple or branched, 15-30 cm long; floral bracts broad, coriaceous, subacute, 0.6 cm long. ***Flowers*** large, pink, purple-red, flesh-coloured or white, lip darker coloured than other segments. ***Sepals*** up to 3.8 x 1.1 cm, narrowly elliptic-lanceolate, acute. ***Petals*** up to 3.9 x 2.2 cm, orbicular-obovate, obtuse, margins undulate. ***Lip*** obscurely 3-lobed, 4 x 3.5 cm; side lobes incurved around column, rounded; mid-lobe subquadrate, margins crisped, deeply emarginate at the apex; disc with 3 lamellate nerves. ***Column*** 1.5 cm long, narrowly winged; pollinia 8 in two groups of 4. Plate 1F & 2A.

HABITAT AND ECOLOGY: On rocks and often beside rivers and streams; roadsides; usually in open habitats. Alt. sea level to 1600 m. Flowering throughout the year.

DISTRIBUTION IN BORNEO: KALIMANTAN: widely distributed. SABAH: Tambunan District; Sook Plain; Mt. Kinabalu, etc., widespread. SARAWAK: Bidi, etc., widespread.

GENERAL DISTRIBUTION: Widespread in South and South-East Asia from Sri Lanka, India and China through Indonesia east to Borneo and Sulawesi, Taiwan and the Philippines; Naturalised in the Pacific islands.

NOTES: A white flowered peloric form (*Lamb* SAN 87141) from Mt. Alab has also been observed and on the Tambunan to Ranau Road in Sabah. It also occurs together with populations of the highland form on Mt. Kinabalu. The highland forms all seem to have purplish green leaves and smaller flowers with white sepals. The lowland form has pure green leaves and pinkish sepals and petals and can be seen in abundance along the Kuching to Serian Road in Sarawak.

The following collections, from Sarawak, of a dwarf narrow-leaved variety with smaller but otherwise identical flowers have been seen: *Dransfield & Chai* S.36486 (K, L, SAR, SING); *Hansen* 405 & 473 (C, K), *Beccari* 3839 (FI, K) & *Tinggi* S.29572 (K, L, SAR, etc.). These agree well with the type material of *A. revoluta* Hook.f. from Peninsular Malaysia. This variant is itself more closely allied to the widespread *A. graminifolia* than to the Sri Lankan endemic *A. minor* Lindl. with which Hooker compared it and the necessary transfer is therefore made below:

A. graminifolia *(D.Don) Hochr.* var. **revoluta** *(Hook.f.) A. Lamb* **comb. et stat. nov.**

A. revoluta Hook.f., Fl. Brit. Ind. 5: 858 (1890). Types: Peninsular Malaysia, Perak, *Scortechini* 1504, *Wray* 1979 & *King's Collector* (syntypes K).

Plants to c. 50 cm tall. ***Leaves*** very narrow, 5 mm or less across at base. ***Flowers*** small, with sepals and petals up to c. 2.5 cm long.

DERIVATION OF NAME: The reed-like stems and grassy leaves have resulted in the popular name of "Bamboo orchid". However, it was the stems that suggested to Blume the generic name from the Greek word *arundo*, a reed. The specific epithet is derived from the Latin *gramini*, grass-like, and *folia*, leaf, referring to the grass-like leaves.

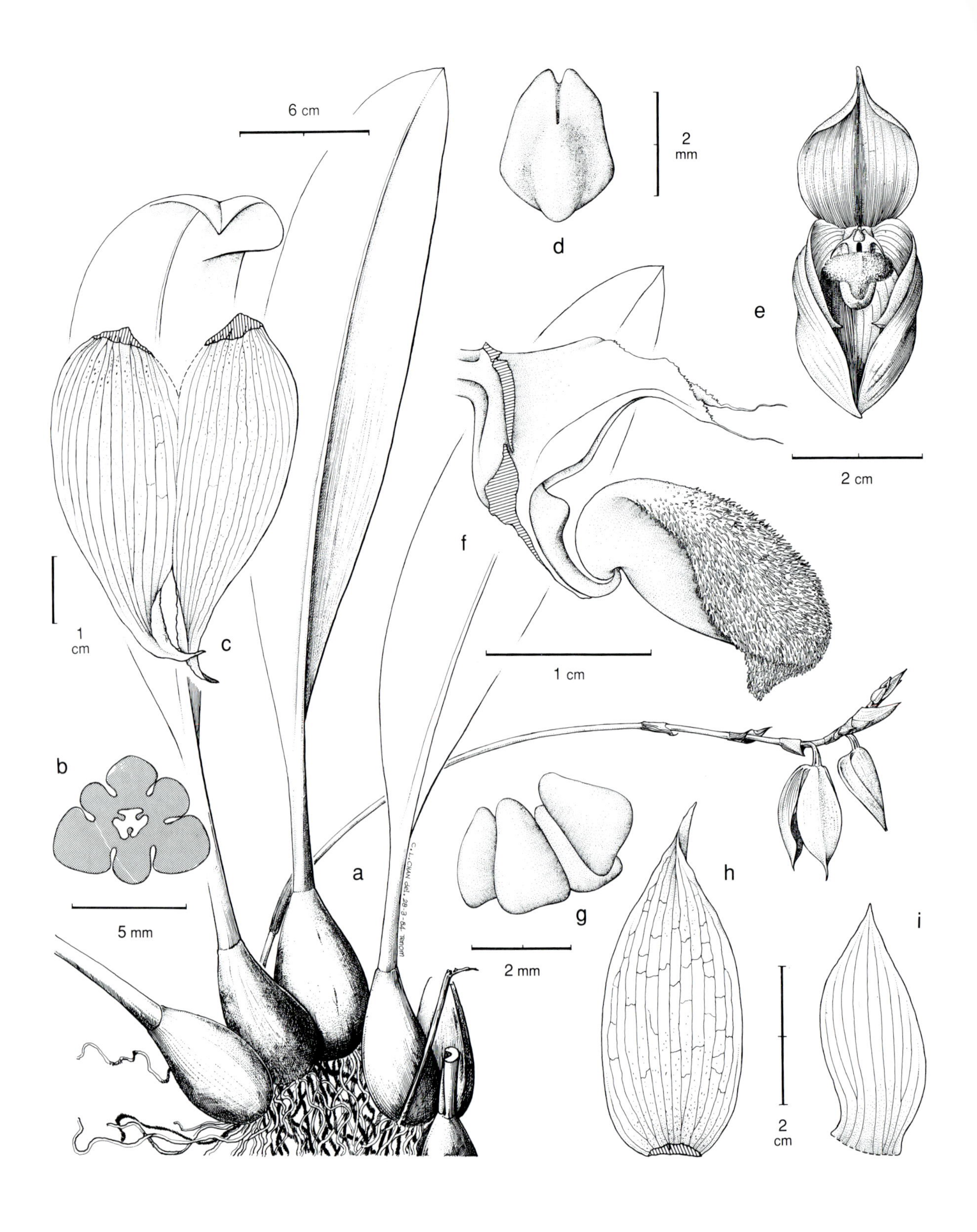

Figure 7. Bulbophyllum mandibulare Rchb. f. - A: plant. - B: ovary, transverse section. - C: lateral sepals. - D: anther. - E: flower. - F: column and lip, lateral view. - G: pollinia. - H: dorsal sepal. - I: petal. All drawn from cult. TOC by Chan Chew

7. BULBOPHYLLUM MANDIBULARE Rchb.f.

Bulbophyllum mandibulare *Rchb.f.* in Gard. Chron. n.s. 17: 366 (1882). Type: Borneo, Sabah, *Burbidge* s.n., cult. *Veitch* (holotype W).

Medium-sized epiphytic ***herb.*** ***Rhizome*** short, stout, up to 0.8 cm in diameter. ***Roots*** wiry, densely clustered on the rhizome. ***Pseudobulbs*** up to 6 cm long, 4.5 cm in diameter, clustered, ovoid, unifoliate, bilaterally compressed and slightly angled. ***Leaf*** up to 40 x 8.2 cm, erect, oblong-oblanceolate, obtuse, petiole up to 9 cm long. ***Inflorescence*** erect, laxly several-flowered; peduncle stout, terete, up to 30 cm long, bearing 2 sheathing sterile bracts; fertile bracts ovate, acute to acuminate, up to 2 cm long, sheathing at base. ***Flowers*** borne in succession, large, fleshy; sepals olive-green or brown with purple veins on the dorsal and a purple-brown base to the synsepal; petals rose-purple; lip yellow, heavily spotted with maroon; column whitish marked with purple at the base. ***Pedicel*** with ***ovary*** 2-3.6 cm long, dark green. ***Dorsal sepal*** lanceolate, acute to acuminate, suberect, 3-5 x 1-2.2 cm. ***Lateral sepals*** connate, free towards the apex and cucullate, 3.3-5 x 0.9-2.3 cm; mentum incurved. ***Petals*** 2.8-3.3 x 0.8-1 cm, falcate, oblong-ovate, acuminate. ***Lip*** 1.5-1.7 x 0.7-0.8 cm, very fleshy, slightly recurved, ovate, obtuse, echinate-papillate; callus of two fleshy ridges on upper surface. ***Column*** 1-1.2 cm long, with two elongate slightly downcurved acuminate apical stelidia; pollinia 4. ***Fruit capsule*** 6 x 2 cm. Plate 2B.

HABITAT AND ECOLOGY: Riverine forest. Alt. 300 to 1000 m.

DISTRIBUTION IN BORNEO: SABAH: Mt. Kinabalu; Tambunan District, etc.

GENERAL DISTRIBUTION: Endemic to Borneo (Sabah only).

NOTES: This species has been placed in section *Intervallatae* Ridl., most species of which occur in New Guinea. They have many flowers per inflorescence with only one to two opening at a time. The plants flower twice a year, each lasting for a long period.

DERIVATION OF NAME: The generic name is derived from the Greek *bulbos*, bulb, and *phyllon*, leaf, referring to the leafy pseudobulbs of most species. The specific epithet is derived from the Latin *mandibulare* as the lip has a shape resembling the lower jaw.

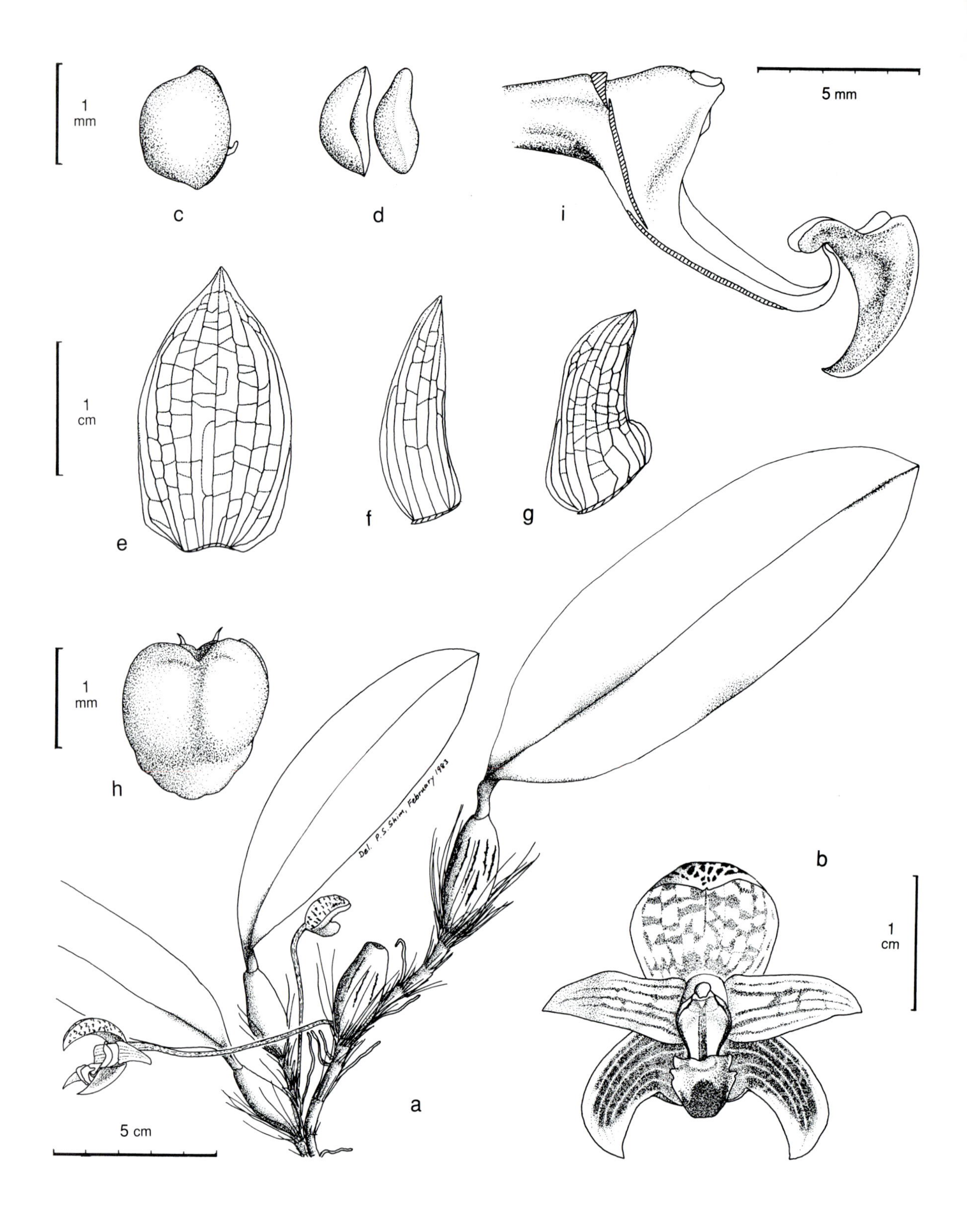

Figure 8. Bulbophyllum microglossum Ridl. - A: plant. - B: flower. C: anther, lateral view. - D: pollinia. - E: dorsal sepal. - F: petal. - G: lateral sepal. - H: anther. - I: column and lip, lateral view. All drawn from *Shim* s.n. by Shim Phyau Soon.

8. BULBOPHYLLUM MICROGLOSSUM Ridl.

Bulbophyllum microglossum *Ridl.* in J. Linn. Soc., Bot. 38: 325 (1908). Type: Peninsular Malaysia, Pahang, Gunung Tahan, *Wray & Robinson* 5397 (holotype BM, isotype K).

Epiphytic ***herb***. ***Rhizome*** long-creeping, 3 mm thick. ***Pseudobulbs*** 6 cm or more apart, to 2 x 1.3 cm, ovoid. ***Leaf*** 8-16 x (2.3-)3.5-4.5(-7) cm, elliptic or oblong-elliptic, obtuse to acute, stiff-textured, petiole 0.5-1.5 cm long. ***Inflorescence*** 1-flowered, peduncle and pedicel 4.5-7.5 cm long. ***Flower*** widely opening; sepals pale yellow-green to ochre-yellow, with red-purple spots and tessellation; petals veined purple; lip purple-pink with an orange patch; column yellow, column-foot speckled pale purple. ***Dorsal sepal*** 1.5-2.5 x 0.9-1.2 cm, ovate, acute. ***Lateral sepals*** 1.5-2.5 x 1.2 cm, triangular-ovate, falcate, acute; mentum 0.8 cm long. ***Petals*** 0.9-1.7 x 0.3-0.6 cm, oblong-elliptic to narrowly-elliptic, acute. ***Lip*** 0.6 x 0.5-0.6 cm, ovate, acute, strongly curved. ***Column*** with only obscure stelidia; pollinia 4 in two groups. Plate 2C.

HABITAT AND ECOLOGY: Lower montane dipterocarp and oak-laurel forest. Alt. 900 to 2200 m. Flowering observed in February, May, June and July.

DISTRIBUTION IN BORNEO: SABAH: Mt. Kinabalu; Mt. Lumaku; Maliau Basin. SARAWAK: Mt. Mulu National Park.

GENERAL DISTRIBUTION: Peninsular Malaysia and Borneo.

NOTES: The status of this species remains uncertain. It appears to be somewhat intermediate between *B. dearei* Rchb.f. and *B. lobbii* Lindl. and may be of hybrid origin. However, the smaller flowers and more creeping habit separate this species from both *B. dearei* and *B. lobbii*. The species has been observed growing high up on branches of dipterocarps at 1000 m in the Crocker Range, Sabah.

DERIVATION OF NAME: From the Greek *micro*, small or little, and *glossa*, tongue, in reference to the small lip.

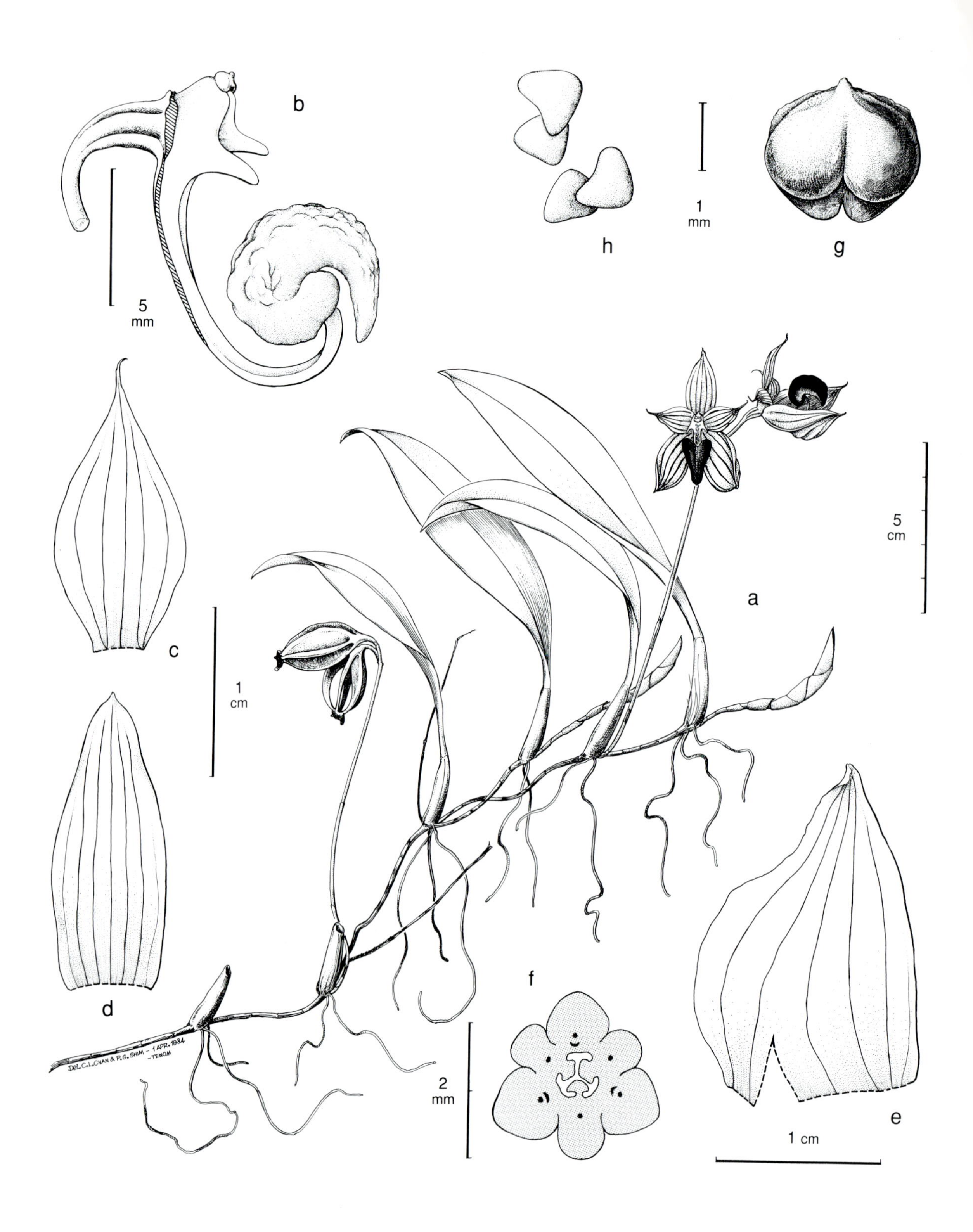

Figure 9. Bulbophyllum nabawanense J.J. Wood & A. Lamb. - A: plant. - B: flower, with sepals and petals removed, lateral view. - C: petal. - D: dorsal sepal. - E: lateral sepal. - F: ovary. - G: anther. - H: pollinia. All from living plant collected from Nabawan. Drawn by Chan Chew Lun and Shim Phyau Soon.

9. BULBOPHYLLUM NABAWANENSE J.J.Wood & A.Lamb

Bulbophyllum nabawanense *J.J. Wood et A. Lamb* **sp. nov.** (Sect. Sestochilo) *B. subumbellato* Ridl. species malayanae et borneensi affine, sed pseudobulbis aliquantum brevioribus proxime dispositi, foliis minoribus, inflorescentia saepe solitaria, floribus eburneis purpureo-nervosis, petalis acuminatis minus caudatis, labello latiore et stelidiis columnae anguste acuminatis distinguenda. Typus: Borneo, Sabah, Keningau District, near Nabawan, 22 September 1983, *Lamb* AL 131/83 (holotypus K).

Terrestrial or occasionally epiphytic ***herb.*** ***Stems*** creeping, up to 30 cm or more long, 2 mm wide; internodes 1.5-5 cm long, enclosed in pale brown sheaths. ***Pseudobulbs*** (1-)1.9- 2.7 x 0.6-0.7 cm, cylindrical, somewhat flattened, 6-ridged, enclosed in a grey sheath which becomes fibrous when mature. ***Leaves*** 7.5-11.1 x 1.5-3.4 cm, narrowly elliptic, acute to acuminate, attenuate to a slender petiole 1-1.5 cm long, thin-textured. ***Inflorescence*** 1- or 2- flowered; peduncle 8-10 cm long, slender, emerging from a basal sheath; sterile bracts 2 or 3, 6-7 mm long, remote, sheathing; floral bracts 1 or 2, 3-6 mm long, ovate, acute. ***Flowers*** opening for up to one week; sepals and petals translucent cream with 5 bright purple nerves; lip purple-violet, paler at the centre; column cream, darker at the apex, flushed purple at the centre. ***Pedicel*** with ***ovary*** 1-1.3 cm long, narrowly clavate. ***Dorsal sepal*** 1.7-1.9 x 0.7-1 cm, oblong-ovate, obtuse and mucronate, nerves prominent and raised on exterior. ***Lateral sepals*** 1.9-2 x 1.3 cm, obliquely ovate-elliptic, acute, concave below, adnate to column-foot, nerves prominent and raised on exterior, mid-nerve carinate at apex. ***Petals*** 1.5-1.7 x 0.8 cm, elliptic, acuminate. ***Lip*** 1 x 0.7-0.8 cm, ovate-cordate, acute, strongly recurved, sulcate, rugulose, glabrous, versatile. ***Column*** 2 mm long, oblong, with pointed, slightly decurved 2 mm long, stelidia, column-foot 1.5 cm long, curved; anther-cap 1.5 x 2 mm, cucullate, papillose; pollinia 4, ovate-triangular, somewhat flattened. Plate 2D.

HABITAT AND ECOLOGY: In *Dacrydium, Eugenia, Garcinia* podsol forest with an understorey of *Rhododendron malayanum*, rattans, etc. and a field layer including *Nepenthes ampullaria, Dendrochilum simplex, Eria* spp., ferns, etc. in shade. Alt. c. 490 m.

DISTRIBUTION IN BORNEO: SABAH: Keningau District.

GENERAL DISTRIBUTION: Endemic to Borneo (Sabah only).

NOTES: The section Sestochilus is primarily montane and includes several familiar species including the widespread and variable *B. lobbii* Lindl., distributed from Assam and Burma to Borneo and the Philippines. A second Bornean representative is *B. uniflorum* (Blume) Hassk., also found in Peninsular Malaysia, Java and Sumatra.

B. nabawanense is a constituent of the dense field layer, including orchids such as *Bromheadia finlaysoniana, Epigeneium speculum* and *Eria* spp., that is a feature of the unusual and very localised acidic podsol forest near Nabawan.

It appears to be related to *B. subumbellatum* Ridl., an epiphyte reported from the lowlands of Johor in Peninsular Malaysia and Sarawak. *B. nabawanense* is a smaller plant with only one or two widely-opening creamy flowers with bold purple nerves and a violet lip. The petals lack the 4 mm long tail of *B. subumbellatum* and the column has two distinct stelidia. *B. subumbellatum* has more distinctly spaced pseudobulbs, larger leaves, and an inflorescence with at least two or three olive-green flowers with lines or red spots on the sepals. The dorsal sepal is hooded, the petals are smaller and tailed, the lip is narrower and the column lacks stelidia.

DERIVATION OF NAME: The specific epithet is derived from Nabawan, the village near where the type locality is situated.

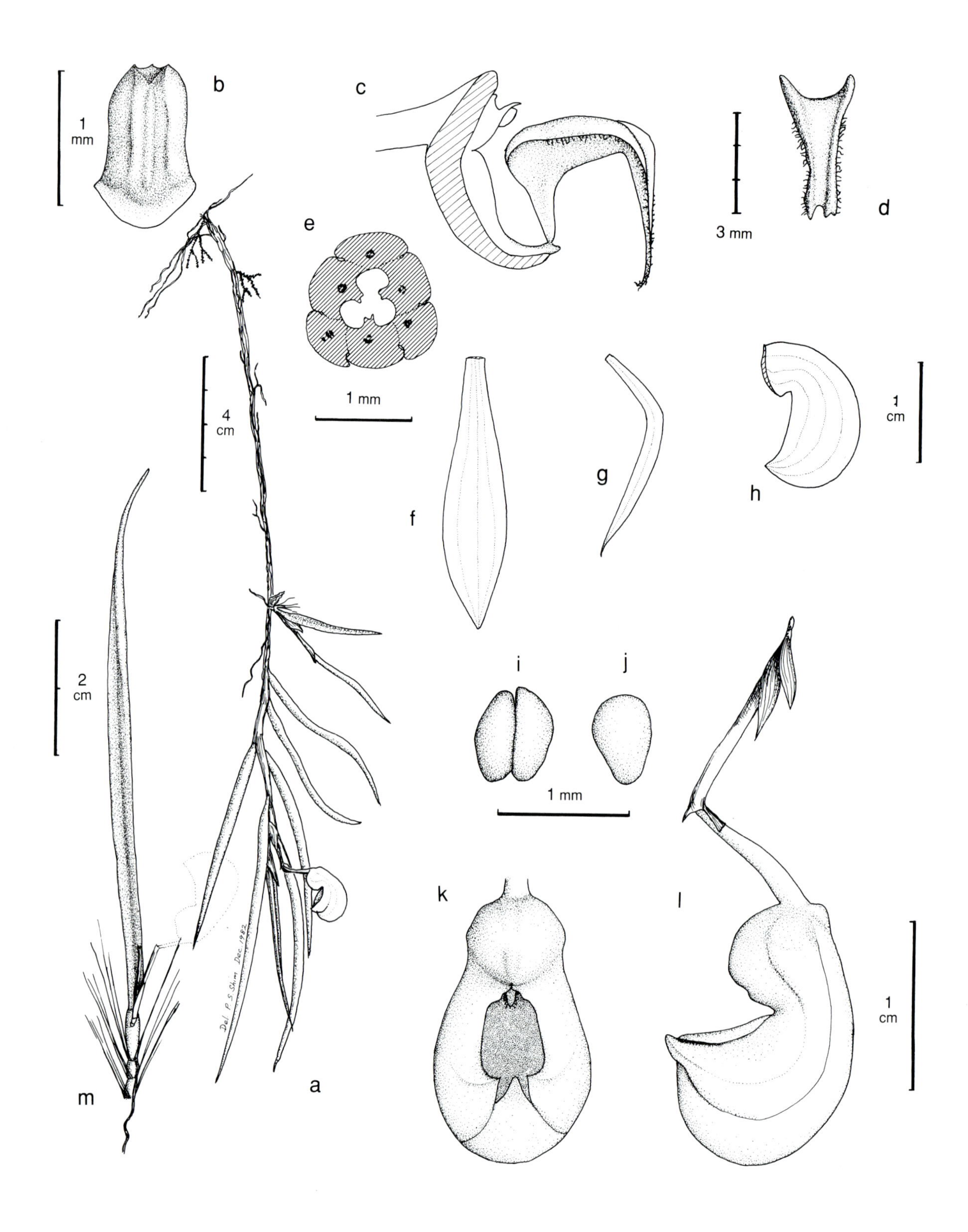

Figure 10. Bulbophyllum pugilanthum J.J. Wood. - A: plant. - B: anther. - C: column and lip, lateral view. - D: lip, view from above. - E: ovary, transverse section. - F: dorsal sepal. - G: petal. - H: lateral sepal. - I: pollinia. - J: pollinium. - K: flower, front view. - L: flower, lateral view. - M: pseudobulb. All drawn from *Shim* s.n. by Shim Phyau Soon.

10. BULBOPHYLLUM PUGILANTHUM J.J.Wood

Bulbophyllum pugilanthum *J.J.Wood* **sp. nov.** in sectio Aphanobulbo Schltr. ponenda. Herba epiphytica pendula; radices e basi pseudobulborum editae, secundum totum caulem percurrentes; pseudobulbi cylindrici, monophylli, in vaginis compluribus scariosis inclusi; folia semiteretia, sulcata, in sectione transversa depresse v-formia, acuta, carnosa; inflorescentiae laterales, a latere pseudobulbi sub apicem ipsum gestae; flores resupinati, textura tenera, formae caestus subsimiles; sepala et petala conniventia, galeam curvatum efficientia; petala falcata; labium trianguli-ovatum, acuminatum, integrum, ecallosum, medio deflexum, margine ciliato-pilosum, versatile; columna brevis, pes longioris, pollinia 2, nuda. Typus: Borneo, Sabah, Mt. Kinabalu, Park Headquarters, May 1983, *Lamb* AL 56/83 (holotypus K).

Pendulous epiphytic ***herb*** to 24 cm long. ***Roots*** long and slender, several produced from base of pseudobulbs and running along the length of and usually covering the stem and pseudobulbs below. ***Pseudobulbs*** 5-7 x 1-2 mm, cylindrical, 1-leaved, borne 5 mm apart on stem, enclosed by several pale brown scarious, elliptic, acute sheaths which become fibrous with age, thereby exposing the older pseudobulbs. ***Leaves*** 6-8.1 x 0.3-0.4 cm, semi-terete, sulcate, appearing shallowly V-shaped in cross-section, acute, fleshy, dark green, straight or gently curved. ***Inflorescence*** with up to 8 flowers appearing one at a time in succession, pseudapical, borne on the side of and partly adnate to the pseudobulb just below the apex and emerging from the enveloping scarious sheaths; sterile bracts 3, up to 5 mm long, elliptic, acute, scarious, overlapping and enclosing all but the uppermost portion of the pedicel; floral bract to 6 mm long, solitary, elliptic, acuminate. ***Flowers*** 1-1.5 cm long, resupinate, thin-textured, pale translucent orange-yellow with a pale yellow lip, or greenish-white, shaped rather like a boxing-glove and reminiscent of some species of *Pterostylis* or *Corybas*. ***Pedicel*** distinct from ovary, narrow, to 1 cm long. ***Ovary*** 5-7 mm long, narrowly clavate, curved. ***Sepals*** and ***petals*** connivent, forming a curved hood. ***Dorsal sepal*** 2.5-2.7 x 0.6-0.7 cm, narrowly elliptic to oblanceolate, acute, curved, cucullate, 3-nerved. ***Lateral sepals*** 1.5-2 x 0.7 cm, broadly falcate, elliptic, acute, adnate to column-foot, 3-nerved, forming a rounded, incurved mentum 4-5 mm long. ***Petals*** 2.1-2.3 x 0.2 cm, linear-ligulate, falcate, acute, 1-nerved. ***Lip*** 8 mm long, 2.5-3 mm broad at base, triangular-ovate, cuneate at base, acute to acuminate, entire, ecallose, geniculately deflexed at middle, versatile, 3-nerved, margin ciliate- hairy, except towards base, hairs longest towards apex, sides erect and becoming reflexed along margin except towards apex. ***Column*** 1 mm long, with 2 small subulate apical stelidia, column-foot curved, 4-5 mm long; anther-cap cucullate, 1 mm long; pollinia 2, ovoid, waxy, naked, viscidia absent. Plate 2E.

HABITAT AND ECOLOGY: Lower montane, upper montane and riverine forest, sometimes on ultramafic substrate, growing low down on trunks of larger trees or on small horizontal branches and mossy saplings two to three metres above ground level, in moderate to deep shade. Alt. 1260 to 2400 m. Flowering observed in May and October.

DISTRIBUTION IN BORNEO: SABAH: Mt. Kinabalu; Sipitang District.

GENERAL DISTRIBUTION: Endemic to Borneo (Sabah only).

NOTES: This interesting plant, allied to *B. Ceratostylis* J.J.Sm. from Sumatra and Borneo, was first believed to belong to a new genus related to *Bulbophyllum.* Although it has petals and a lip typical of *Bulbophyllum,* other characters such as inflorescence type, connivent sepals and two pollinia suggest separate status. However, all of these characters have been found to occur in *Bulbophyllum* sensu lato (Vermeulen, pers. comm.).

Contrary to much of what is stated in the literature, many species of *Bulbophyllum* have two instead of four pollinia. Sections widespread in Malaysia and Indonesia with species having only two pollinia, include *Aphanobulbon* Schltr. (to which *B. pugilanthum* belongs), *Monilibulbus* J.J.Sm., *Micromonanthe* Schltr. and the mainly New Guinean *Polyblepharon* Schltr. Connivent sepals and petals are also found, for example, in the Sumatran *B. subclausum* J.J.Sm., recently recorded from Mt. Kinabalu. Semi-terete or terete leaves occur in several very different sections, including *B. teres* Ridl. in section *Aphanobulbon.*

The inflorescence of both *B. ceratostylis* and *P. pugilanthum* is pseudapical, ie. arising from just below the apex of the pseudobulb. It is actually sunken into the pseudobulb and partly adnate to it. This type of inflorescence can also be found, for example, in *B. monilibulbum* Carr from Peninsular Malaysia and *B. mutabile* (Blume) Lindl. from Malaysia and Indonesia.

It is interesting that *B. ceratostylis* and *B. pugilanthum* should have been collected from Kinabalu Park in the north of Sabah and from Sipitang District in the southwest close to the Sarawak border. Both are no doubt more widespread than the few collections to hand would suggest.

DERIVATION OF NAME: The specific epithet is derived from the Latin *pugillus,* a handful, and the Greek *anthe,* flower, in reference to the fist-shaped or boxing-glove-shaped flower.

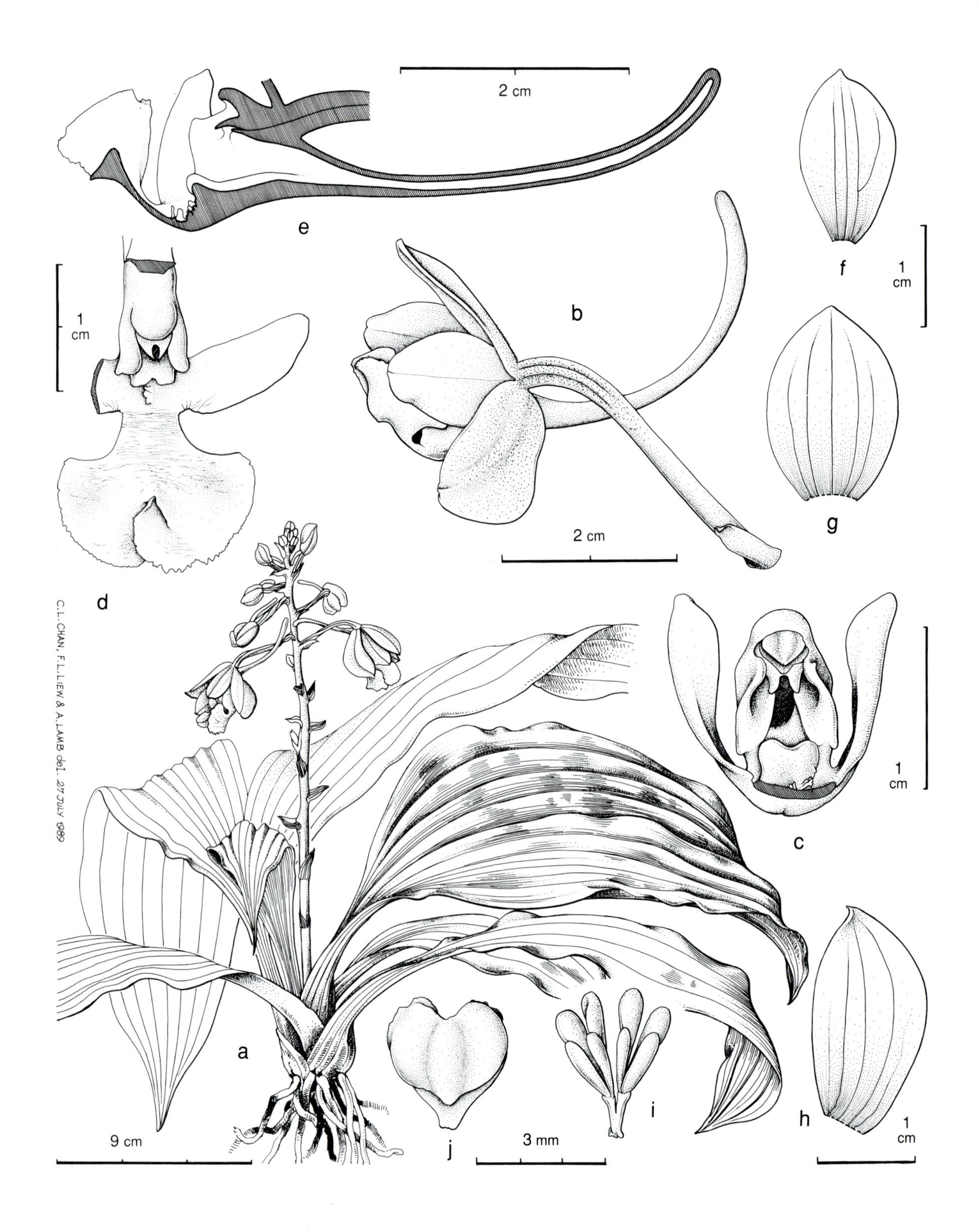

Figure 11. Calanthe crenulata J.J. Sm. - A: plant. - B: flower, lateral view. - C: column and lip, with mid-lobe removed, front view. - D: column and lip, spread out. - E: column and lip, longitudinal section. - F: petal. - G: dorsal sepal. - H: lateral sepal. - I: pollinia. - J: anther. All drawn from cult. TOC by Chan Chew Lun, Anthony Lamb and Liew Fui Ling.

11. CALANTHE CRENULATA J.J. Sm.

Calanthe crenulata *J.J. Sm.* in Bot. Jahrb. Syst. 48: 97 (1912). Type: Borneo, Kalimantan, between Muara Uja and Kundim Baru, *Winkler* 2676 (holotype L, isotypes BO, P).

Terrestrial ***herb*** with one or two growths on a short rhizome. ***Roots*** elongate, spreading, 1.5 mm in diameter. ***Stems*** up to 2.5 cm long, hidden by the leaf bases, pseudobulbous, ovoid, 3- to 4-noded, up to 2.5 cm long. ***Leaves*** 25-36 x 4.5-9.5 cm, strongly plicate, glabrous, suberect to arcuate, elliptic-obovate, acute or subacute; petiole slender but slightly dilated at the base, up to 12 cm long. ***Inflorescence*** terminal, erect, subdensely 6- to 20-flowered, up to 50 cm long; peduncle up to 35 cm long, shortly pubescent; rachis shortly pubescent. ***Flowers*** showy; sepals and petals white; lip white with pale yellow side lobes . ***Pedicel*** with ***ovary*** 3-3.6 cm long, pubescent. ***Dorsal sepal*** 1.5-2.3 x 1-1.3 cm, erect, elliptic, obtuse or very shortly acuminate. ***Lateral sepals*** 1.9-2.4 x 1.2-1.5 cm, spreading, elliptic-ovate, obtuse. ***Lip*** adnate to the column at base, 3-lobed in basal part, 1.4-1.6 x 1.9- 2.4 cm, papillose; side lobes 0.8-1.2 x 0.5 cm, erect, linear-elliptic, rounded at the apex; mid- lobe 1-1.5 x 1-1.5 cm, shortly clawed, transversely reniform-circular, emarginate, with an erosulate margin; callus basal, of three short verrucose broken lines; spur 2.5-3.6 cm long, pointing backwards, sinuous, slightly pubescent, white. ***Column*** 0.7-0.8 cm long, fleshy, shortly clavate; pollinia 8, clavate, mealy. Plate 2F.

HABITAT & ECOLOGY: In Kalimantan it is recorded from dense primary rain forest on deep alluvial clay soil. On Mt Kinabalu in Sabah it has been collected from lower montane forest on ultramafic substrate. It has been found on sandstone on ridges elsewhere on the Crocker Range. Alt. 80 to 1200 m.

DISTRIBUTION IN BORNEO: KALIMANTAN SELATAN: Djaro Dam area. SABAH: Mt Kinabalu; Crocker Range.

GENERAL DISTRIBUTION: Endemic to Borneo.

NOTES: Plants from the Crocker Range in Sabah flower two to three times a year in cultivation. The side lobes of the lip can vary from white to deep yellow.

DERIVATION OF NAME: The generic epithet is derived from the Greek *kalos*, beautiful, and *anthe*, flower, in praise of the beautiful flowers of many species. The specific epithet is derived from the Latin *crenulatus*, with a small notch, referring to the shape of the lip.

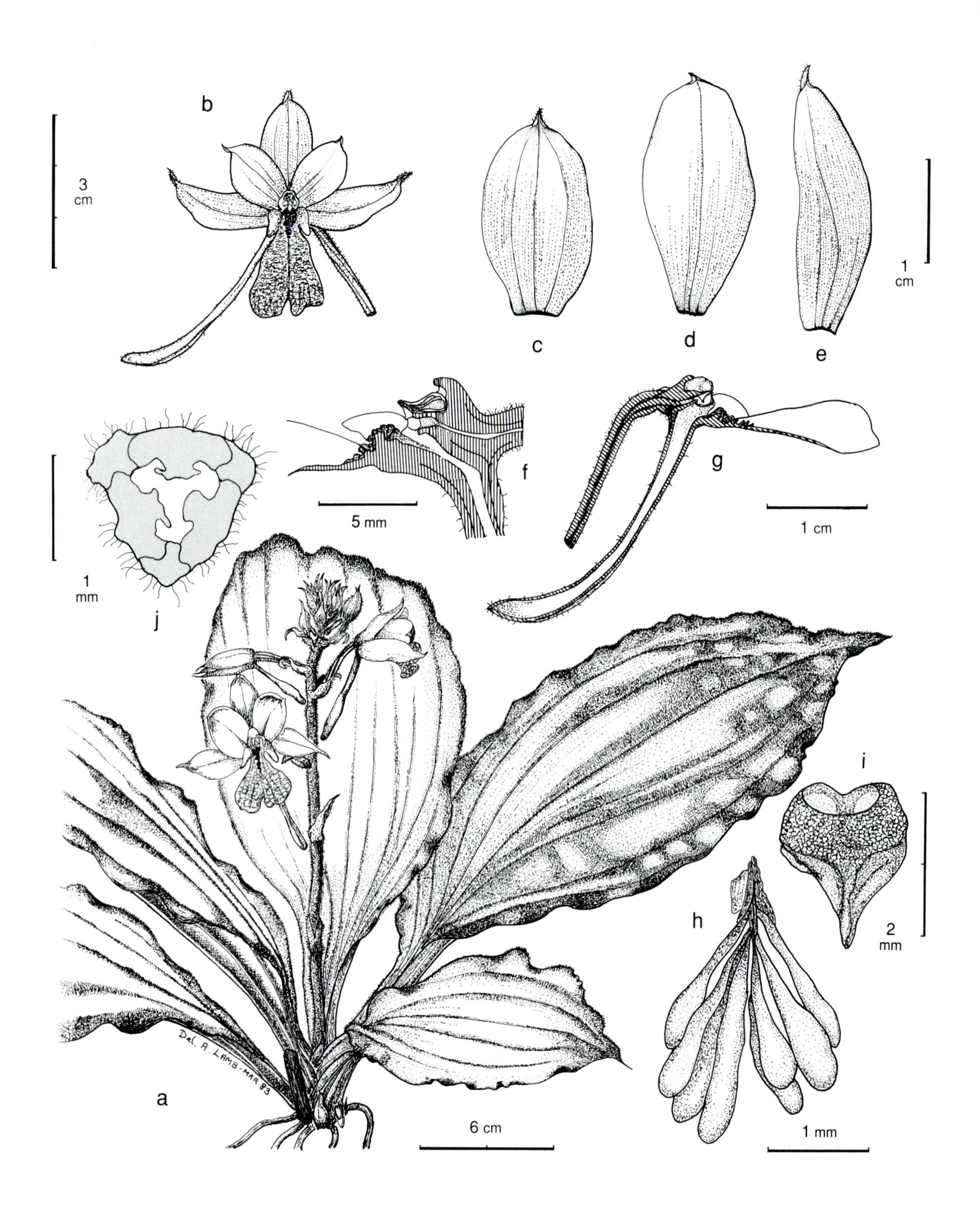

Figure 12. Calanthe sylvatica (Thouars) Lindl. - A: plant. - B: flower. - C: dorsal sepal. - D: petal. - E: lateral sepal. - F: section through column and lip. - G: column, pedicel and lip, longitudinal section. - H: pollinia. - I: anther. - J: ovary, transverse section. All drawn from *Lamb* AL 85/83 by Anthony Lamb.

12. CALANTHE SYLVATICA (Thouars) Lindl.

Calanthe sylvatica *(Thouars) Lindl.,* Gen. Sp. Orch. Pl.: 250 (1833). Type: Mascarene Islands, *du Petit Thouars* s.n. (holotype P).

Centrosis sylvatica Thouars, Orch. Iles. Aust. Afr.: t. 35, 36 (1822).

Amblyglottis emarginata Blume, Bijdr.: 370 (1825). Type: Java, *Blume* s.n.(holotype L).

Bletia masuca D. Don, Prodr. Fl. Nepal.: 30 (1825). Type: Nepal, *Hamilton* s.n.(holotype not located).

Calanthe masuca (D. Don) Lindl., Gen. Sp. Orch. Pl.: 249 (1833).

Calanthe emarginata (Blume) Lindl., loc. cit.: 249 (1833).

For further synonomy see Cribb in Flora of Tropical East Africa, Orchidaceae, part 2: 282 (1984).

Terrestrial ***herb. Pseudobulbs*** 2-5 x 1.2 cm, cylindrical to conical, concealed by sheathing leaf-bases. ***Leaves*** 3-5(-8), 13-25 x 5-13 cm, lanceolate to elliptic-oblong, acute or acuminate, softly pubescent above and below, petiole 4-10 cm long, sulcate. ***Inflorescence*** 2-12 cm long, erect, few- to many-flowered; peduncle, rachis and bracts softly pubescent; floral bracts 1-2 cm long, lanceolate, acute or acuminate, suberect to recurved. ***Flowers*** large, showy; sepals and petals pale to dark mauve or whitish pink; lip usually paler; callus purple-red or orange. ***Pedicel*** with ***ovary*** 2.3-3 cm long, pubescent. ***Dorsal sepal*** 2-2.4 x 1 cm, narrowly elliptic or oblong-elliptic, acute or acuminate, hirsute. ***Lateral sepals*** 2.6 x 1cm, similar but oblique. ***Petals*** 2-2.2 x 1.1 cm, elliptic to oblanceolate, obtuse and mucronate. ***Lip*** 3-lobed, adnate at base to column; side lobes 4.5 x 2.5 mm, oblong, auriculate; mid-lobe 2.1 x 2.2 cm, flabellate to obovate, emarginate, with or without a minute apical tooth; callus 3-5 mm long, bilobed at base, verrucose; spur 2.8-4.5cm long, narrowly clavate, obtuse, shortly pubescent, incurved. ***Column*** 5 mm long; pollinia 8. ***Fruit*** to 4 cm long. Plate 3A.

HABITAT AND ECOLOGY: Primary mixed dipterocarp forest on sandstone soils, also recorded from limestone and basalt, in deep shade. Alt. 400 to 1500 m.

DISTRIBUTION IN BORNEO: KALIMANTAN: Mt Medadem. SABAH: Tambunan District; Mt Lumaku area. SARAWAK: Mt Batu Lawi; Mt Murud; Baram District, etc.

GENERAL DISTRIBUTION: Widespread in Africa, Madagascar, the Mascarene Islands and tropical Asia from India and Sri Lanka east to Borneo.

NOTES: *Calanthe sylvatica* is a relatively recent addition to the Bornean orchid flora. In mainland Asia it is widely known by the later synonym *C. masuca* (D. Don) Lindl. It was first collected by Dr. Andrew Bacon in the 1970s from near Iburu in Sipitang District. Living material was sent to Edinburgh Botanic Garden in Scotland where it was identified as *C. masuca*. *C. sylvatica* probably has the widest distribution of any species represented in Borneo. Variation is extreme throughout its extensive range whether it be stature, flower size, the degree of development of the lip side lobes or the curvature of the spur. Three trends in variation can be recognised in East Africa alone. Plants from Borneo generally have a lip with small, auriculate side lobes and correspond with Javan and Sumatran populations previously assigned to *C. emarginata* (Blume) Lindl. The lack of any obvious discontinuities in this pattern of variation does not allow more than a single species to be recognised. Populations in Sabah and Sarawak have pale to dark mauve or purple to white or pink sepals and petals. The lip is usually paler with a red, brick red or orange callus. This was one of the first orchids used in artificial hybridisation.

DERIVATION OF NAME: The specific epithet is derived from the Latin *sylvaticus*, from the forest, or dwelling in the forest.

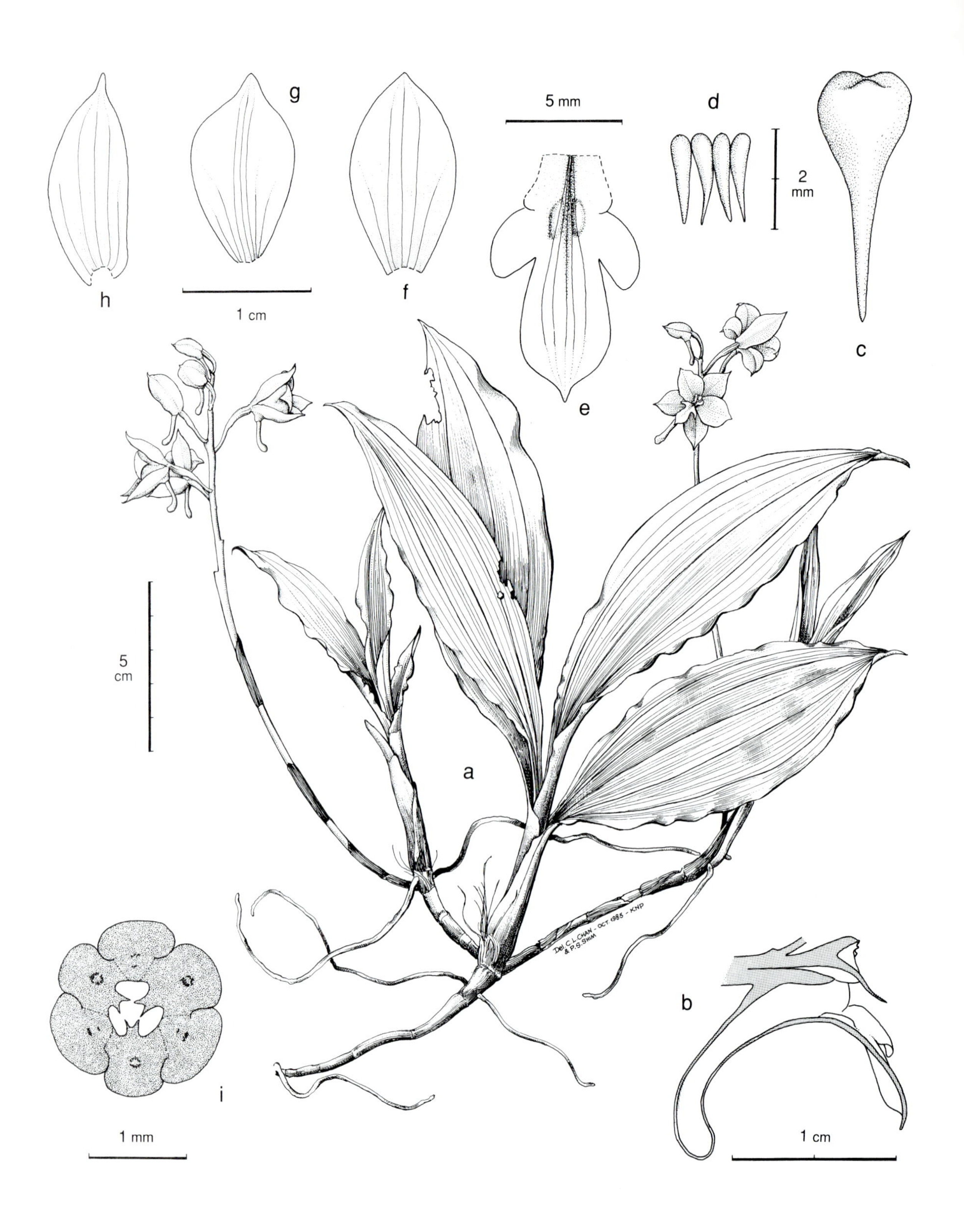

Figure 13. Calanthe truncicola Schltr. - A: plant. - B: column and lip, longitudinal section. - C: anther. - D: pollinia. - E: lip. - F: dorsal sepal. - G: petal. - H: lateral sepal. - I: ovary, transverse section. All drawn from *Chan, Kiat Tan & Phillipps* s.n. by Chan Chew Lun and Shim Phyau Soon.

13. CALANTHE TRUNCICOLA Schltr.

Calanthe truncicola *Schltr.* in Bot. Jahrb. Syst. 45, Beibl. 104: 26 (1911). Type: Sumatra, Bukit Djarat, *Schlechter* 16001 (holotype B destroyed, isotype K).

Terrestrial or occasionally epiphytic ***herb.*** ***Rhizome*** to 40 cm. or more long, narrow and creeping, rooting and producing new growths at the nodes, internodes 7-8 cm long, bases of old leaf sheaths persistant. ***Roots*** villose. ***Pseudobulbs*** 1-2 cm long, enclosed by sheathing leaf bases. ***Leaves*** 3 or 4 per node, glabrous, (7-)14-18 x (2-)3-4.8 cm, narrowly elliptic, acuminate; petiole 2-3 cm long; sheathing base 3-4 cm long. ***Inflorescence*** lax, oblong to ovate, racemose, up to 10-flowered; peduncle 18-25 cm long, with 3-5 sterile bracts, mostly at the base; rachis 4-6 cm long; floral bracts 4 cm long, ovate-elliptic, acuminate, deciduous. ***Flowers*** deep orange. ***Pedicel*** with ***ovary*** 1.7 cm long, narrowly clavate. ***Dorsal sepal*** 1.3-1.4 x 0.7 cm, oblong to oblong-elliptic, subacute, concave. ***Lateral sepals*** 1.3-1.4 x 0.6 cm, oblong, subacute. ***Petals*** 1.3 x 0.8 cm, obliquely elliptic, obtuse. ***Lip*** 0.8 cm long, 0.75 cm wide across side lobes; side lobes 0.25 x 0.2-0.3 cm, auriculate, obtuse; mid-lobe 0.4-0.5 x 0.45-0.5 cm, oblong, obtuse, sometimes slightly retuse; callus reduced to 2 small basal swellings; spur 1.2-1.3 cm long, cylindrical, gently curved, apex swollen and narrowly clavate. ***Column*** 0.5 cm long; pollinia 8. Plate 3B.

HABITAT AND ECOLOGY: Lower montane forest, amongst bamboo and scrub. Alt. 1400 to 1500 m. Recorded from ultramafic substrate at Marai Parai and Mamut on Mt. Kinabalu. Populations can also be found in abundance along the Silau Silau Trail near Park Headquarters. Here they grow in deep shade on mixed conglomerate of granitic and sandstone soils in mixed lower montane forest.

DISTRIBUTION IN BORNEO: SABAH: Mt. Kinabalu; Lahad Datu District, Mt. Nicola.

GENERAL DISTRIBUTION: Sumatra and Borneo.

NOTES: *Calanthe truncicola* was originally described from collections made by Rudolf Schlechter on Bukit Djarat, Sumatra, in 1907. The recent collection from Mt. Kinabalu represents a new record for Borneo. The Kinabalu material differs from the Sumatran plant in a few minor details, particularly the fewer flowered inflorescence and slightly larger, less retuse lip mid-lobe. Differences of this type commonly occur between different island populations and are mostly of little significance. The natural variation within Sumatran populations is unknown.

C. angustifolia (Blume) Lindl., a widespread species from West Malaysia to the Philippines, is closely related, but has longer, usually narrower leaves and white flowers with a deep yellow callus and emarginate mid-lobe to the lip. Plants with deep cream or yellowish flowers with an orange-yellow lip, named var. *flava* Ridl., could be confused with *C. truncicola*.

DERIVATION OF NAME: The specific epithet is derived from the Latin *truncatus*, cut off at the end, referring to the side lobes of the lip.

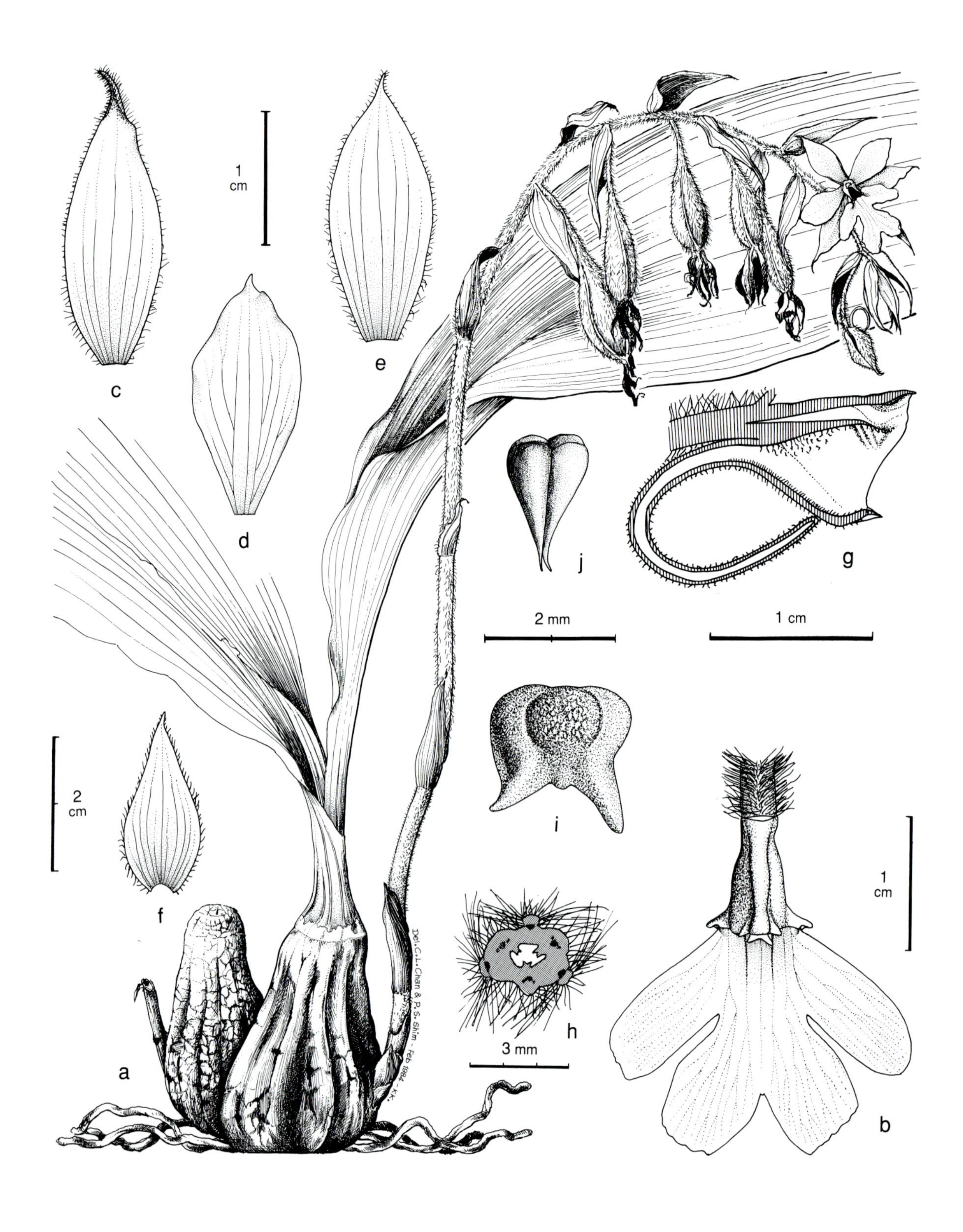

Figure 14. Calanthe vestita Lindl. - A: plant. - B: column and lip. - C: lateral sepal. - D: petal. - E: dorsal sepal. - F: floral bract. - G: column and spur, longitudinal section. - H: ovary, transverse section. - I: anther. - J: pollinia. All from cult. *Bacon*. (*Bacon* in *Lamb* AL 169/84) Drawn by Chan Chew Lun and Shim Phyau Soon.

14. CALANTHE VESTITA Lindl.

Calanthe vestita *Lindl.*, Gen. Sp. Orch. Pl.: 250 (1833). Type: Burma, Tavoy, *Wallich* s.n. (holotype K).

Cytheris griffithii Wight, Ic. 5: t. 1751 (1852). Type: Burma, Mergui, *Griffith* s.n. (holotype K).

Preptanthe vestita (Lindl.) Rchb.f. in Fl. des Serres 8: 245 (1852).

Amblyglottis pilosa de Vriese ex Lindl., Fol. Orch., Cal.: 11 (1854). Type: Sumatra, *de Vriese* s.n.(holotype ?L).

Calanthe pilosa (de Vriese ex Lindl.) Miq., Fl. Bat. 3: 711 (1859).

C. regnieri Rchb.f. in Gard. Chron. n.s. 19: 274 (1883). Type: Cochin China, cult. *Veitch*, *Regnier* s.n.(holotype W).

C. turneri Rchb.f. in Gard. Chron. n.s. 19: 274 (1883). Type: Java, cult. *Veitch* (holotype W).

C. grandiflora Rolfe in Orchid Rev. 9: 141 (1901). Type: Borneo, cult. *Lawrence* (holotype K).

C. padangensis Schltr. ex Mansf. in Feddes Repert. Beih. 74: t. 54, fig. 214 (1934) nom. nud.

Terrestrial, rarely epiphytic ***herb.*** ***Pseudobulbs*** 4-10 cm long, forming a cluster, ovoid, conical, angled. ***Leaves*** 40-60 x 12-20 cm, lanceolate, shortly petiolate, plicate, glabrous, deciduous, young leaves produced with inflorescence in Borneo, elsewhere flowering when leafless during the dry season, following new leaves. ***Inflorescence*** 60 cm long, erect, arching at apex, hairy, bearing up to 12 large, well-spaced flowers; peduncle green, hairy, with 4-6 sterile persistent bracts; rachis green, hairy; floral bracts 1.5-2.5 x 1.2 cm, ovate, persistent, broad. ***Flowers*** with white sepals; lip white with a dark yellow to brownish yellow patch at base of lip in Bornean populations (elsewhere often rose-coloured); spur cream to pale yellow, sometimes greenish at base. ***Pedicel*** with ***ovary*** 1.5-3.5 cm, pale green, hairy. ***Dorsal sepal*** 2-2.2 x 0.7-0.8 cm, shorter than laterals, hairy on reverse, elliptic, acuminate. ***Lateral sepals*** 2-2.5 cm, similar, hairy on reverse, lanceolate, acuminate. ***Petals*** 1.8-2 x 0.8 cm, glabrous on reverse, oblong-elliptic, apiculate. **Lip** 1.8-2 x 2-2.3 cm; side lobes 1 x 0.5 cm wide; mid-lobe 1.2 x 1.5-1.8 cm, widening from base, forming 2 rounded lobes; spur 2-2.5 cm long, slender, recurved, shortly hairy. ***Column*** 0.8-1 cm long, footless, joined to the claw of the lip along its lower length; anther-cap roughly 2-celled, bicornute; pollinia 8 in two groups of 4. Plate 3C.

HABITAT AND ECOLOGY: In Sarawak and elsewhere mainly terrestrial, often on limestone but occasionally epiphytic. In Sabah it has been found on dead tree stumps, or among fallen rotting tree trunks at 600 to 700 m in lower montane forest, on ultramafic substrate.

DISTRIBUTION IN BORNEO: KALIMANTAN. SABAH: Mt. Kinabalu.

GENERAL DISTRIBUTION: Burma, Thailand, Vietnam, Peninsular Malaysia, Java, Sulawesi, Seram.

NOTES: The Bornean populations are probably at the limit of their distribution and out of phase with their normal climatic environment, since there is only a short dry season, hence flowering spikes are often produced with the unfolding leaves, and they are partly deciduous. This very showy species is often cultivated, and some attractive hybrids have been produced.

DERIVATION OF NAME: The specific epithet is derived from the Latin *vestitus*, clothed, referring to the hairy coat covering the inflorescence.

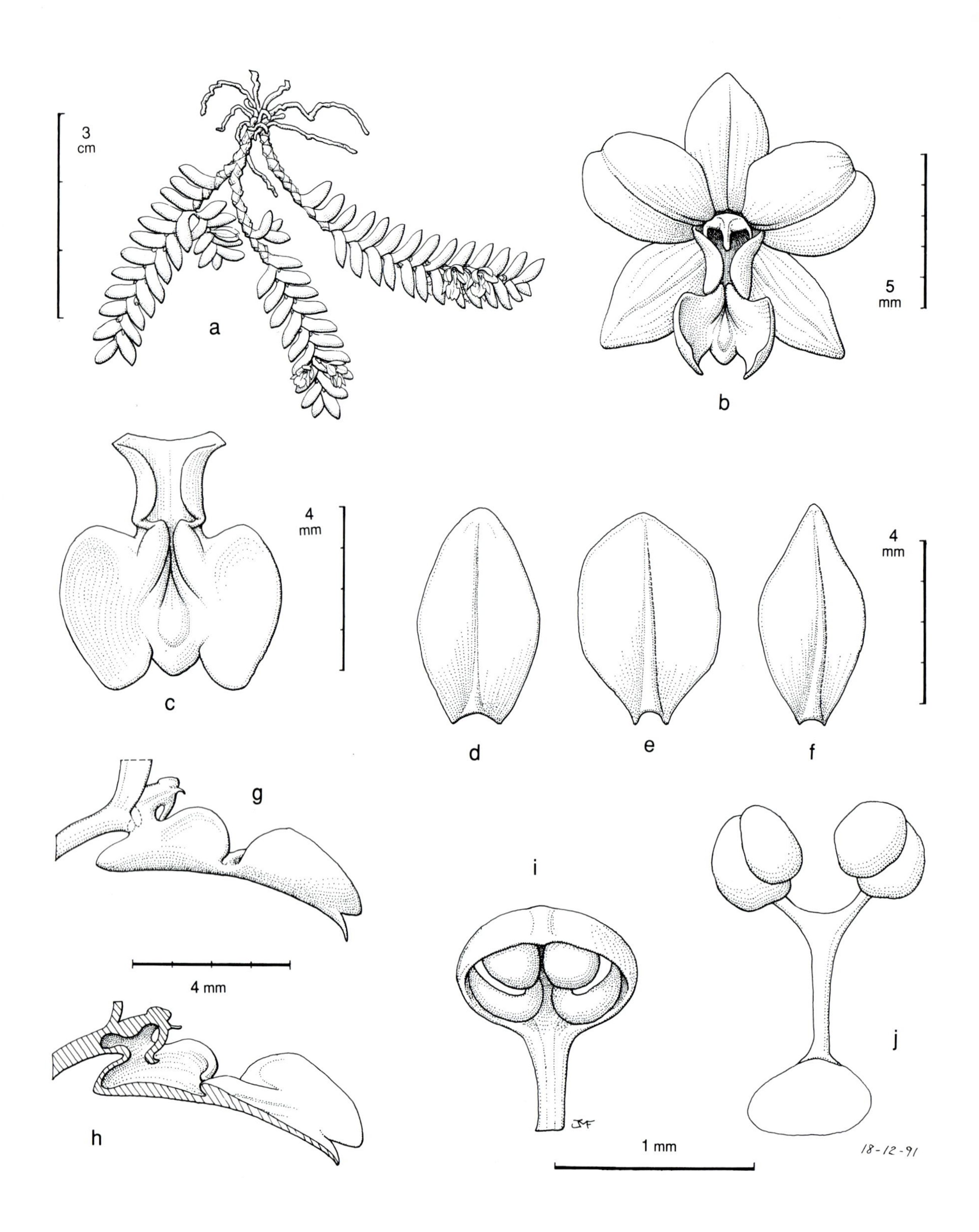

Figure 15. Ceratochilus jiewhoei J.J. Wood & Shim. - A: plant. - B: flower, front view. - C: dorsal sepal. - D: lateral sepal. - E: petal. - F: lip, front view. - G: column and lip, lateral view - H: column and lip, longitudinal section. - J: anther-cap with pollinia, front view. - K: pollinarium. - A from *Lamb* AL 611/86, B-K from *Beaman* 10673 and *Lamb* AL 58/83. Drawn by Mark Fothergill.

15. CERATOCHILUS JIEWHOEI J.J. Wood & Shim

Ceratochilus jiewhoei *J.J. Wood et Shim* **sp. nov.** a *C. biglanduloso* (Kuntze) Blume speciei javensi, floribus minoribus, petalis multo latioribus, labii lobis lateralibus distinctis lobo medio conspicuo late oblongo-elliptico, calcari breviore pilis in ore deficientibus. Typus: Borneo, Sabah, Mt. Kinabalu, Pinosuk Plateau, January 1983, *Lamb* AL 58/83 (holotypus K).

Epiphytic ***herb*** 3-4 cm long with the habit of *Microsaccus griffithii*. ***Stems*** simple, leafy, rooting at the base. ***Leaves*** 0.7-1 x 0.2-0.28 cm, distichous, ensiform, acute, flattened, hard and fleshy, with sheathing bases. ***Inflorescences*** lateral, axillary, from upper portion of stem, 1-flowered, subtended by two brownish, ovate bracts, each c. 1 mm long. ***Flower*** 1 cm in diameter, pure white, sometimes with a small yellow spot in centre of lip, translucent. ***Pedicel*** with ***ovary*** c. 0.3 cm long. ***Sepals*** and ***petals*** spreading, 1-nerved. ***Dorsal sepal*** 0.5-0.6 x 0.3-0.35 cm, elliptic, acute. ***Lateral sepals*** 0.6 x 0.3 cm, elliptic, acute, slightly carinate at apex. ***Petals*** 0.6 x 0.42 cm, broadly elliptic, obtuse. ***Lip*** 0.4-0.5 cm long, 0.4-0.45 cm broad (width measured across mid-lobe), 3-lobed, shortly spurred, somewhat fleshy; side lobes rounded, erect, clasping column at base; mid-lobe broadly oblong-elliptic, obscurely thickened at base, apex emarginate, with a broadly triangular tooth in the sinus; spur 1-1.5 mm long, obtuse. ***Column*** 0.5 mm long; anther-cap 1 mm long, broadly spathulate; pollinia 4, ovoid, unequal, on a linear, clavate stipes; viscidium elliptic. Plate 3D.

HABITAT AND ECOLOGY: Lower montane oak-laurel forest, growing near the tips of thin branches of small saplings, in shade. Alt. 900 to 1800 m. Flowering observed in March, May, July, September and November.

DISTRIBUTION IN BORNEO: SABAH: Mt. Kinabalu; Mt. Alab.

GENERAL DISTRIBUTION: Endemic to Borneo (Sabah only).

NOTES: *C. jiewhoei* differs from the only other species, the Javan *C. biglandulosus*, in having smaller flowers with much broader petals, a lip with distinct side lobes and a conspicuous broadly oblong-elliptic mid-lobe. The spur is shorter and lacks any hairs at its mouth.

DERIVATION OF NAME: The generic name is derived from the Greek *keras*, *kerato*, meaning a horn and *cheilos*, a lip, and alludes to the rather obscure 'horn-like' swellings at the mouth of the lip of *C. biglandulosus*. These swellings, which are more flange-like than horn- like, probably represent rudimentary side lobes and are developed as such in *C. jiewhoei*. The specific epithet is named in honour of Mr Tan Jiew Hoe of Singapore for his services towards conservation and his generosity towards the Sabah Society without which the publication of this volume would not have been possible.

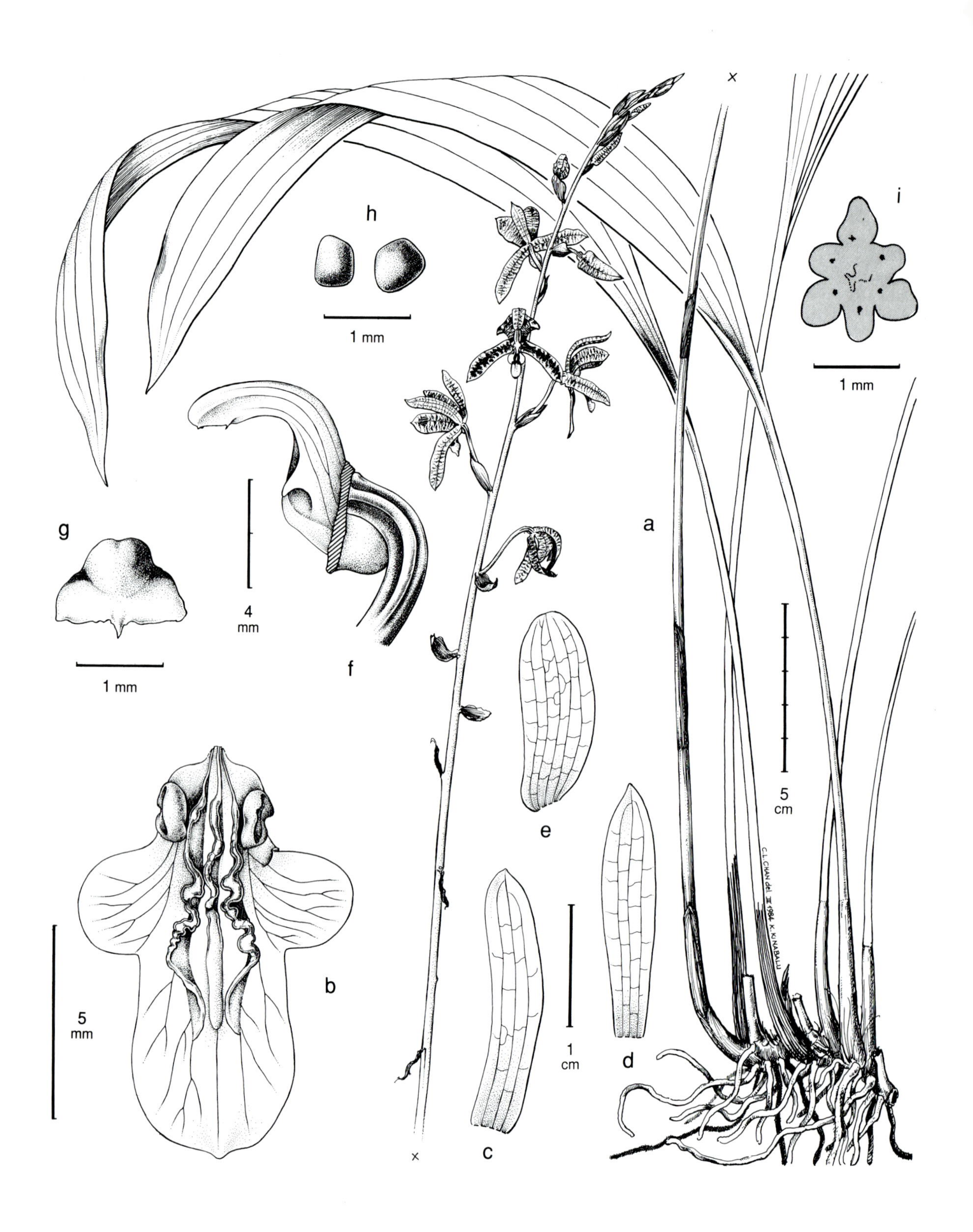

Figure 16. Chrysoglossum reticulatum Carr. - A: plant. - B: lip, spread out. - C: lateral sepal. - D: dorsal sepal. - E: petal. - F: column and ovary, lateral view. - G: anther. - H: pollinia. - I: ovary, transverse section. All drawn from *Beaman* 8953 by Chan Chew Lun.

16. CHRYSOGLOSSUM RETICULATUM Carr

Chrysoglossum reticulatum *Carr* in Gard. Bull. Straits Settlem. 8: 197 (1935). Types: Borneo, Sabah, Mt. Kinabalu, main spur west of Tenompok Pass, *Carr* 3314 (syntype SING, isosyntype K); Tenompok/Tomis, 1600 m, *J. & M.S. Clemens* s.n. (syntype BM).

Terrestrial ***herb.*** ***Rhizome*** to 9 cm long. ***Pseudobulbs*** 1-3.8 x 0.5 cm, 1-leafed, terete to cylindrical, tapering at the top, creeping. ***Leaves*** 15-23 x 2.4-4.5 cm, linear-lanceolate, acuminate; petiole 7-17 cm long. ***Inflorescence*** erect, racemose, 15- to 20-flowered; peduncle 6-56 cm long; rachis to 28 cm long; floral bracts 9-10 x 5.5-6 mm, ovate, acuminate, spreading. ***Flowers*** resupinate; sepals and petals yellow or greenish to whitish, with many transverse pale purple bars and spots; lip white with a short brown or purple streak at the base of the white or yellowish side lobes, mid-lobe spotted dark lilac, keels white; column white, yellow or orange, anther-cap yellow. ***Pedicel*** with ***ovary*** 0.9-1.6 cm long. ***Dorsal sepal*** 1.5-1.9 x 0.24-0.4 cm, linear-lanceolate, acute. ***Lateral sepals*** 0.9-2 x 0.3-0.4 cm, lanceolate, slightly falcate. ***Petals*** 1.3-1.6 x 0.4-0.6 cm, ovate-oblong, slightly falcate. ***Lip*** 3-lobed, mobile, fleshy, 0.85-1 cm long; side lobes 4.5-5 x 5-5.5 mm, parallel with column, claw pleated, velutinous, sides obliquely semi-orbicular to ligulate; mid-lobe 4-4.4 x 4-4.5 mm, obovate in outline, slightly emarginate, obtuse; disc with 3 slightly undulate keels, the outer extending from near base and ending halfway along mid-lobe, median keel extending from base and continuing as a low ridge ending on basal half of mid-lobe. ***Column*** 0.75-0.8 cm long, slender, apex curved forward, with two white lateral wings on each side of 2 fleshy keels near base; column-foot cleft, 2 x 1 mm. with a narrow entrance to a small spur 1.5-2 x 1.5-2 mm which is formed by the column-foot; anther-cap 1.5 x 1.5 mm; pollinia 2. Plate 3E & F.

HABITAT AND ECOLOGY: Lower and upper montane forest, often in moss or cloud forest. Alt. 1300 to 1800 m. Flowering observed from February to August, and December.

DISTRIBUTION IN BORNEO: SABAH: Mt. Kinabalu; Mt. Alab; Kimanis Road. SARAWAK: Bario District.

GENERAL DISTRIBUTION: Endemic to Borneo.

DERIVATION OF NAME: The generic epithet is derived from the Greek *chrysous*, golden, and *glossa*, tongue, referring to the colour of the lip in the type species. The specific epithet is derived from the Latin *reticulatus*, netted, a reference to the markings on the sepals and petals.

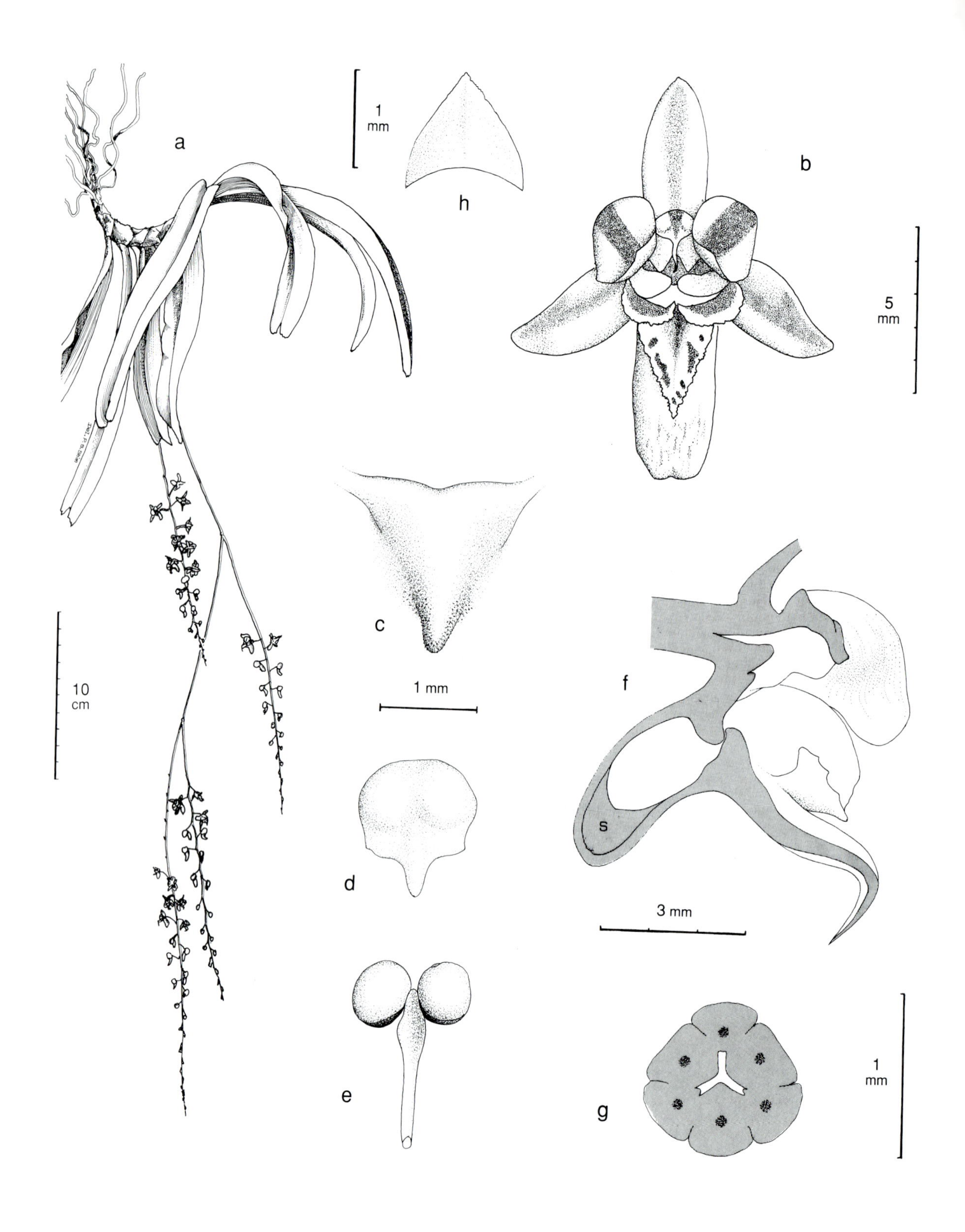

Figure 17. Cleisostoma discolor Lindl. - A: plant. - B: flower. - C: callus at spur entrance. - D: anther. - E: pollinarium. - F: column and lip, longitudinal section. - G: ovary, transverse section. - H: floral bract. All drawn from *Lamb* AL 75/83 by Shim Phyau Soon.

17. CLEISOSTOMA DISCOLOR Lindl.

Cleisostoma discolor *Lindl.*, in Bot. Reg. 31: misc. 59 (1845). Type: India, cult. *Loddiges* s.n. (holotype K).

Sarcanthus termissus Rchb.f. in Hamburger Garten-Blumenzeitung 16: 15 (1860). Type: Java, cult. *Stange* (holotype W).

S. macrodon Rchb.f. in Gard. Chron. 1872: 1555 (1872). Type: India, Madras, *Veitch* ex *Benson* s.n. (holotype W).

Saccolabium rostellatum Hook.f., Fl. Brit. Ind. 6: 59 (1890). Type: India, Sikkim, *Gamble* s.n. (holotype CAL).

Sarcanthus auriculatus Rolfe in Kew Bull. 1895: 9 (1895). Type: cult. *O'Brien* (holotype K).

S. discolor (Lindl.) J.J. Sm. in Natuurk. Tijdschr. Ned.-Indië 72: 85 (1912).

S. josephii J.J. Sm. in Bull. Jard. Bot. Buitenzorg, ser. 2, 9: 103 (1913). Type: Java, Gede, cult. *Joseph* (holotype BO).

S. angkorensis Guillaumin in Bull. Soc. Bot. France 77: 328 (1930). Type: Cambodia, Angkor, *Thorel* s.n. (holotype P).

Cleisostoma auriculatum (Rolfe) Garay in Bot. Mus. Leafl. 23 (4): 169 (1972).

C. termissus (Rchb.f.) Garay, loc. cit.: 175 (1972).

Pendent, epiphytic ***herb***. ***Stem*** 2-25 cm long, up to 1 cm in diameter, leafy in upper part. ***Roots*** elongate, unbranched, 1-2.5 cm in diameter. ***Leaves*** 5-20 x 1-2.2 cm, coriaceous, falcate, linear, unequally roundly bilobed at apex, apical lobe erose. ***Inflorescence*** up to 60 cm long, simple or branched, pendent, densely many-flowered; peduncle terete, up to 15 cm long; floral bracts triangular, acute, c. 1 mm long. ***Flowers*** fleshy; sepals purple, edged with yellow or green; petals yellow with purple central vein; lip yellow with purple spotting; spur yellow or white. ***Pedicel*** with ***ovary*** 0.3-0.5 cm long, purple. ***Dorsal sepal*** 0.5-0.55 x 0.2 cm, erect, elliptic, rounded at apex. ***Lateral sepals*** 0.5-0.6 x 0.2-0.25 cm, reflexed-recurved, elliptic-obovate, obtuse. ***Lip*** three-lobed in basal half, 0.6 x 0.5 cm; side lobes erect, each with a fleshy longitudinal ridge-like callus, obliquely oblong, erose; mid-lobe oblong to elliptic, obtuse or rounded, erose on margin; spur 3-5 mm long, decurved in middle, cylindric, with a dorsal and ventral fleshy callus almost blocking the mouth. ***Column*** 2 mm long, fleshy; pollinia 4, attached by an elastic strap-like stipes to a small circular viscidium. Plate 4A.

HABITAT AND ECOLOGY: Lowland and lower montane mixed dipterocarp forest, often near to streams. Alt. sea level to 600m.

DISTRIBUTION IN BORNEO: KALIMANTAN SELATAN: Banjarmasin area. SABAH: Mt. Kinabalu; Tambunan District. SARAWAK: Bau area.

GENERAL DISTRIBUTION: India, Thailand, Cambodia, Peninsular Malaysia, Sumatra, Java and Borneo.

NOTES: This species appears to be widespread in Sabah in the interior and Crocker Range, but few collections have been made. It usually flowers in the monsoon seasons.

DERIVATION OF NAME: The generic name is derived from the Greek *kleis(t)os*, closed, and *stoma*, mouth, referring to the mouth of the inflated spur which is blocked by two calli. The specific epithet is derived from the Latin *discolor*, of different colour, referring to the colour of the flower.

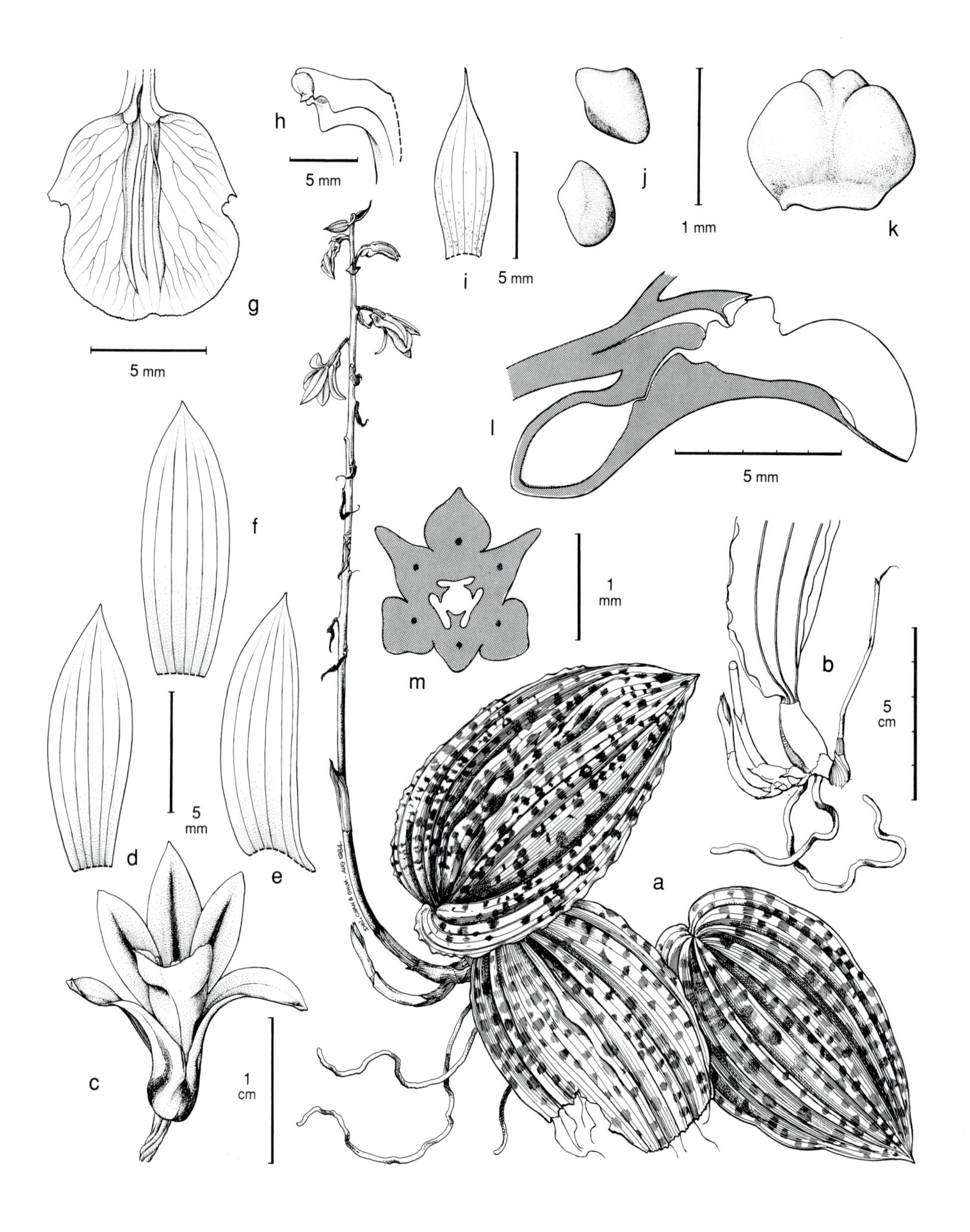

Figure 18. Collabium bicameratum (J.J. Sm.) J.J. Wood - A: plant. - B: pseudobulb, lateral view, showing base of inflorescence. - C: flower. - D: petal. - E: lateral sepal. - F: dorsal sepal. - G: lip. - H: column. - I: floral bract. - J: pollinia. - K: anther. - L: column and lip, longitudinal section. - M: ovary, transverse section. All drawn from *Lamb* AL 240/84 by Chan Chew Lun and Shim Phyau Soon.

18. COLLABIUM BICAMERATUM (J.J. Sm.) J.J. Wood

Collabium bicameratum *(J.J. Sm.) J.J. Wood*, **comb. nov.**

Chrysoglossum bicameratum J.J. Sm. in Bull Jard. Bot. Buitenzorg, ser. 3, 11: 91 (1931). Type: Borneo, Kalimantan, West Koetai, Gunung Kemal (Kemoel), *Endert* 4351 (lectotype L).

Terrestrial ***herb. Rhizome*** 7-8 cm long, creeping. ***Pseudobulbs*** 0.9-1.5 cm long, 1.5-2.3 cm apart, narrowly conical, 1-leafed. ***Leaves*** 6.5-10 x 3.2-4 cm, elliptic or ovate-elliptic, acute or shortly acuminate, margin undulate, texture papery, nerves prominent; petiole 3-5 mm long, sulcate, brownish green, blotched and spotted grey-green. ***Inflorescence*** erect, laxly 5-flowered; peduncle 8-15 cm long, with 3 acute sheaths at base; rachis 3-5 cm long; floral bracts 9 x 3 mm, narrowly elliptic, acuminate. ***Flowers*** c. 1.4 x 0.9 cm; pedicel purplish pink; ovary green; sepals and petals yellowish green stained pink to rose-violet; lip white with yellow disc and keels; column and anther-cap white. ***Pedicel*** with ***ovary*** 0.7-0.8 cm long, 6-winged, 3 broad, 3 narrow. ***Dorsal sepal*** 1-1.15 x 0.4 cm, oblong, acute. ***Lateral sepals*** 0.8-1.1 x 0.34 cm, obliquely oblong, subfalcate, acute, connate at base, adnate to column-foot forming a 3.5-4 mm long, obtuse mentum. ***Petals*** 1-1.05 x 0.35 cm, oblong or oblong-elliptic, acute. ***Lip*** 0.8-0.9 cm long, obscurely 3-lobed, shortly clawed and narrowed at base; side lobes 1-2 mm long, triangular, irregularly toothed, erect; mid-lobe 4-5 x 8 mm, narrowly semi-orbicular, truncate, retuse, sometimes erose denticulate, gently curved; disc with 3 parallel ridges, the outer 2 prominent and swollen at the base, the inner narrower, slightly shorter and not swollen at the base. ***Column*** 0.4-0.5 cm long, slightly curved, shortly winged, column-foot 3.5-4 mm long; anther-cap ovate, retuse; pollinia 2. Plate 4B.

HABITAT AND ECOLOGY: Lower montane forest. alt. 1600 to 1800 m.

DISTRIBUTION IN BORNEO: KALIMANTAN TIMUR: Mt. Kemal (Kemoel). SABAH: Tenom District; Keningau District. SARAWAK: Bario District.

GENERAL DISTRIBUTION: Endemic to Borneo.

NOTES: *Collabium* is a small genus of eleven often rare and poorly collected species distributed from China and Taiwan through Indochina east to New Guinea, Vanuatu and Fiji. It is distinguished from the closely related *Chrysoglossum* by the long column-foot which, together with the connate bases of the lateral sepals, forms a slender nectary or mentum. *Chrysoglossum* has a short column-foot with a saccate nectary and free lateral sepals.

DERIVATION OF NAME: The generic name is derived from the Latin *collum*, a neck, and *labium*, a lip, referring to the base of the lip which embraces the column like a collar. The specific epithet is derived from the Latin *bi*, two, and *camera*, meaning an arch or vault, and refers to the two swollen bases of the two outer ridges on the lip.

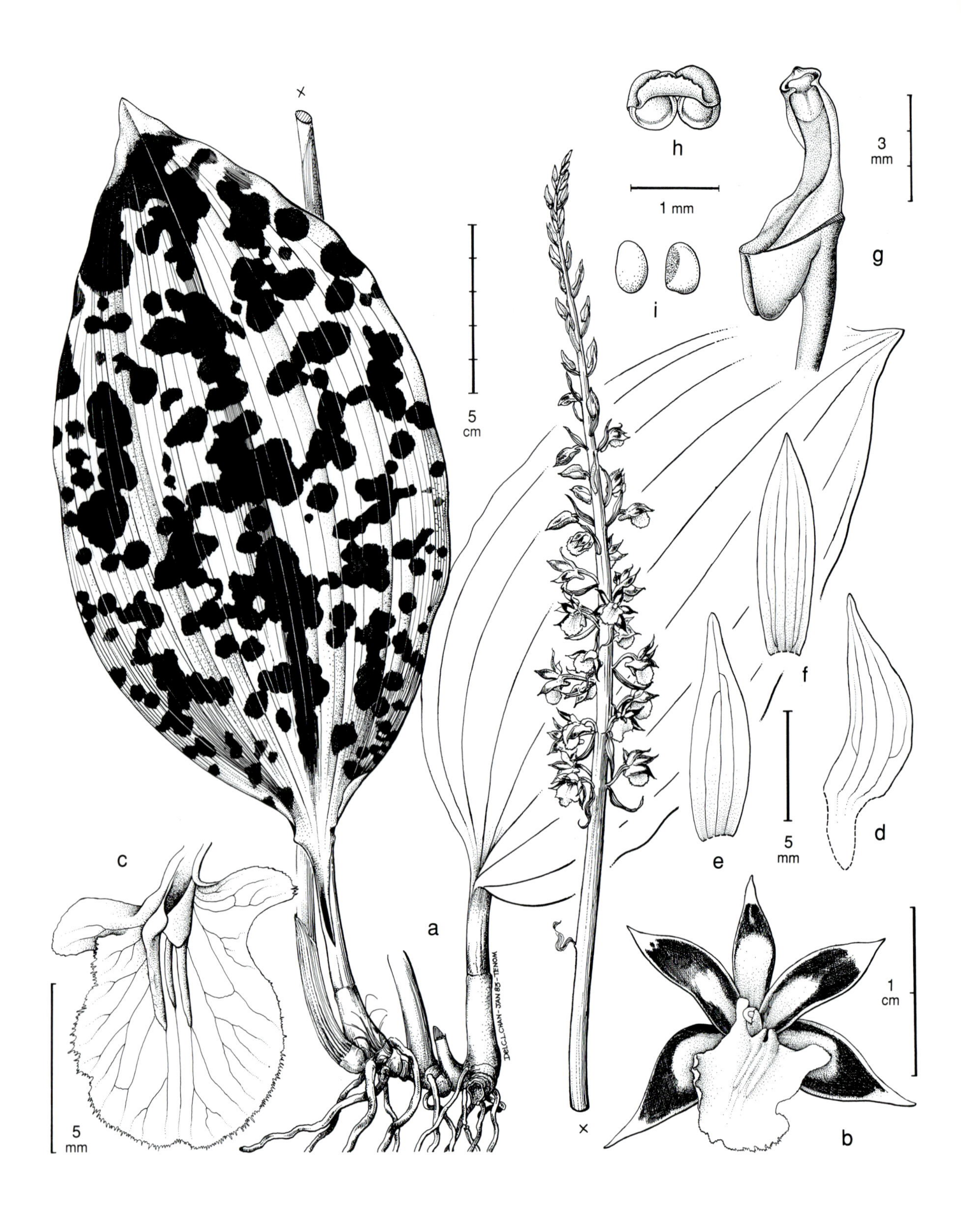

Figure 19. Collabium simplex Rchb. f. - A: plant. - B: flower. - C: lip. - D: lateral sepal. - E: petal.. - F: dorsal sepal. - G: column. - H: anther, adaxially. - I: pollinia. A from *Lamb* AL 59/83, B-I from *Phillipps* s.n. Drawn by Chan Chew Lun.

19. COLLABIUM SIMPLEX Rchb.f.

Collabium simplex *Rchb.f.* in Gard. Chron. n.s. 15: 462 (1881). Types: Borneo, cult. *Messrs. Veitch & Sons* 288 & *Bull* (syntypes W).

Chrysoglossum simplex (Rchb.f.) J.J. Sm., Orch. Java: 177 (1905).

Terrestrial ***herb.*** ***Rhizome*** to 5 cm long. ***Pseudobulbs*** 2-3 x 0.5-0.7 cm, 0.5-1.5 cm apart, quadrangular, tapering towards apex, 1-leafed, purple. ***Leaves*** (9-)21-37.5 x 6.5-15 cm, broadly lanceolate, acuminate, attenuate at base, margin slightly undulate, nerves prominent below; petiole 1.5-6 cm long, sulcate, blade purple when young, turning pale green mottled dark green above and blue green below. ***Inflorescence*** erect, densely 20- to 30-flowered; peduncle 22-30 cm long, purple, with usually 3 acute 2-5 cm long sheaths; rachis 10-28 cm long, purple; floral bracts 1-1.65 x 0.15-0.26 cm, linear, acute, reflexed, minutely hairy. ***Flowers*** with greenish yellow sepals and petals, stained with purple-red; mentum yellow; lip white flushed purple red near keels; column white, bright purple at base. ***Pedicel*** with ***ovary*** 0.6-0.9 cm long, sulcate, purple. ***Dorsal sepal*** 0.85-0.96 x 0.18-0.27 cm, narrowly elliptic, acute, slightly falcate. ***Lateral sepals*** 0.75-0.9 x 0.2-0.28 cm, narrowly elliptic, acute to acuminate, falcate, adnate to column-foot forming a 1.8-2.3 mm long, saccate, rather flattened and slightly retuse mentum. ***Petals*** 0.75-0.95 x 0.18-0.25 cm, narrowly elliptic, acute to acuminate. ***Lip*** 3-lobed, rather fleshy, 0.7-0.85 cm long, clawed at base; claw 1.5-2 x 0.7-1.6 mm; side lobes 1.6-2.3 x 0.8-1.5 mm, oblong-ligulate, obtuse to acute, margin slightly crenate-undulate; mid-lobe 4-4.7 x 3.5-5.5 mm, obovate to almost rectangular, obtuse to retuse, slightly recurved, margin irregularly denticulate to lacerate, undulate, sometimes slightly incurved; disc with 3 parallel fleshy keels, the outer extending from the base of the claw, becoming erect and rounded, 1-1.5 mm high between the side lobes, diminishing in height and ending about the centre of the mid-lobe, central keel low, extending from between the side lobes to near or at the centre of the mid-lobe. ***Column*** 0.4-0.5 cm x 1- 1.2 mm, slender, truncate, wings obscure, slightly curved, twisted to one side; column-foot 1.8-2.3 mm long; anther-cap reniform, retuse. Plate 4C & D.

HABITAT AND ECOLOGY: Lower montane forest. Alt. 1100 m. Flowering observed in January and February.

DISTRIBUTION IN BORNEO: SABAH: Mt. Alab.

GENERAL DISTRIBUTION: Peninsular Malaysia, Sumatra, Java & Borneo.

DERIVATION OF NAME: The Latin specific epithet *simplex* means simple and it is not certain to which feature this applies.

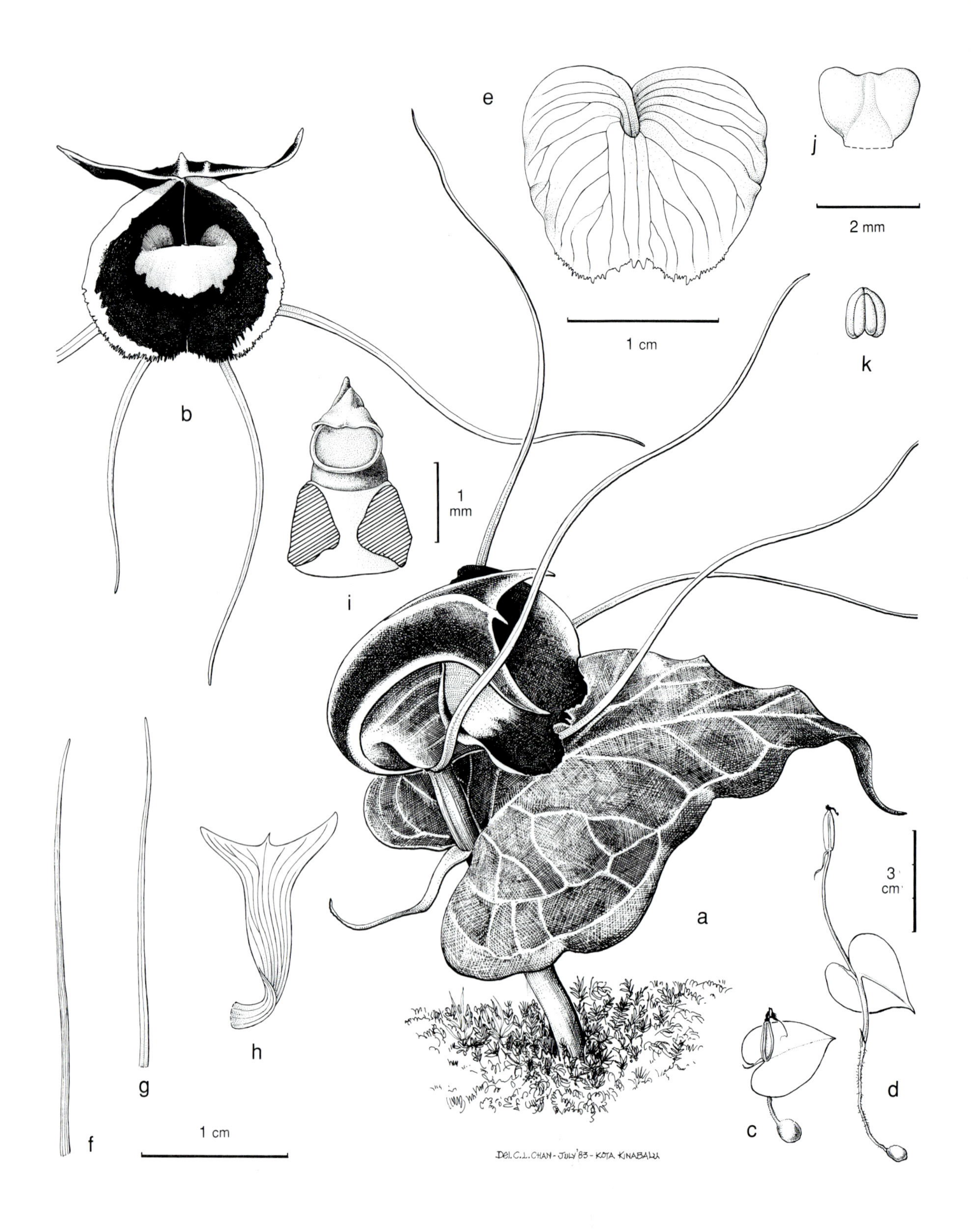

Figure 20. Corybas pictus (Blume) Rchb. f. - A: plant. - B: flower. - C & D: plants in fruit showing underground tubers. - E: lip. - F: petal. - G: lateral sepal. - H: dorsal sepal. - I: column with anther. - J: anther. - K: pollinia. All drawn from *Lamb* SAN 91526 by Chan Chew Lun and Jaap J. Vermeulen.

20. CORYBAS PICTUS (Blume) Rchb.f.

Corybas pictus *(Blume) Rchb.f.*, Beitr. Syst. Pflanz.: 67 (1871). Type: Java, Gunung Salak, *Blume* s.n. (holotype L).

Calcearia picta Blume, Bijdr.: 418, t.33 (1925).

Corysanthes picta (Blume) Lindl., Gen. Sp. Orch. Pl.: 394 (1840).

Corysanthes mucronata Blume, Fl. Javae, ser. 2, 1 Orch. (Coll. Orch. Arch. Ind.): 147, fig. 66 (1859). Type: Java, Gunung Pangrango, *Blume* s.n. (not traced).

Corysanthes limbata Hook.f. in Bot. Mag. 89: t. 5357 (1863). Type: cult. *Bull* (not traced).

Corybas limbatus (Hook.f.) Rchb.f., Beitr. Syst. Pflanz.: 67 (1871).

Corysanthes picta (Blume) Lindl. var. *karangensis* J.J. Smith in Bull. Jard. Bot. Buitzenorg, ser. 3, 3: 230 (1921). Type: Java, Gunung Karang, *Winckel* 554B (holotype BO, isotypes K, L).

Corybas mucronatus (Blume) Schltr. in Feddes Repert. 19: 20 (1924).

Corybas pictus (Blume) Rchb.f. var. *dorowatiensis* J.J.Sm. ex Backer & Bakh.f., Fl. Java 3: 256 (1968). Type: Java, Gunung Dorowati-Kukusan, *van Steenis* 2585 (holotype BO, isotype L).

Corybas crenulatus sensu Carr, non (J.J. Sm.) Schltr., in Gard. Bull. Straits Settlem. 8: 173 (1935).

Tiny terrestrial ***herb*** up to 7 cm tall, often growing in colonies. ***Tuber*** small, globose, c. 0.5 cm in diameter, pubescent. ***Stem*** short, 2-4.5 cm long, unifoliate at apex. ***Leaf*** 1.5-3 x 1-2 cm, borne horizontally just above substrate, heart-shaped, acute to apiculate, dark green veined with white or pink. ***Inflorescence*** one-flowered, very short; bract linear-tapering, to 1.2 cm long. ***Flower*** large for plant, maroon marked with white in throat of lip and with translucent white and purple sepals and petals. ***Pedicel*** with ***ovary*** 0.5-1 cm long, pedicel elongates to 3 cm or more before the fruit capsule dehisces. ***Dorsal sepal*** 1-1.3 x 0.7-1.2 cm, curving forwards over the lip, oblanceolate with an abruptly dilated apex, truncate and shortly mucronate at the apex to appear 3-lobed. ***Lateral sepals*** and ***petals*** 2.5-3.5 cm long, spreading, filiform. ***Lip*** c. 1 cm long, 1.5 cm wide when flattened, strongly recurved in the middle, entire, somewhat funnel-shaped, front margin crenulate, ecallose, with two short conical spurs at the base, each 0.5-0.6 cm long. ***Column*** 0.25 cm long; pollinia 2, 2-lobed, granulate. Plate4 E & F.

HABITAT AND ECOLOGY: On mossy rocks and mossy banks and on the mossy boles of tree ferns and trees, in deep shade in hill- and lower montane forest on ultramafic or sandstone soils. Alt. 700 to 1800 m. Flowering has generally been observed during the rainy seasons in May, June and from October to December.

DISTRIBUTION IN BORNEO: SABAH: Mt. Kinabalu; Mt. Alab; Mt. Tavai.

GENERAL DISTRIBUTION: Sumatra, Java and Borneo.

NOTES: The plants seen on ultramafic substrate often have deep pink venation on the leaves. Plants previously ascribed to *C. crenulatus* by Carr on Mt. Kinabalu are *C. pictus*. The flowers reportedly mimic fungi and produce a fungus-like scent to attract fungus flies. During the dry season the plants often die back, when no leaves can be observed.

DERIVATION OF NAME: The generic name is derived from the Greek *korybas*, drunken man, priest of Cybele, which describes the dorsal sepal which simulates a veiled drooping head, alluding to a priest's head-dress or to the nodding of a drunken man. The specific epithet is derived from the Latin *pictus*, coloured or painted, referring to either the silver or red venation of the leaves, or the flower colours.

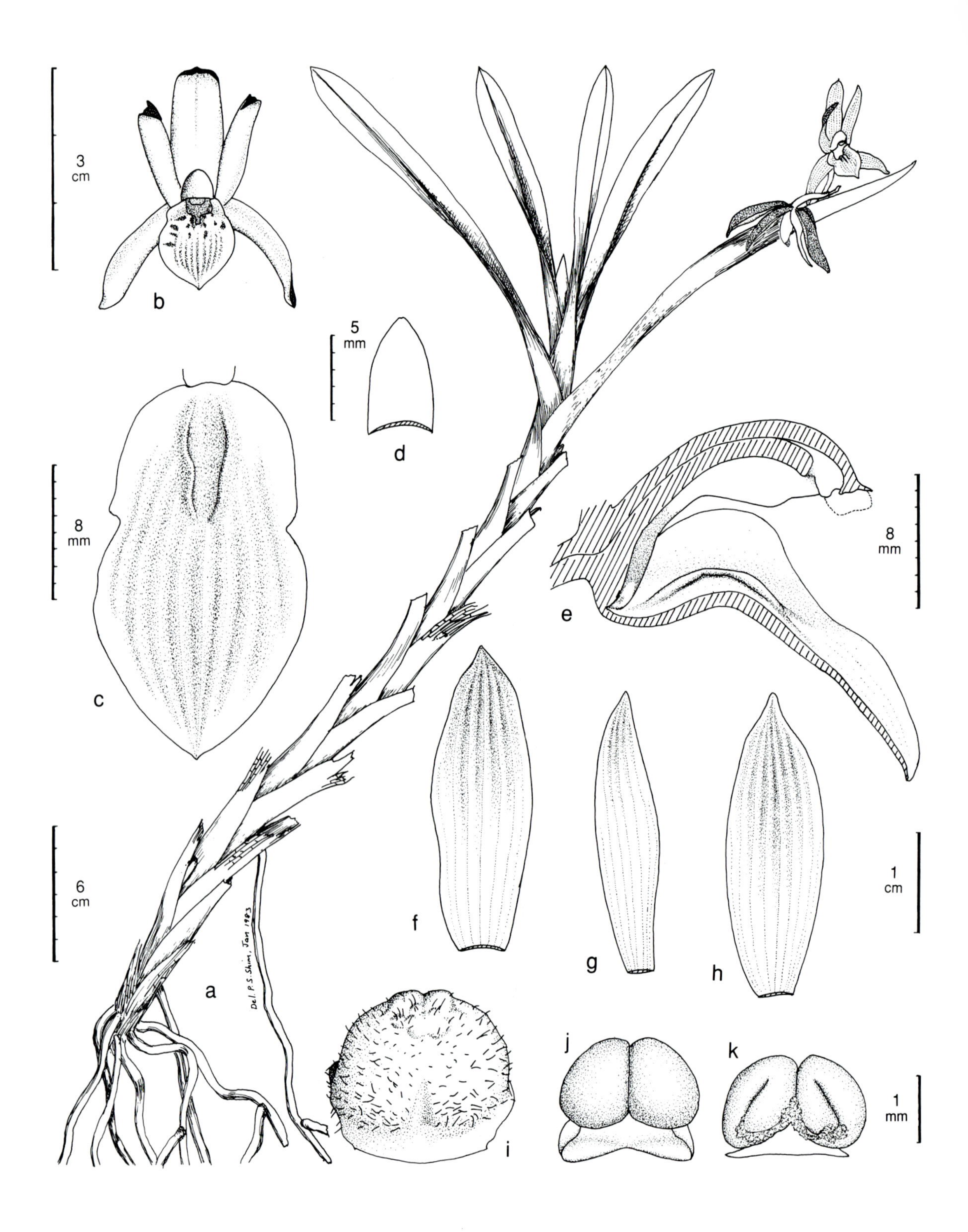

Figure 21. Cymbidium elongatum J.J. Wood, DuPuy & Shim. - A: plant. - B: flower. - C: lip. - D: floral bract. - E: column and lip, longitudinal section. - F: dorsal sepal. - G: petal. - H: lateral sepal - I: anther - J: pollinia, abaxially. - K: pollinia, adaxially. All from living plant collected from Marai Parai Spur. Drawn by Shim Phyau Soon.

21. CYMBIDIUM ELONGATUM J.J.Wood, DuPuy & Shim

Cymbidium elongatum *J.J. Wood, DuPuy & Shim* in DuPuy & Cribb, The Genus Cymbidium: 103, fig. 22.1, photos 76 & 77 (1988). Type: Borneo, Sabah, Mt. Kinabalu, Marai Parai, *Collenette* A 47 (holotype BM, isotype K).

Robust, glabrous, terrestrial ***herb***, the young sterile growth resembling certain *Phreatia* and the mature growth certain *Agrostophyllum* and *Dipodium* in habit. ***Roots*** produced from within the leaf sheaths opposite the leaf blades along the mid to lower portion of the stem. ***Stems*** simple, the immature erect, the mature becoming much elongated and reclinate, usually about 30-130 cm or more long, leafy at the apex, entirely covered in persistent sheathing leaf bases, the lowermost of which become fibrous. ***Leaves*** 4-8, distichous, blade 10-19 x 1.2-2.3 cm, broadly linear, ensiform or ligulate, conduplicate, gently curving, obtuse and mucronate, usually very slightly oblique, somewhat cucullate, coriaceous, articulated at a sheathing base, eventually deciduous; sheaths (3.5-)6-7 cm long, with scarious, sometimes dark brown margins up to 1 mm broad. ***Inflorescence*** a lax 1- to 4- (perhaps more) flowered axillary raceme; peduncle naked above, with about 6 narrowly elliptic, acute to acuminate, scarious basal sheaths each 0.4-3.5 cm long, which are normally concealed within the leaf blade; rachis 1-6.5 cm long; floral bracts 4-7 mm long, ovate or triangular-ovate, acute, scarious. ***Flowers*** c. 4 cm in diameter, slightly scented; sepals and petals spreading, purplish red outside, olive-green, reddish olive green or creamy white inside; lip off-white or yellowish green, usually spotted crimson or purple-red, callus yellow; column greenish yellow or reddish, flushed crimson or purple at base. ***Pedicel*** with ***ovary*** 2-3.5(-4) cm long, slender, curving. ***Dorsal sepal*** 2.6-3.6 x 1-1.2 cm, oblong-elliptic, acute. ***Lateral sepals*** 1.4-3 x 0.9-1 cm, sometimes slightly oblique, oblong-elliptic, apex slightly cucullate and carinate, mucronate. ***Petals*** 2.2-3.1 x 0.6-0.8 cm, narrowly elliptic, acute. ***Lip*** obscurely 3-lobed, free, gently recurved, 2-2.4 x 1.1-1.3 cm, disc c. 7 mm across; side lobes 5.6 mm broad, rounded, erect, not clasping the column; mid-lobe 1.2-1.4 cm long, ovate, entire, acute; callus composed of two low ridges extending from base of lip to base of mid-lobe. ***Column*** 1.5-1.6 cm long, with a short foot, curved, the wings broadened towards the apex; anther-cap cucullate, hirsute; pollinia 2, cleft. Plate 5A.

HABITAT AND ECOLOGY: Marshy areas in open, scrubby woodland of stunted trees, often rooted at the base of *Leptospermum* or amongst rattans, sedges, Ericaceae and *Begonia*, on ultramafic serpentine rock. Alt. 1200 to 2300 m.

DISTRIBUTION IN BORNEO: SABAH: Mt. Kinabalu; Mt. Monkobo; Mt. Tembuyuken. SARAWAK: Mt. Murud; Mt. Mulu National Park.

GENERAL DISTRIBUTION: Endemic to Borneo.

NOTES: *C. elongatum* is one of the most unusual species in the genus. It has a monopodial rather than sympodial habit, with indeterminately growing stems. These tend to lean on and scramble over the surrounding vegetation as they elongate. This atypical habit is reminiscent of climbing species of *Dipodium*. In Sarawak it has been found to occur occasionally as an epiphyte.

DERIVATION OF NAME: The generic name is derived from the Greek *kymbes*, a boat-shaped cup, in allusion to the lip shape in some species. The specific epithet is derived from the Latin *elongatus*, elongated, referring to the habit.

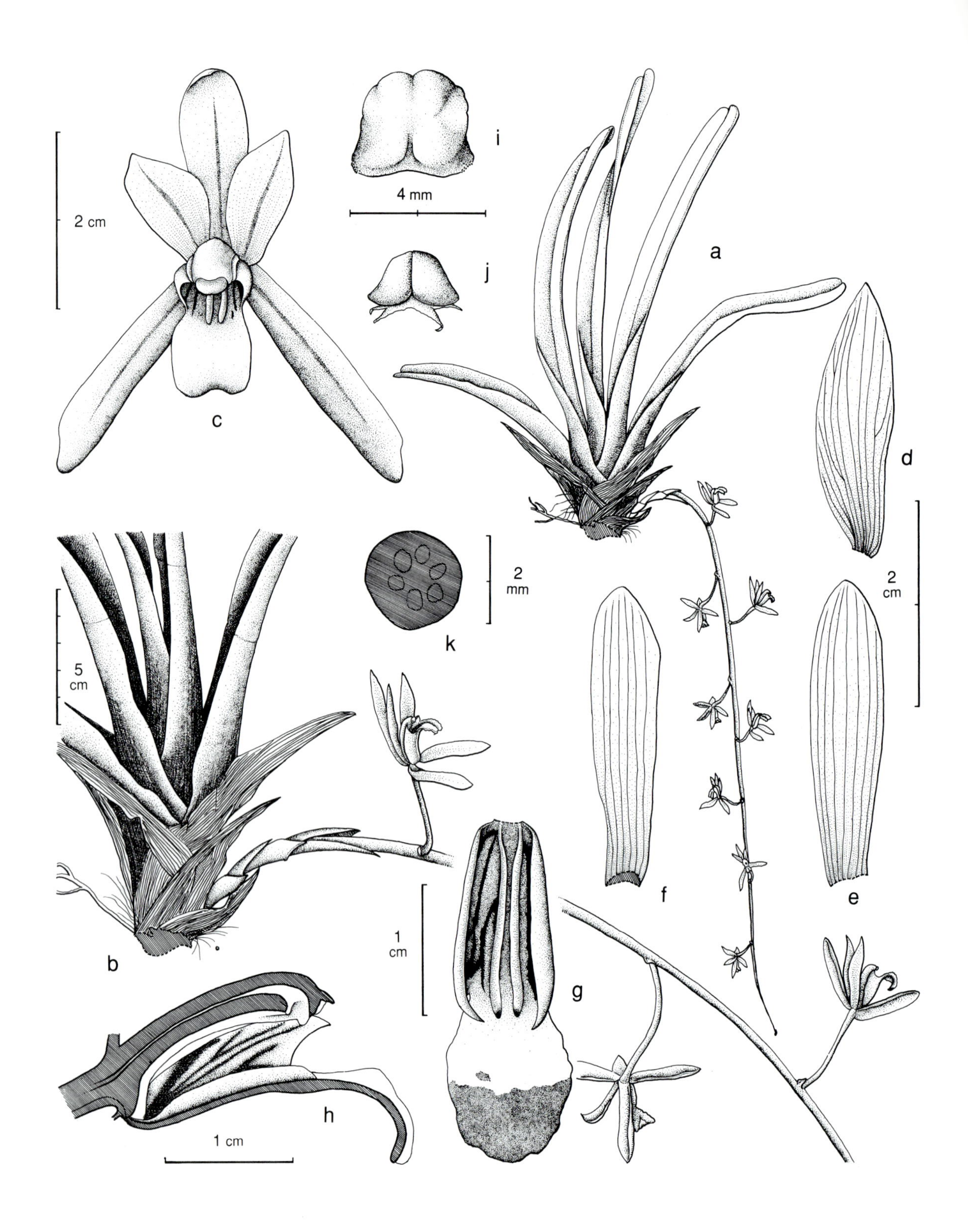

Figure 22. Cymbidium finlaysonianum Lindl. - A & B - plant. - C: flower. - D: petal. - E: dorsal sepal. - F: lateral sepal. - G: lip. - H: column and lip, longitudinal section - I: anther - J: pollinia. - K: ovary, transverse section. All drawn from living plant by Liew Fui Ling.

22. CYMBIDIUM FINLAYSONIANUM Lindl.

Cymbidium finlaysonianum *Lindl.*, Gen. Sp. Orch. Pl.: 164 (1833). Type: Vietnam, Turon (Da Nang) Bay, *Finlayson* in *Wallich* 7358 (holotype K).

Large epiphytic or lithophytic ***herb.*** ***Pseudobulbs*** up to 8 x 5 cm, ovoid, bilaterally flattened, usually obscure, enclosed in the persistent leaf-bases and 5 to 8 cataphylls. ***Leaves*** distichous, 4 to 7 per pseudobulb, 36-100 x 2.7-6 cm, ligulate, obtuse to emarginate and unequally bilobed at apex, coriaceous and rigid, almost erect, articulated to a broadly sheathing base up to 8-16 cm long. ***Inflorescence*** scape 20-140 cm long, with 7-62 well-spaced flowers; peduncle short, 5-12 cm, covered by 6-8 overlapping, cymbose, acute, spreading sheaths 3.7-8 cm long; rachis about 25-110 cm long, usually becoming slender and fractiflex towards the apex. ***Flowers*** 4-5.7 cm across; sepals and petals dull green to straw-yellow, usually suffused with red brown; lip white, side lobes suffused and strongly veined with purple, mid-lobe yellow with purple and red blotches; callus-ridges bright yellow with purple stains; column purple at the tip, becoming yellowish towards the base. ***Pedicel*** with ***ovary*** 1.4-4.5 cm long. ***Dorsal sepal*** 2.5-3.3 x 0.7-1.1 cm, narrowly ligulate-elliptic, obtuse, erect, margins revolute. ***Lateral sepals*** similar, slightly oblique, spreading. ***Petals*** 2.4-3 x 0.7-1.1 cm, narrowly elliptic to ovate, obtuse to subacute, margins weakly revolute, 7-9-veined. ***Lip*** 2.4-2.8 x 1.4-1.8 cm when flattened, 3-lobed, papillose or with some minute hairs; side lobes erect, upper margins involute, tips triangular, acute to acuminate, porrect; mid-lobe large, 1.1-1.4 x 0.9-1.4 cm, broadly elliptic, obtuse to emarginate, mucronate, recurved, margin undulate; callus of 2 parallel, strongly raised and well-defined ridges which terminate abruptly at the base of the mid-lobe. ***Column*** 1.5-1.8 cm long, arching, slightly winged; pollinia about 2 mm long, triangular, deeply cleft. Plate 5B.

HABITAT AND ECOLOGY: Found on cliffs and rocky limestone outcrops where humus has collected. It can withstand direct sunlight and, in Sabah, is found on rocky coastal shorelines. It can equally tolerate light shade and is common in old rubber plantations, secondary forest as well as lowland primary forest to hill dipterocarp forest. Alt. sea level to 1200 m.

DISTRIBUTION IN BORNEO: KALIMANTAN TIMUR: Apokayan. SABAH: Mt. Kinabalu; Tenom District; Papar Beach. SARAWAK: 1st Division; Bako National Park; Miri River; 4th Division.

GENERAL DISTRIBUTION: S. Vietnam, Cambodia, S. Thailand, Peninsular Malaysia, Sumatra, Java, Borneo, Sulawesi and the Philippines.

NOTES: This species can form very large clumps on larger trees, or form colonies on rocks and cliff ledges. In Sabah it has been noted that plants growing along the coast have golden coloured sepals and petals whereas inland they are more often pale greenish with a reddish mid-vein. Large bees *(Apis dorsata)* visit the flowers and good fruiting results. *C. finlaysonianum* is closely related to *C. aloifolium* (L.) Sw., a species having smaller flowers. *C. atropurpureum* (Lindl.) Rolfe is distinguished by its much narrower leaves, with deep burgundy coloured flowers with a scent of desiccated coconut. Lamb has observed plants on the offshore Sabah islands of

Bodgaya and Boheydulang (Semporna), and Gaya and Manukan (Kota Kinabalu).

DERIVATION OF NAME: The specific epithet honours George Finlayson, a surgeon in the service of the East India Company, who first collected the species in Vietnam, at what is now Da Nang Bay.

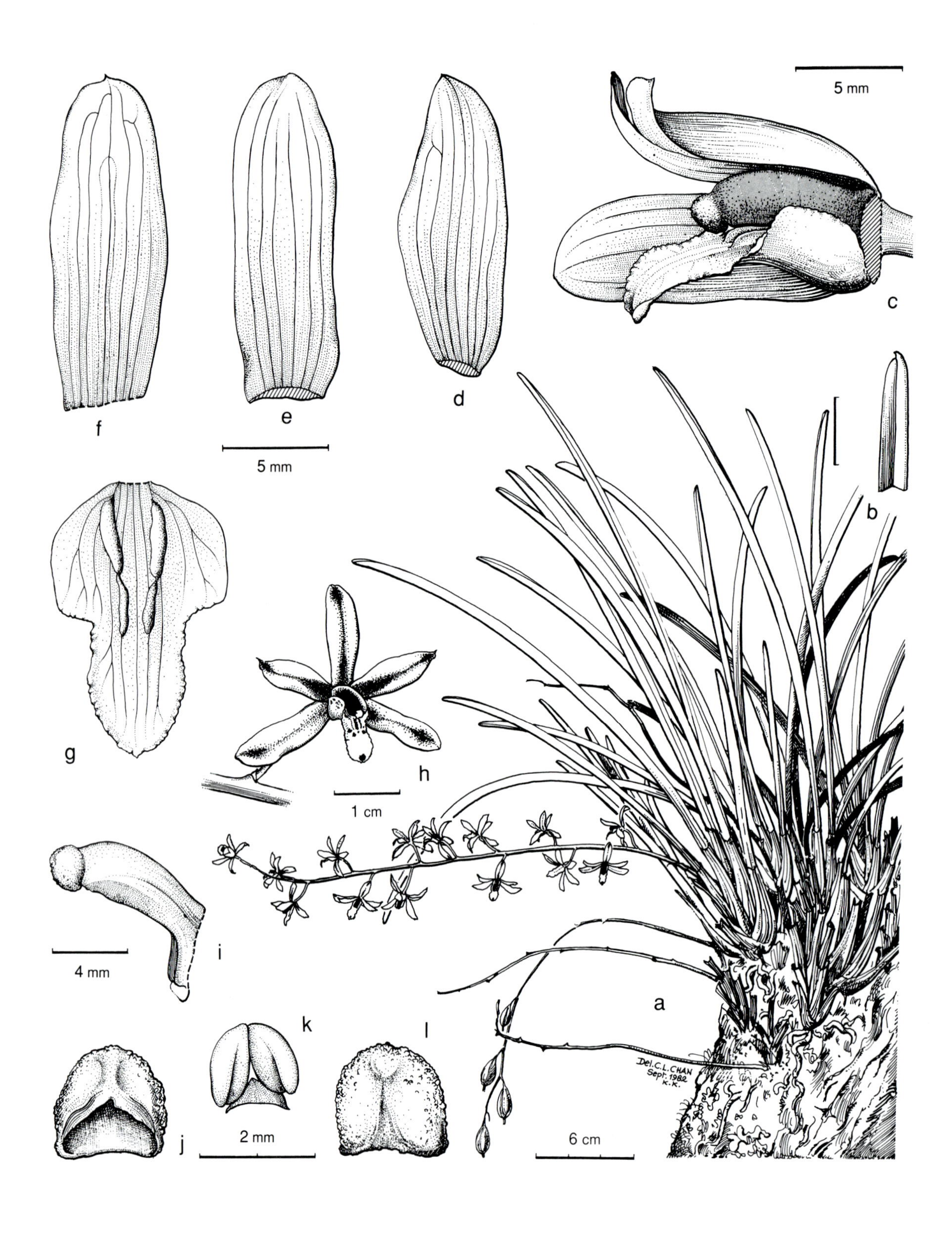

Figure 23. Cymbidium rectum Ridl. - A: plant. - B: leaf tip. - C: flower, lateral view, with a petal and a lateral sepal removed. - D: petal. - E: dorsal sepal - F: lateral sepal. - G: lip. - H: flower. - I: column, lateral view. - J: anther, abaxially. - K: pollinia. - L: anther, abaxially. All drawn from *Bacon* s.n. by Chan Chew Lun.

23. CYMBIDIUM RECTUM Ridl.

Cymbidium rectum *Ridl.* in J. Straits Branch Roy. Asiat. Soc. 82: 198 (1920). Type: Peninsular Malaysia, Negeri Sembilan, *Genyns-Williams* s.n. (holotype SING).

Medium-sized epiphytic ***herb.*** ***Pseudobulbs*** not strongly inflated, about 2-5 cm long, elongate-ovoid, enclosed in the sheathing leaf-bases and 2 to 3 cataphylls. ***Leaves*** 7-9 per pseudobulb, up to 60 x 0.8-1.4 cm, narrowly ligulate, unequally bilobed to oblique at the tip, strongly V-shaped in section, coriaceous, very stiff, arching, articulated 1-8 cm from the pseudobulb. ***Inflorescence*** up to 40 cm long, suberect to horizontal, often pendulous in fruit, with up to 17 flowers; peduncle short, up to 9 cm, covered towards the base by about 5 sheaths; sheaths up to 4 cm long, cymbose, overlapping, spreading; floral bracts 2-3 mm long, triangular. ***Flowers*** 3-4 cm across, usually slightly sweet-scented; sepals and petals pale yellow or cream with a broad central stripe of maroon-brown extending to the tip; lip white with a pale yellow patch at the base, side lobes lightly spotted maroon, mid-lobe lightly spotted maroon at the base, with a broad primrose yellow central band and a single maroon spot at the apex; callus ridges white with yellow; column cream, strongly stained maroon above, white and speckled maroon below. ***Pedicel*** with ***ovary*** 1.5-3 cm long. ***Dorsal sepal*** 1.7-2 x 0.7-0.8 cm, narrowly oblong to elliptic, obtuse or weakly mucronate, erect, margins lightly revolute. ***Lateral sepals*** similar, slightly oblique, spreading. ***Petals*** 1.7-1.8 x 0.6 cm, narrowly elliptic, acute, weakly porrect, but not covering the column. ***Lip*** 1.5 cm long, 3-lobed; side lobes erect, obtusely angled and appearing truncated at the apex; mid-lobe 0.8-0.9 x 0.5-0.55 cm, ligulate, acute, minutely papillose, margin undulate towards the base; callus of two entire sigmoid ridges. ***Column*** short, 8 mm long, weakly winged at the apex, with a very short column-foot; pollinia 2, deeply cleft. ***Fruit capsule*** about 3.5 cm long, beaked. Plate 5C.

HABITAT AND ECOLOGY: On *Baeckea frutescens* and other heath forest trees in stunted kerangas forest on podsolic soils and in swamp forest, often 6 metres or less from the ground. Alt. 400 to 500 m. Flowering sporadic from September to February.

DISTRIBUTION IN BORNEO: KALIMANTAN TIMUR: near Balikpapan. SABAH: Sook Plain: Nabawan area.

GENERAL DISTRIBUTION: Peninsular Malaysia and Borneo.

NOTES: This species appears to be very rare in Sabah, as most of the Sook Plain has been cleared and burnt, and only a few remnant strips of podsolic kerangas (heath) and swamp forest are left near Nabawan. The species has potential for breeding smaller pot-plant hybrid cymbidiums on account of its small size and erect inflorescence.

DERIVATION OF NAME: The specific epithet is a corruption of the Latin *erectum*, erect, referring to the erect inflorescence.

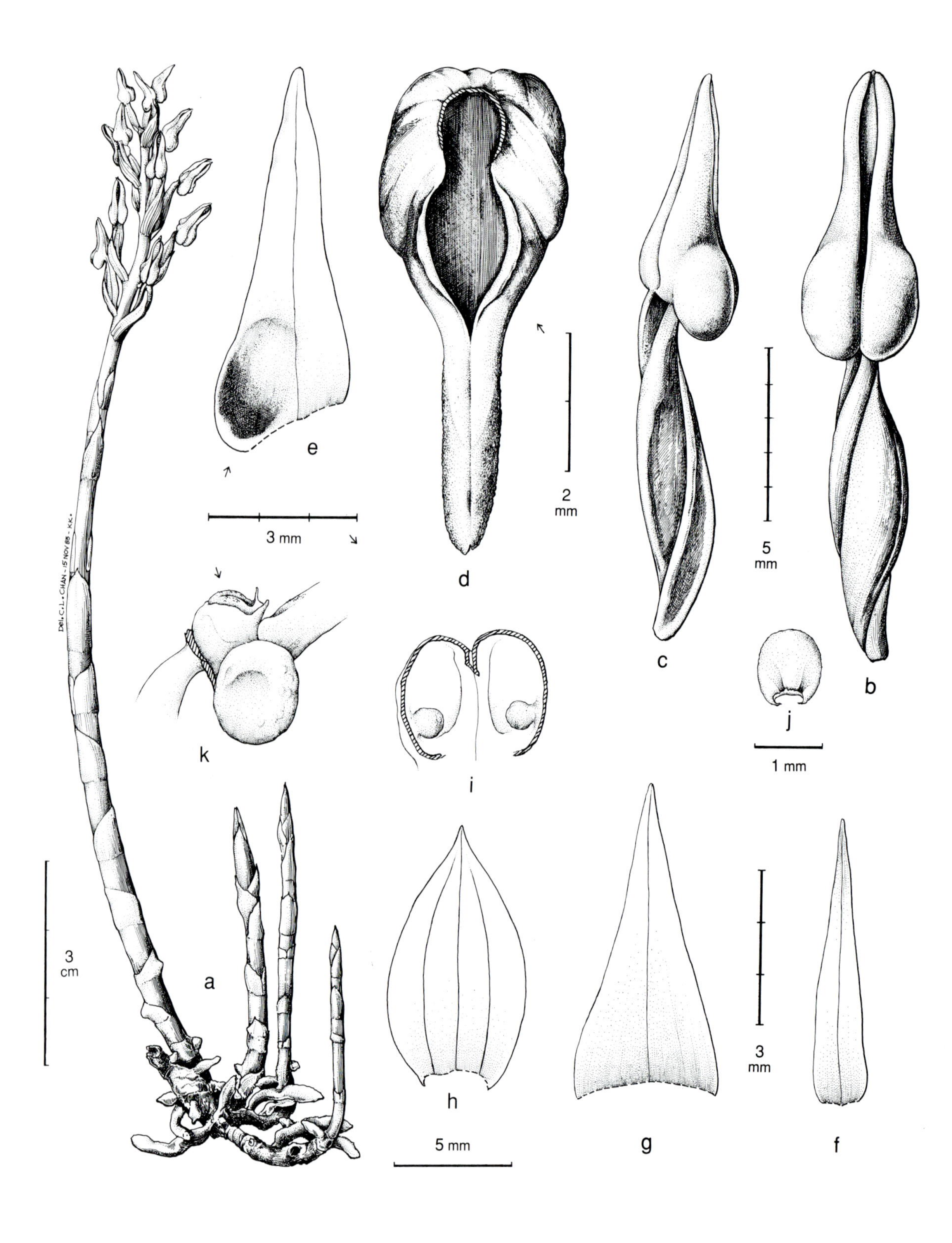

Figure 24. Cystorchis aphylla Ridl. - A: plant. - B: flower, ventral view. - C: flower, lateral view. - D: lip. - E: lateral sepal. - F: petal. - G: dorsal sepal. - H: floral bract. - I: lip, showing two glands inside. - J: anther. - K: column and base of lip, lateral view. All drawn from *Beaman* 7409a by Chan Chew Lun.

24. CYSTORCHIS APHYLLA Ridl.

Cystorchis aphylla *Ridl.* in J. Linn. Soc., Bot. 32: 400 (1896). Type: Peninsular Malaysia, *Ridley* s.n. (holotype SING).

Slender pale buff to reddish terrestrial saprophytic ***herb***, with a branching underground rhizome 2-4 mm in diameter. ***Stems*** 1 to few, erect, 10-35 cm long, bearing many overlapping ovate, acute, pink sterile bracts 0.8-1.5 cm long. ***Inflorescence*** terminal, erect, 2- to 14-flowered; floral bracts pale brown, lanceolate, acuminate, 0.5-1 x 0.6 cm. ***Flowers*** not opening widely, pink to reddish with a paler base and white tips to sepals, suberect, subtubular. ***Pedicel*** with ***ovary*** 0.4-1 cm long, glabrous. ***Dorsal sepal*** 0.4-0.6 x 0.15 cm, concave at base, porrect, oblong-lanceolate, rounded at the apex, forming a hood with the petals. ***Lateral sepals*** 0.5-0.8 x 0.25 cm, enclosing the lip, porrect, obliquely oblong-lanceolate, obtuse, forming at the base a bilobulate saccate mentum, 0.15 cm deep. ***Petals*** 0.4-0.6 x 0.1 cm, linear, obtuse, adnate to the dorsal sepal. ***Lip*** very fleshy, tripartite, 0.5-0.6 x 0.2 cm; hypochile saccate, bilobulate, with a fleshy spherical callus and sessile glands on each side; mesochile oblong with strongly incurved sides that meet along their margins; epichile recurved, ovate, obtuse. ***Column*** 0.2 cm long, fleshy, pointed at the apex; pollinia 2, mealy; viscidium small, terminal. Plate 5D.

HABITAT AND ECOLOGY: In deep shade in mixed hill-dipterocarp forest and lower montane forest on steep slopes on sandstone and ultramafic substrate. Alt. 200 to 1300 m.

DISTRIBUTION IN BORNEO: SABAH: Mt. Kinabalu; Lahad Datu District. SARAWAK: Mt. Mulu; Niah.

GENERAL DISTRIBUTION: Peninsular Malaysia, Thailand, Sumatra, Java, Borneo, Buru and the Philippines.

NOTES: Seidenfaden (1978) depicts a Thai specimen (*Larsen* 30724 (AAU)) as having a hairy rachis with scattered hairs on the floral bracts whereas Bornean and Javan specimens are glabrous. It would appear that this species is self-fertilised, since with a suppressed rostellum and upturned stigma, the pollen readily falls onto the stigma and pollination occurs.

DERIVATION OF NAME: The generic name is derived from the Greek *kystis*, a bladder, and *orchis*, orchid, in reference to the inflated or bladder-like base to the lip hypochile. The specific epithet is derived from the Latin *aphyllus*, meaning leafless, a feature of all the saprophytes.

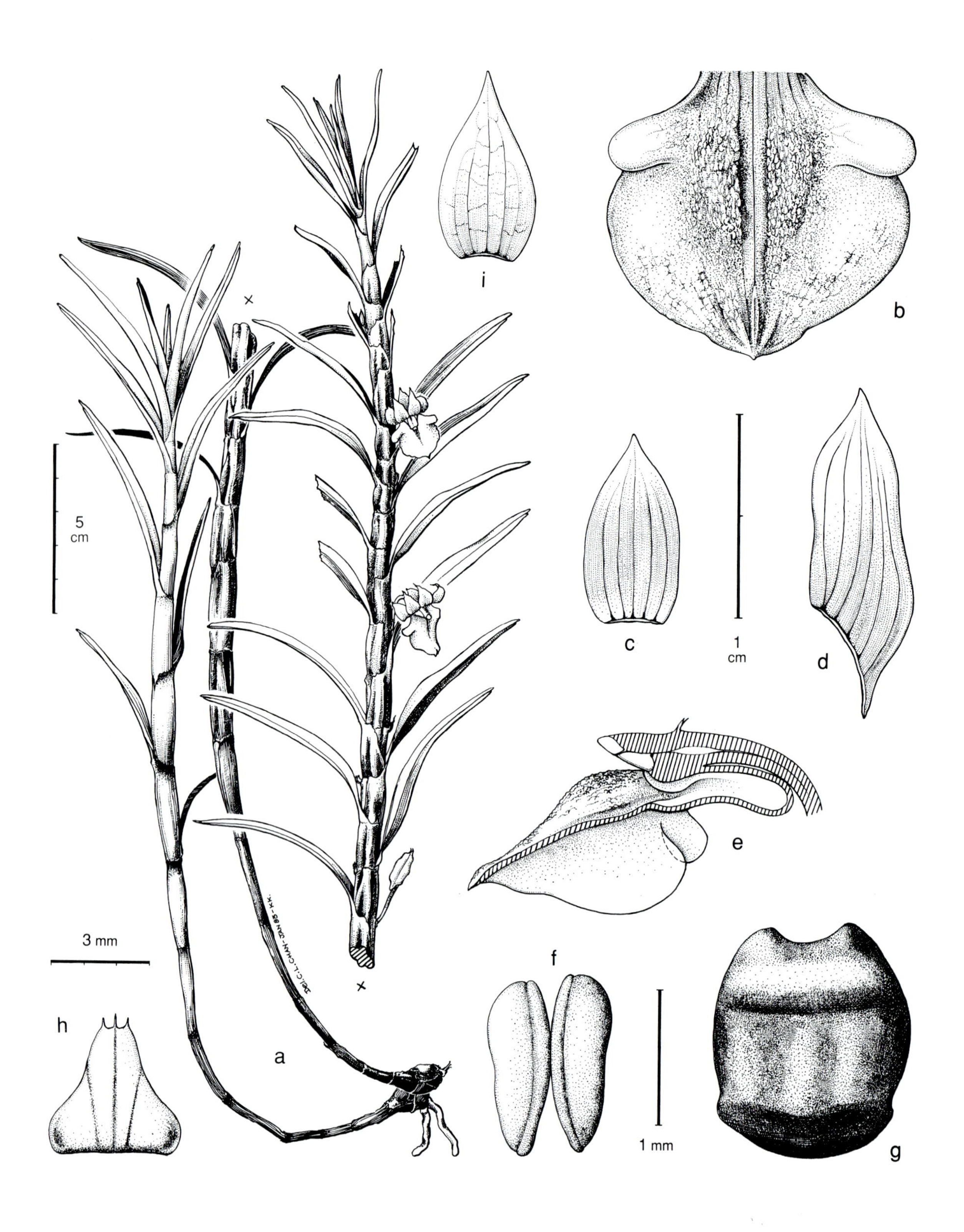

Figure 25. Dendrobium maraiparense J.J. Wood & C.L. Chan. - A: plant. - B: lip, spread out. - C: dorsal sepal. - D: lateral sepal. - E: column and lip, longitudinal section. - F: pollinia. - G: anther. - H: column, dorsal view. - I: petal. All drawn from *Lamb* SAN 93368 by Chan Chew Lun.

25. DENDROBIUM MARAIPARENSE J.J.Wood & C.L.Chan

Dendrobium maraiparense *J.J.Wood et C.L.Chan* **sp. nov.** (sect. *Distichophyllo*) *D. unifloro* Griff. affinis sed foliis angustioribus numerosioribus acute bilobatis, sepalis petalisque reflexis, labello leviter plicato, disco obscure papilloso-verruculoso lamellis centralibus 2 ornato distinguenda. Typus: Borneo, Sabah, Mt. Kinabalu, Bembangan River, 20 April 1964, *Chew & Corner* RSNB 4962 (holotypus K, isotypus SING).

Stiffly erect epiphytic ***herb.*** ***Stems*** 24-45(-50) cm high, entirely covered by persistent leaf-sheaths. ***Leaves*** 3.5-6.5 x 0.5-0.8 cm, narrowly elliptic, apex acutely unequally bilobed, distichous, articulate, stiff and leathery, spreading, sheaths 0.8-2(-3) cm long, nigrohirsute when young. ***Inflorescence*** 1-flowered; floral bracts 2-3 mm long, ovate, acute. ***Flowers*** white, ageing to orange-apricot. ***Pedicel*** with ***ovary*** 2-2.4 cm long. ***Sepals*** and ***petals*** reflexed. ***Dorsal sepal*** 0.9-1 x 0.5 cm, oblong-elliptic, acute. ***Lateral sepals*** obliquely oblong, acute, free portion 0.9-1 x 0.5 cm, lower portion adnate with column-foot to form an obtuse mentum 0.5 cm long. ***Petals*** 0.9 x 0.5 cm, ovate-elliptic, acute. ***Lip*** 1.3 cm long, 3-lobed, gently folded, minutely papillose-verruculose; side lobes 0.5-0.6 x 0.3 cm, narrowly oblong, obtuse, overlapping base of mid-lobe; mid-lobe 0.6-0.7 x 1.4-1.5 cm, broadly ovate, obtuse, with a minute apical mucro, disc with 2 obscure papillose-verruculose central ridges. ***Column*** 0.3-0.35 x 0.2-0.25 cm, broadly triangular-oblong, foot 0.5 cm long; anther-cap c. 2 mm long, cucullate. Plate 5E.

HABITAT AND ECOLOGY: Lower montane *Leptospermum* forest on ultramafic substrate; lower montane oak forest. Alt. 1200 to 2000 m.

DISTRIBUTION IN BORNEO: SABAH: Mt. Kinabalu.

GENERAL DISTRIBUTION: Endemic to Sabah (Mt. Kinabalu only).

NOTES: D. *maraiparense* is related to *D. uniflorum* Griff. from Peninsular Malaysia, Thailand, Vietnam, Borneo (Sabah and Sarawak) and the Philippines. *D. maraiparense* is at once distinguished by the narrower, more acutely bilobed leaves, reflexed sepals and petals and gently folded lip having only two obscure central ridges. *D. uniflorum* has broader, obtusely bilobed leaves, spreading sepals and petals and the lip has two distinct outer ridges and a short but prominent central apical keel.

Another related species, *D. bifarium* Lindl., occurs on Mt. Kinabalu, including Marai Parai Spur. This is distinguished from *D. maraiparense* by its smaller, broader, obtusely bilobed leaves and smaller flowers having an entire lip.

DERIVATION OF NAME: The generic name is derived from the Greek *dendron*, tree, and *bios*, life, referring to the epiphytic habit. The specific epithet refers to the type locality.

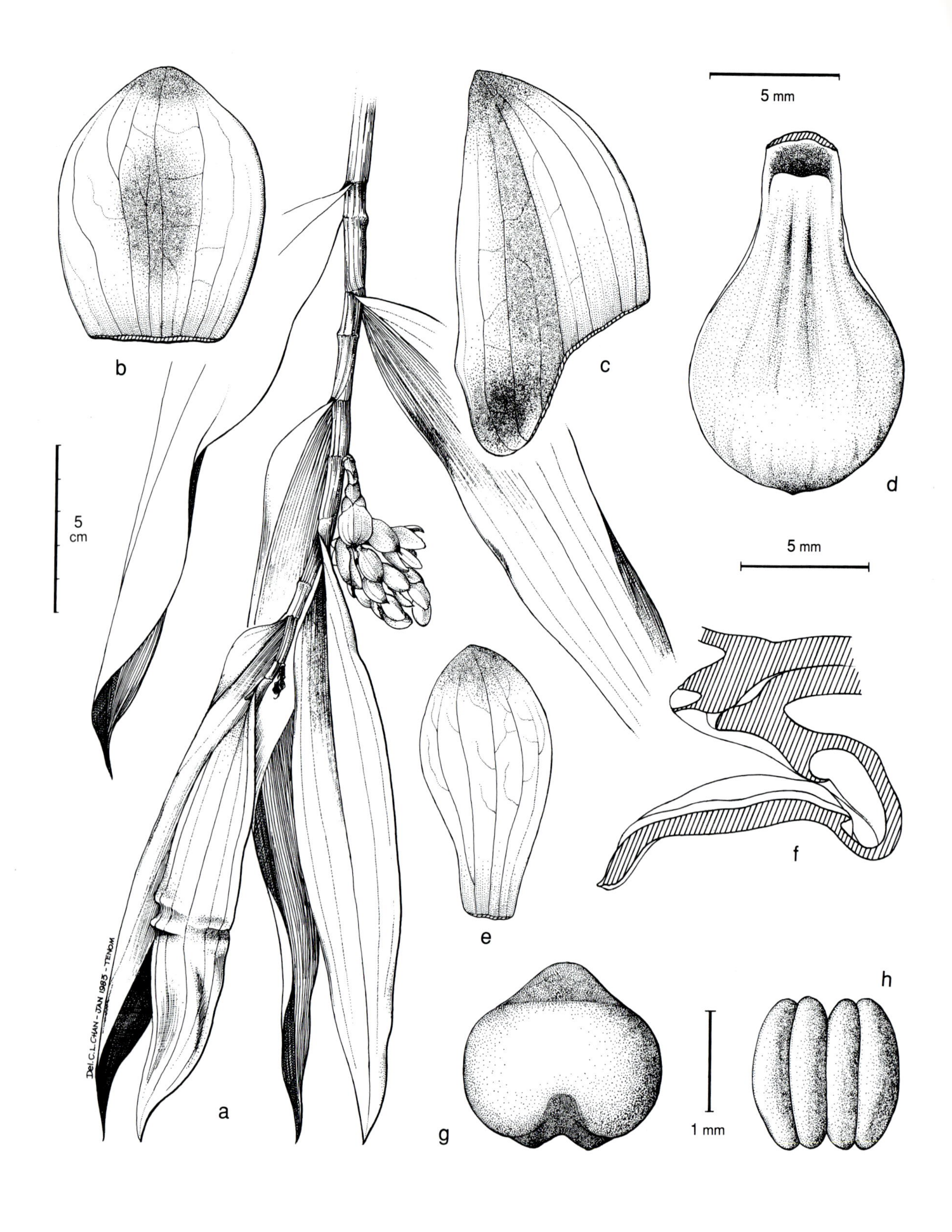

Figure 26. Dendrobium microglaphys Rchb. f. - A: plant. - B: dorsal sepal. - C: lateral sepal. - D: lip. - E: petal. - F: column and lip, longitudinal section. - G: anther. - H: pollinia. All drawn from *Lamb* AL 61/83 by Chan Chew Lun.

26. DENDROBIUM MICROGLAPHYS Rchb.f.

Dendrobium microglaphys *Rchb.f.* in Gard. Chron. 1868: 1014 (1868). Type: Borneo, *Low* s.n., cult. *Bullen* (holotype W).

D. callibotrys Ridl. in J. Linn. Soc., Bot. 32: 258 (1896). Type: Singapore, Toas, Sungai Mora, *Ridley* s.n. (holotype SING).

Epiphytic or terrestrial ***herb***. ***Pseudobulbs*** (20-)28-44 cm long, 5-8 mm broad, slender, cane-like, strongly sulcate, erect or pendent when epiphytic, leafy above, enclosed in the fibrous remains of the old leaf-sheaths below. ***Leaves*** (2-)6-10, (8-)15-19 x 2.7-4.5 cm, apical, narrowly elliptic, obliquely acute to acuminate, thin-textured, coriaceous, conduplicate at base, upper surface sparsely brown furfuraceous-ramentaceous towards the base, otherwise ± glabrous, lower surface similar, particularly towards the base and along the midrib; sheaths 1-2.5 cm long, sulcate, green turning pale brown, distinctly furfuraceous-ramentaceous, appearing brown-speckled. ***Inflorescence*** a rather dense, (6)8- to 12-flowered raceme (the Kinabalu form has up to 20 flowers borne on an upper node at the base of the leaf sheath); peduncle 0.5-1.5 cm long, almost entirely enclosed by 4-6 scarious, slightly furfuraceous-ramentaceous, green to pale brown ovate-elliptic, acute, imbricate sheaths each 4-8 mm long; rachis 1-2.5 cm long; floral bracts ovate-elliptic to narrowly elliptic, obtuse to acute, concave, scarious, pale fawn, slightly furfuraceous-ramentaceous to ± glabrous, distinctly nerved, equalling or longer than pedicel with ovary, 0.9-1.3 x 0.4-0.8 cm. ***Flowers*** with white sepals and petals; lip white, tip pale yellow, or yellow, with 5 or more orange-red striations, sweetly scented (the Mt. Kinabalu form has mauve striations on the lip). ***Pedicel*** with ***ovary*** 1 cm long, narrowly clavate, curving. ***Sepals*** glabrous. ***Dorsal sepal*** 0.9 x 0.7-0.75 cm, oblong to oblong-elliptic, obtuse. ***Lateral sepals*** 1.4 cm long, 1.1 cm broad at base, obliquely ovate-elliptic, acute, adnate to column-foot; mentum 7 mm long, obtuse and rounded. ***Petals*** 1 x 0.4 cm, obovate-elliptic, acute and minutely papillose at apex. ***Lip*** 1.2 x 0.7 cm, oblong-spathulate, entire, obtuse, hard and fleshy, margins somewhat upturned, disc rough and papillose, provided with a truncate shelving ledge-like appendage near the base. ***Column*** 0.2 cm long, with 2 short apical stelidia, column-foot 0.7 cm long; anther-cap ovate, obtuse, 1 x 1.5 mm; pollinia 4 in 2 adpressed pairs, slightly unequal. Plate 5F.

HABITAT AND ECOLOGY: Podsol forest on sandstone composed of *Dacrydium* etc., with a *Rhododendron malayanum* understorey on sandstone; low, rather open swampy forest on ultramafic substrate; coastal *Agathis* forest; usually found growing low down on tree trunks near the forest floor or sometimes as a terrestrial. Reported occurring on mangrove by Holttum (1964). Alt. sea level to 800 m. Flowering sporadic, but often in November.

DISTRIBUTION IN BORNEO: KALIMANTAN TENGAH: Sampit area. KALIMANTAN TIMUR: Balikpapan area. SABAH: Pamol area; Nabawan area; Mt. Tawai; Mt. Kinabalu. SARAWAK: Baram District; Kuching area.

GENERAL DISTRIBUTION: Southern Peninsular Malaysia, Singapore and Borneo.

NOTES: Reichenbach (1868) originally thought *D. microglaphys* to be allied to *D. aduncum* Wall. ex Lindl. (section Breviflores) native to China and from the Himalayan region to Thailand. Ridley (1896) placed his *D. callibotrys* in Hooker's section Breviflores (as "Breviflorae"), but Holttum (1964) subsequently included it, along with *D. linguella* Rchb.f. in section Eugenanthe Schltr., now renamed section Dendrobium (Seidenfaden 1985). *D. linguella*, also native to Borneo, is now included in the quite distinct Breviflores. Further study of *D. microglaphys*, however, shows it to belong to section Amblyanthus established by Schlechter in 1905 and hitherto only recorded from New Guinea and the Solomon Islands.

Section Amblyanthus is characterised by the short or much abbreviated inflorescence provided with conspicuous scarious sterile basal sheaths and floral bracts. The lip, which may be entire or obscurely 3-lobed, is somewhat obovate and always has a distinctive fleshy flange-like basal appendage. The individual species appear to show little variation and are rather uniform in habit. Schlechter (1912) records nine species from New Guinea, all except two of which inhabit montane "mist forest" where they frequently grow low down on tree trunks. Section Amblyanthus appears to be allied to section Breviflores Hook.f. rather than to Distichophyllum Hook.f. as suggested by Schlechter. *D. microglaphys* differs from the allied New Guinean *D. melanostictum* Schltr. and *D. squamiferum* J.J.Sm. by the longer, broader leaves and papillose rather than barbellate lip with an entire instead of toothed basal appendage. In *D. microglaphys* the lower part of the leaf blade, leaf sheaths and floral bracts are scaly but the ovary and sepals are glabrous. In all other respects, however, it agrees well with the circumscription of the section. Ridley (1896) records that the peduncle and pedicels elongate in fruit.

DERIVATION OF NAME: The specific epithet is derived from the Greek *micro*, small, little, and *glaphy*, meaning a hole or hollow, referring to the base of the lip.

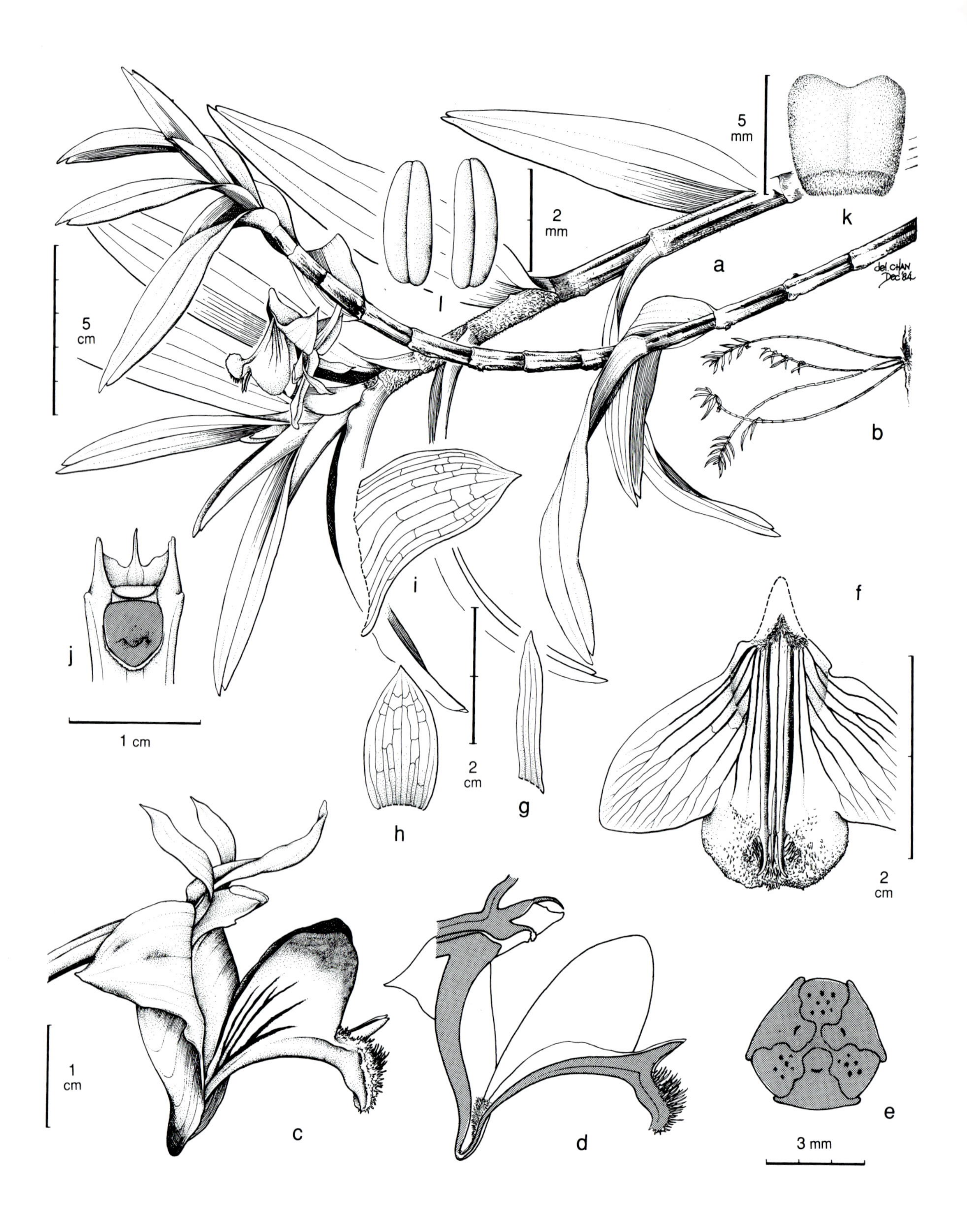

Figure 27. Dendrobium olivaceum J.J. Sm. - A & B: plant. - C: flower, lateral view. - D: column and lip, longitudinal section. - E: ovary, transverse section. - F: lip, spread out. - G: petal. - H: dorsal sepal. - I: lateral sepal. - J: column, ventral view. - K: anther. - L: pollinia. All drawn from *Lamb* AL 298/84 by Chan Chew Lun.

27. DENDROBIUM OLIVACEUM J.J.Sm.

Dendrobium olivaceum *J.J.Sm.* in Bull. Jard. Bot. Buitenzorg, ser. 2, 8: 41 (1912). Type: Borneo, Kalimantan, Liang Gagang, *Hallier* s.n., cult. Bogor under number 540a (holotype BO).

Large epiphytic ***herb*** up to 90 cm high. ***Roots*** white, elongate, branching, up to 3 mm in diameter. ***Stems*** cane-like, to 1 cm thick, spreading, or pendent, leafy in apical half, more or less covered by sheathing leaf bases, yellow when dry. ***Leaves*** 5-11 x 0.8-2 cm, suberect to spreading, coriaceous, subacute and unequally bilobed at apex, articulated to sheathing leaf-bases up to 5 cm long, with a black, scurfy, stellate indumentum. ***Inflorescences*** emerging through leaf-sheaths on upper part of stem; peduncle 3 mm long, 1-flowered; floral bract ovate, 4-5 mm long, dark green. ***Flowers*** very fleshy, non-resupinate, sweetly scented; sepals greenish brown to olive turning orange with age; petals orange-brown; lip white with olive-green to dark brown veins, green keels and orange-green margins to the side lobes; apex of mid-lobe light brown with tufts of white hairs; column white with a green base to the foot. ***Pedicel*** with ***ovary*** up to 2.5 cm long, green. ***Dorsal sepal*** 2-2.5 x 0.7-1 cm, porrect, ovate-lanceolate, acute, slightly upturned at the apex and concave at the base. ***Lateral sepals*** 2-3.2 x 2-2.5 cm, spreading, reflexed in apical half, obliquely oblong-triangular, acute; mentum c. 2-2.5 cm long, conical. ***Petals*** 2-2.2 x 0.5-0.6 cm, spreading, much narrower than sepals, linear, apex acute, slightly reflexed, 2-2.2 x 0.5-0.6 cm. ***Lip*** 3-lobed in apical half, 2.5-4 x 3.5-4 cm, pubescent at base, side lobes erect, obliquely triangular, obtuse or rounded; mid- lobe very fleshy, transversely elliptic-oblong, erose; callus of three longitudinal ridges, the outer two from base almost to the apex of the lip ending in two erect teeth, the central one on mid-lobe only, with a tuft of papillae at the apex of ridges on mid-lobe. ***Column*** 0.9-1 cm long, fleshy, laterally acute at apex, with two obscure teeth on lower margin, column-foot slightly sigmoid, 2-2.5 cm long. Plate 6A.

HABITAT AND ECOLOGY: Lower montane forest, epiphytic on *Phyllocladus*; very low and open podsol forest, recorded as a terrestrial among shrubs on a sandstone road cutting. Alt. 900 to 1600 m. Flowering usually in October.

DISTRIBUTION IN BORNEO: KALIMANTAN: Liang Gagang. SABAH: Mt. Kinabalu; Mt. Alab; Kimanis Road.

GENERAL DISTRIBUTION: Endemic to Borneo.

NOTES: *D. olivaceum* belongs to section Distichophyllum. Plants collected at 1700 m along the Kiau View Trail on Mt. Kinabalu (*Lamb & Phillipps* AL 161/83 (K) and *Lamb* SAN 89671) are very similar to this species but have pendulous stems up to 110 cm long, and very large flowers 5 cm across.

DERIVATION OF NAME: The specific epithet is derived from the Latin *olivaceus*, olive-green, referring to the flower colour.

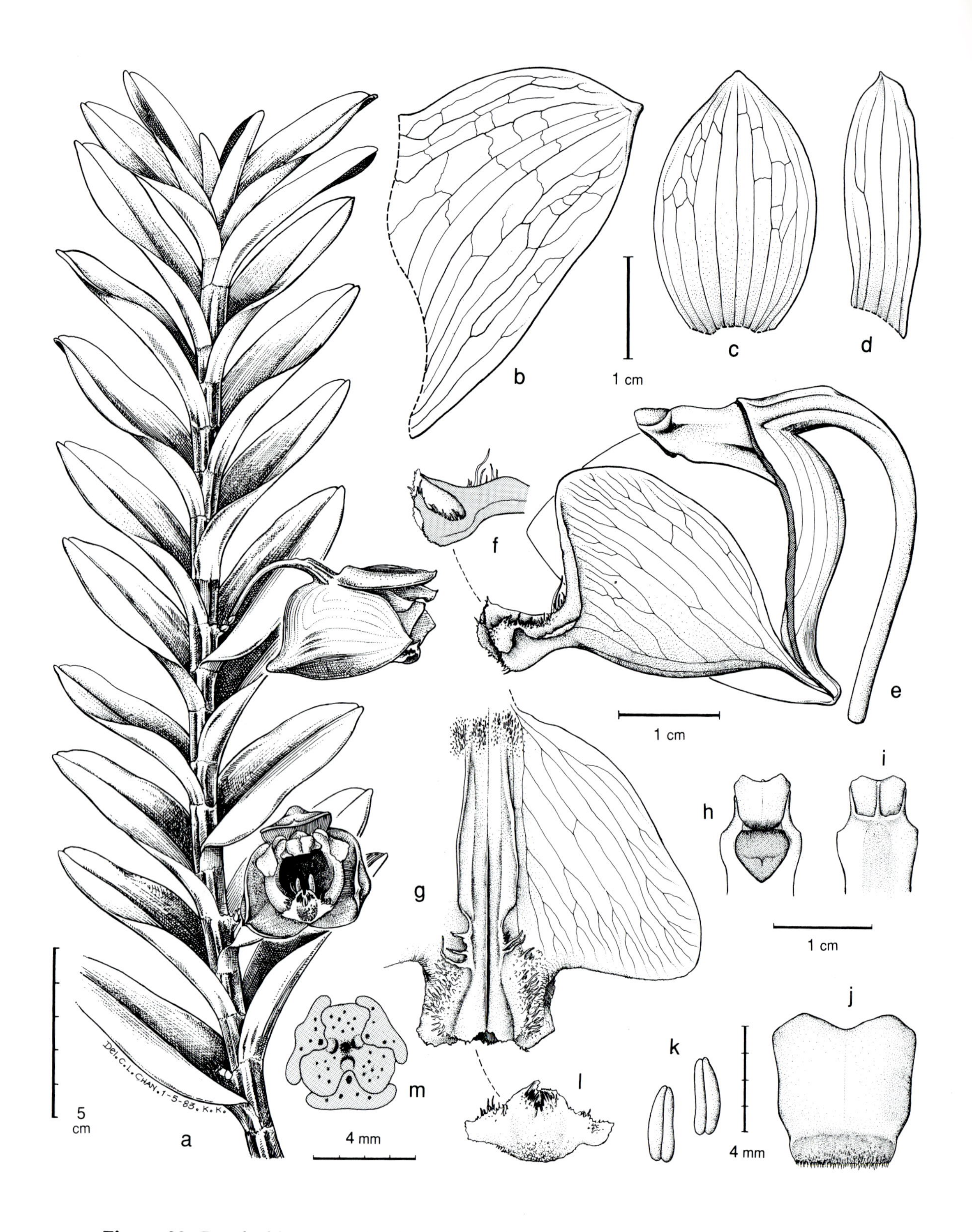

Figure 28. Dendrobium piranha C.L. Chan & P.J. Cribb. - A: plant. - B: lateral sepal. - C: dorsal sepal. - D: petal. - E: flower, lateral view. - F: lip mid-lobe, transverse section. - G: lip, spread out. - H: column, ventral view. - I: column, dorsal view. - J: anther. - K: pollinia. - L: lip mid-lobe, front view. - M: ovary, transverse section. All drawn from *Bailes & Cribb* 815 by Chan Chew Lun.

28. DENDROBIUM PIRANHA C.L.Chan & P.J.Cribb

Dendrobium piranha *C.L.Chan et P.J.Cribb* **sp. nov.** (sect. *Distichophyllo*) *D. olivaceo* J.J.Sm. affinis sed foliis brevioribus et latioribus, floribus resupinatis, petalis oblongis obtusis, sepalis petalisque non reflexis, lobo medio labelli subquadrato carnoso ad apicem excavato, lamellis calli trilaceratis satis distinguenda. Typus: Borneo, Sabah, Mt. Kinabalu, Marai Parai Spur, 1600 m, *Bailes & Cribb* 815 (holotypus K).

Large erect epiphytic or terrestrial ***herb.*** ***Stems*** to 2 m or more long, cane-like, flexuous, leafy. ***Leaves*** up to 8 x 1.8 cm, distichous, narrowly elliptic-lanceolate, unequally obliquely roundly bifid at apex, greyish green, articulated below to a tubular leaf-sheath 1-2.5 cm long, covered with scurfy brown or black hairs when young. ***Inflorescences*** 1-flowered, emerging through leaf-sheath opposite leaf below; peduncle very short; floral bracts elliptic, obtuse, 4 mm long. ***Flowers*** very fleshy, olive-brown to flesh-coloured or salmon, not opening widely. ***Pedicel*** with ***ovary*** up to 3.5 cm long. ***Dorsal sepal*** 1.6-2.6 x 1-1.1 cm, ovate-elliptic to elliptic, acute. ***Lateral sepals*** 2.5-3 x 1.1-2 cm, obliquely ovate-elliptic, subacute, forming with the column-foot a slightly S-shaped, conical, inflated mentum, (1-)2-2.5 cm long. ***Petals*** 2.5-2.6 x 0.8 cm, oblong, obtuse. ***Lip*** 3-lobed, very fleshy, 3.5 x 4 cm when flattened; side lobes erect, obliquely obovate, rounded; mid-lobe very fleshy, subquadrate, excavated in front, papillose-pubescent; callus dark brown, of 3 ridges, the outer two lacerate towards apex, the middle one obscure. ***Column*** c. 1 cm long, with two acute apical stelidia. Plate 6B.

HABITAT AND ECOLOGY: Lower montane forest; low scrubby ridge forest on serpentine; unstable rocky ground on ridges; recorded from ultramafic and sandstone substrates. Alt. 1400 to 2400 m.

DISTRIBUTION IN BORNEO: SABAH: Mt. Kinabalu.

GENERAL DISTRIBUTION: Endemic to Borneo (Sabah only).

NOTES: The largest flowered member of section Distichophyllum. Closely allied to *D. olivaceum* J.J.Smith but with resupinate flowers that do not open widely and have broader obtuse petals and a distinctive lip with a smaller fleshier mid-lobe and a more lacerate callus.

DERIVATION OF NAME: The specific epithet *piranha* refers to the sinister-looking flowers, whose unusual excavated lip gives the appearance of an Amazonian piranha fish with its jaws open, the sharply pointed lacerate apical calli being the fish's teeth. *D. piranha* has become affectionately known by the nickname 'jaws'.

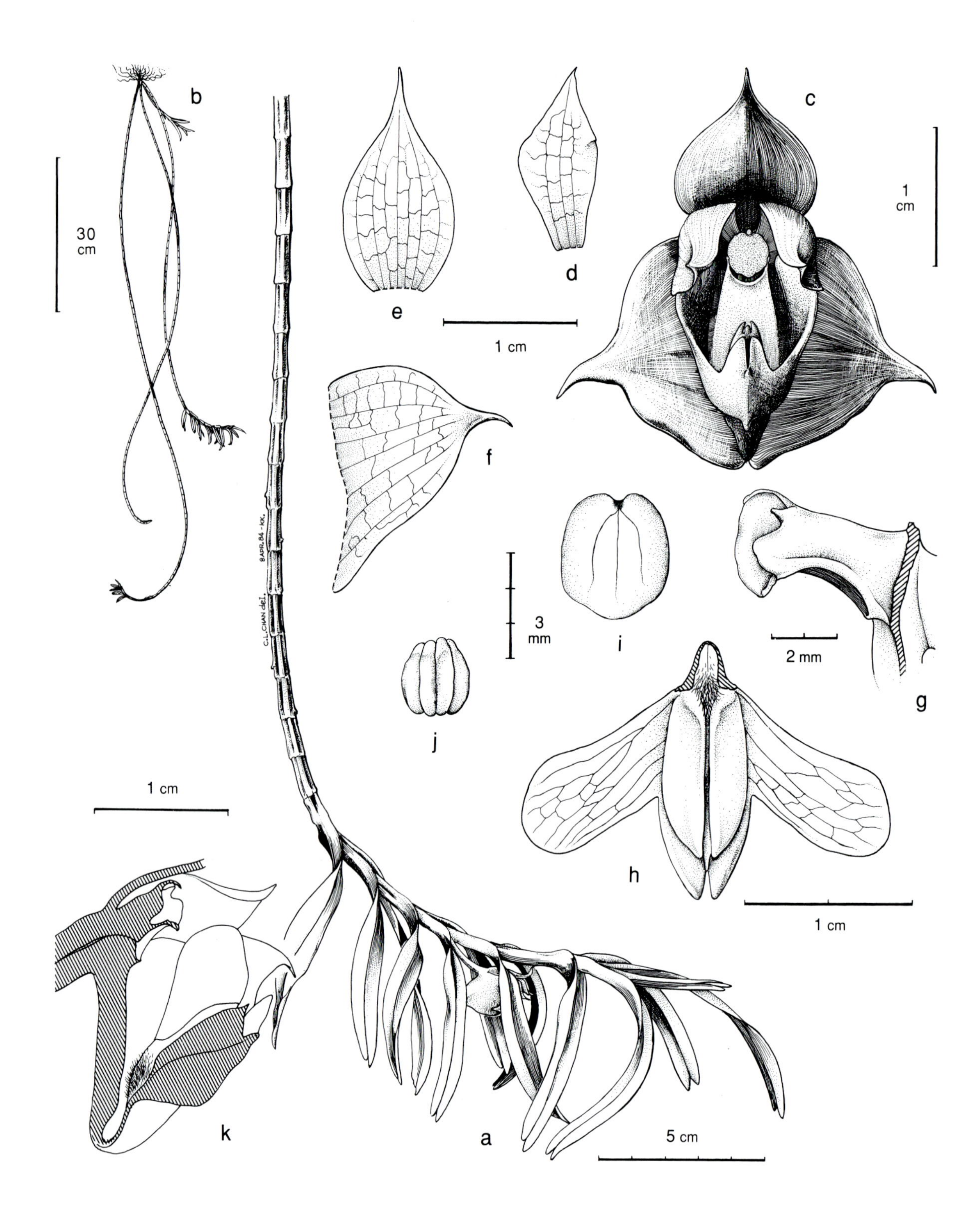

Figure 29. Dendrobium sandsii J.J. Wood & C.L. Chan. - A & B: plant. - C: flower. - D: petal. - E: dorsal sepal. - F: lateral sepal. - G: column, lateral view. - H: lip. - I: anther. - J: pollinia. - K: column and lip, longitudinal section. All drawn from *Sands* 4064 by Chan Chew Lun.

29. DENDROBIUM SANDSII J.J.Wood & C.L.Chan

Dendrobium sandsii *J.J.Wood et C.L.Chan* **sp. nov.** (sect. *Distichophyllo*) *D. olivaceo* J.J.Sm. affinis sed habitu pendulo, floribus resupinatis, petalis oblique anguste-ellipticis, lobis lateralibus labelli angustioribus, lobo medio anguste ovato emarginato glabro, callo trilamellato humili carnoso non lacerato, anthera glabra distinguenda. Typus: Borneo, Sabah, Kinabatangan District, Maliu river valley, ESE of Enteleben & SW of Bukit Tawai summit, 6 April 1984, *Sands* 4064 (holotypus K).

Pendulous epiphytic ***herb.*** ***Stems*** (60-)95-112 cm long, entirely covered by persistent leaf-sheaths, becoming curved towards apex. ***Leaves*** 6-7 x 1-1.2 cm, ligulate, acutely unequally bilobed, articulate, distichous. ***Inflorescence*** 1-flowered; floral bracts 3 mm long, ovate, acute. ***Flowers*** resupinate; sepals dull pale green ageing to dull orange with a greenish apex; petals greenish white ageing to pale orange; lip white with pale olive side lobes and mentum. ***Pedicel*** with ***ovary*** 1.5-1.6 cm long, narrowly clavate. ***Sepals*** spreading. ***Dorsal sepal*** 1.5-1.6 x 0.8-0.9 cm, ovate, acuminate, apex recurved. ***Lateral sepals*** 1.5 x 1.4 cm, obliquely broadly ovate, acuminate, apex somewhat recurved, adnate to column-foot to form an obtuse mentum 1-1.2 cm long. ***Petals*** 1.3 x 0.6 cm, obliquely narrowly elliptic, acute. ***Lip*** 3-lobed, 1.5 cm long x 2 cm wide when flattened, very fleshy, with a hairy tuft at the base; side lobes 1.2-1.3 x 0.5 cm, oblong, rounded, erect; mid-lobe 0.5-0.6 cm, narrowly ovate, apex deeply emarginate, with a small mucro below the sinus on the lower surface; callus of 3 hard, low fleshy keels, the median lower and slightly longer than the outer. ***Column*** 0.5 cm long, with 4 acute apical stelidia, column-foot 1-1.2 cm long; anther-cap 3.5 mm long, oblong, glabrous. Plate 6C.

HABITAT AND ECOLOGY: Mixed hill-forest, found on a rotting trunk in partial shade near a river. Alt. 350 m.

DISTRIBUTION IN BORNEO: SABAH: Kinabatangan District.

GENERAL DISTRIBUTION: Endemic to Borneo (Sabah only).

NOTES: *D. sandsii* is distinguished from *D. olivaceum* J.J.Sm. by its pendulous habit and resupinate flowers with broader, obliquely narrowly elliptic petals, lip with narrower side lobes and a narrowly ovate, emarginate, glabrous mid-lobe. The callus consists of three low fleshy keels and the anther is glabrous. *D. olivaceum* is at once distinguished from *D. sandsii* by its non-resupinate flowers with narrow petals, lip with a broad papillose-verruculose mid-lobe and prominent, erect lacerate keels and hairy anther-cap.

DERIVATION OF NAME: Named in honour of Martin Sands, the *Begonia* specialist at Kew, who collected the type.

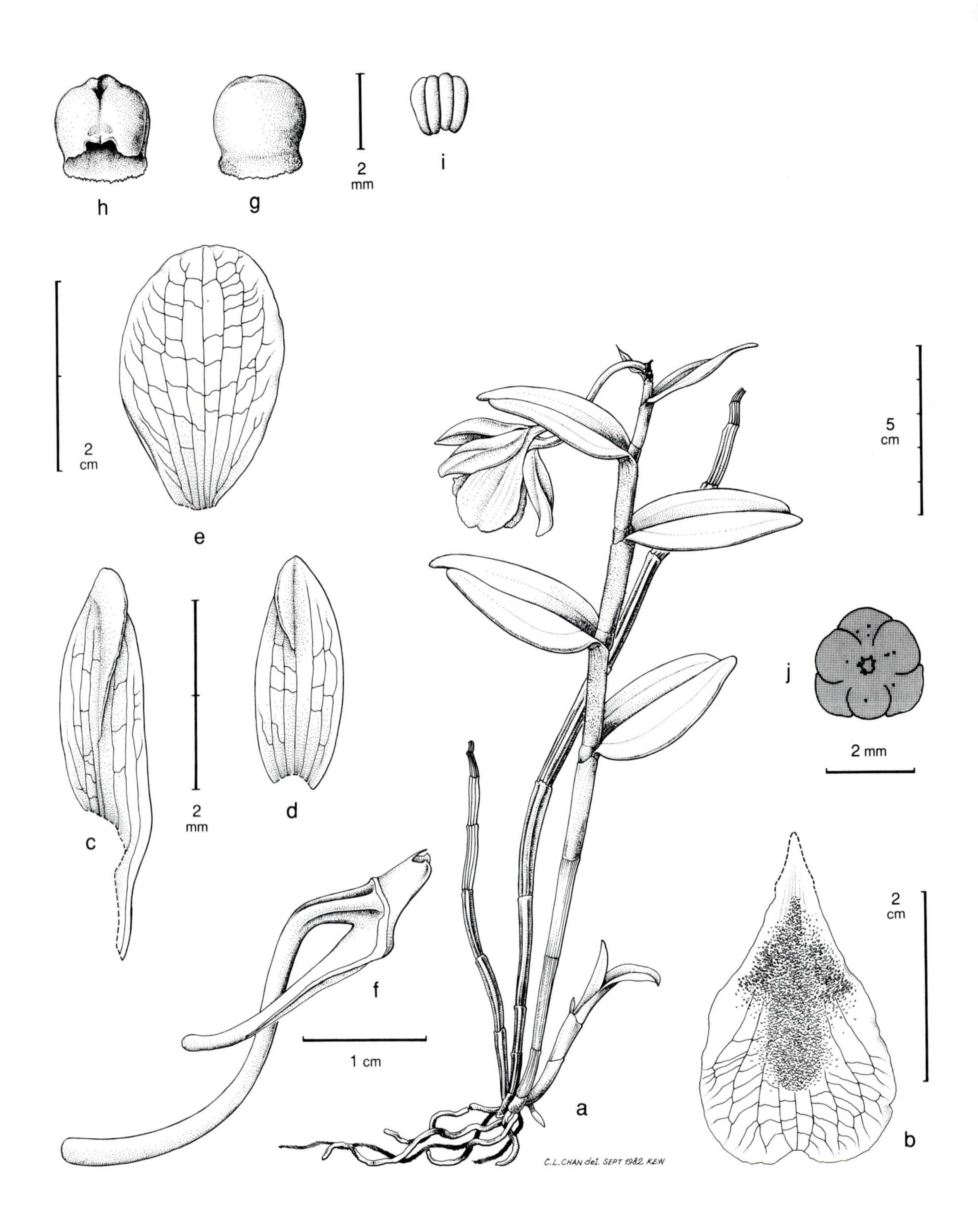

Figure 30. Dendrobium sculptum Rchb. f. - A: plant. - B: lip. - C: lateral sepal. - D: dorsal sepal. - E: petal. - F: column and pedicel with ovary, lateral view. - G: anther, abaxially. - H: anther, adaxially. - I: pollinia. A drawn from living plant cultivated at Kew, B-I from *Lamb* SAN 93491 by Chan Chew Lun.

30. DENDROBIUM SCULPTUM Rchb.f.

Dendrobium sculptum *Rchb.f.* in Bot. Zeitung (Berlin) 21: 128 (1863). Type: Borneo, imported by *Messrs. H. Low & Co.,* cult. *Bullen* (holotype W).

Epiphytic ***herb*** up to 75 cm high. ***Roots*** branching, 1-1.5 mm in diameter, grey-white. ***Stems*** erect to pendulous when old, cane-like, yellow, covered by sheathing leaf-bases, leafy in upper half. ***Leaves*** 3-6 x 1.5-3 cm, elliptic to ovate-elliptic, unequally roundly bilobed at apex, 3-6 x 1.5-3 cm, densely covered in black scurfy hairs above and below, articulated to sheathing leaf-bases; sheaths cylindric, up to 3.5 cm long, densely covered in black hairs. ***Inflorescences*** borne from near the stem apex, emerging through leaf-bases, 1- to 4-flowered; peduncle very short, up to 1.5 cm long; floral bracts up to 2 cm long, lanceolate, acute, with a scurfy black indumentum. ***Flowers*** unscented, pure white with an orange to orange-yellow mark in the middle of the lip and deep orange around the entrance to the spur at the base of the lip; sometimes the disc of the lip is suffused pale yellow. ***Pedicel*** with ***ovary*** 4-5 cm long, white. ***Dorsal sepal*** 2.3-2.7 x 1 cm, suberect, elliptic or elliptic-ovate, subacute or acute, mid-vein apically carinate on outer surface. ***Lateral sepals*** 2.5-2.8 x 1 cm, recurved, obliquely oblong-elliptic, subacute or acute; mentum 1.8-2.5 cm long, spur-like; spur formed mostly from the fused bases of lateral sepals, cylindric, straight but slightly upcurved towards apex. ***Petals*** 2.9-3.5 x 2-2.6 cm, spreading, obovate, obtuse and truncate. ***Lip*** 3-3.5 x 2-2.4 cm, porrect to slightly decurved, 3-lobed and rugulose in basal half, side lobes obscurely obliquely oblong, obtuse; mid-lobe transversely oblong-reniform, fleshy at base. C***olumn*** 0.5 cm long, with two short, upcurved obtuse stelidia, papillate on lower margins, column-foot 0.5 cm long. Plate 6D.

HABITAT AND ECOLOGY: In open, rather low, small-crowned montane forest, on podsolic soil formations over sandstone. This forest approaches highland heath forest, with species of *Gymnostoma, Podocarpus* and *Tristaniopsis* dominant. *D. sculptum* is epiphytic on shrubs about 1 to 2 m above the ground. Alt. 1200 to 1500 m. Flowering observed in April and June.

DISTRIBUTION IN BORNEO: SABAH: Mt. Lotung. SARAWAK: Dulit Range.

GENERAL DISTRIBUTION: Endemic to Borneo.

NOTES: *D. sculptum* belongs to section Formosae.

DERIVATION OF NAME: The specific epithet is derived from the Latin *sculptus*, engraved, carved out, presumably referring to the ornamentation on the lip.

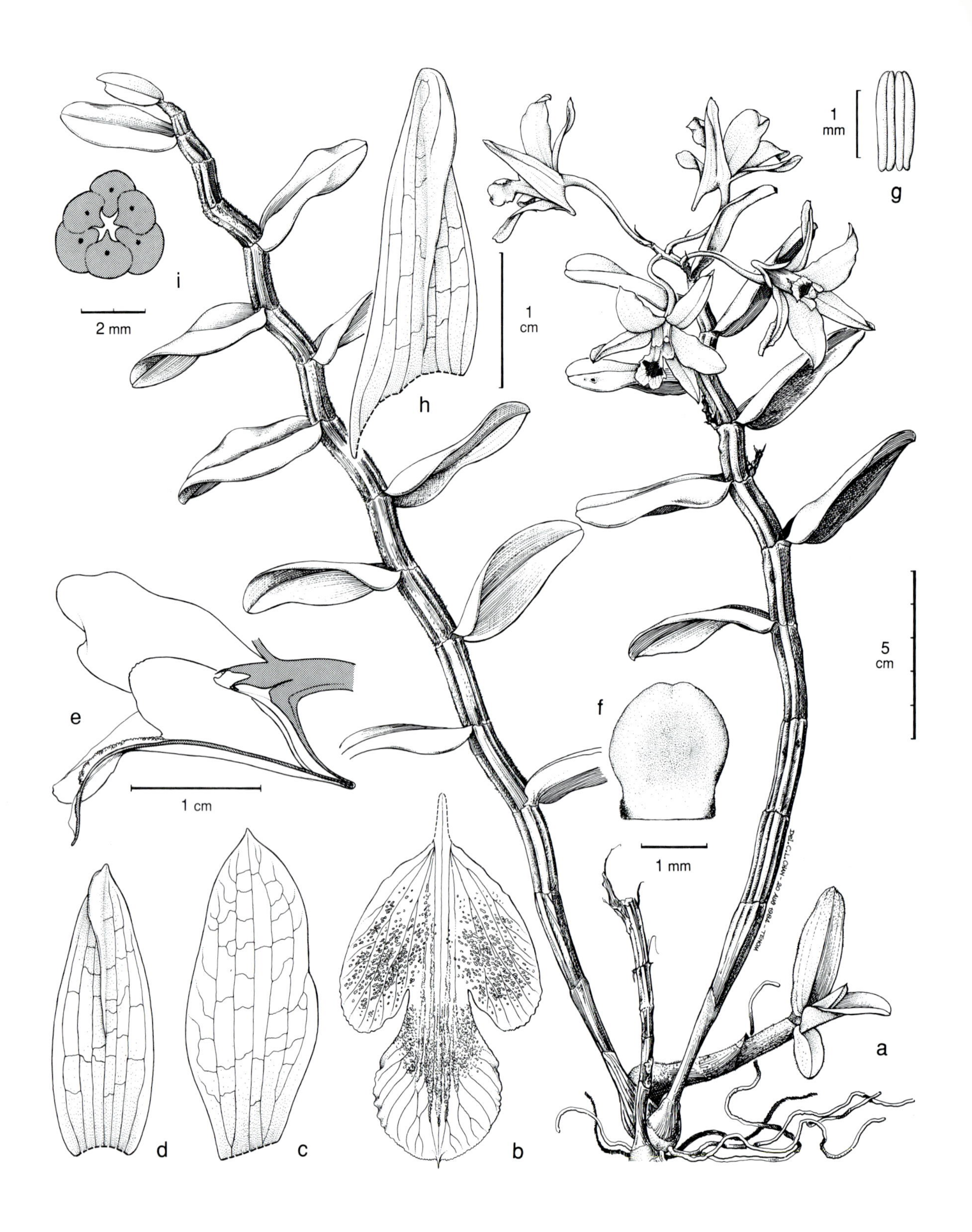

Figure 31. Dendrobium singkawangense J.J. Sm. - A: plant. - B: lip. - C: petal. - D: dorsal sepal. - E: column and lip, longitudinal section. - F: anther. - G: pollinia. - H: lateral sepal. - I: ovary, transverse section. All drawn from living plant collected from Sinsuran Road, Crocker Range by Chan Chew Lun.

31. DENDROBIUM SINGKAWANGENSE J.J. Sm.

Dendrobium singkawangense *J.J. Sm.* in Gard. Bull. Straits Settlem. 9: 91 (1935). Type: Borneo, Kalimantan, Singkawang, 20 October 1934, *Paath* s.n. (holotype BO).

Epiphytic ***herb*** often forming a large clump. ***Stems*** clustered, up to 75 cm long, 0.5-1 cm in diameter, cane-like, covered by loosely inflated sheathing leaf-bases. ***Leaves*** 4-8, up to 11.5 x 2.5-2.7 cm, distichous, oblong-elliptic, apex unequally bilobed, obtuse and rounded, coriaceous; leaf-bases sheathing, 1-4 cm long, minutely black-hairy. ***Inflorescences*** several, emerging opposite leaves at upper nodes of leaf-bearing and leafless stems, 2- to 10-flowered; scape 1-2 cm long; floral bracts c. 1 cm long, linear-lanceolate, acuminate, minutely black-hairy. ***Flowers*** opening widely, 3-4.5 cm across, waxy, unscented, pure white or creamy, with a dark orange mark on the ventral surface of the column and mid-lobe of the lip, pale orange on the side lobes, the disc yellow, the claw flushed orange. ***Pedicel*** with ***ovary*** 2.2-3.5 cm long, ovary borne at an obtuse angle to the pedicel. ***Dorsal sepal*** 1.9-2.5 x 0.6-0.7 cm, lanceolate, acute, mid-nerve carinate on outer surface. ***Lateral sepals*** 2.2-2.7 x 0.7-1 cm, obliquely oblong-lanceolate, acute, mid-nerve broadly carinate on outer surface; mentum 7-10 mm long, conical. ***Petals*** 2-2.5 x 0.7-1 cm, elliptic-lanceolate, acute. ***Lip*** 3-lobed, 2-2.3 x 1.4-1.5 cm, verrucose all over; side lobes enclosing column, obliquely oblong, rounded and erose in front; mid-lobe broadly clawed, oblong-obovate, mucronate, recurved at apex; callus short, obscure, 3-ridged, verrucose, in centre of mid-lobe. ***Column*** 0.6 cm long. Plate 6E.

HABITAT AND ECOLOGY: Riverine hill-forest; open montane mixed oak/chestnut/conifer forest, on sandstone ridges; usually epiphytic on canopy branches. Alt. 500 to 1700 m. Flowering observed in July, August, September and October.

DISTRIBUTION IN BORNEO: KALIMANTAN BARAT: Singkawang area. SABAH: Mt. Kinabalu; Mt. Alab.

GENERAL DISTRIBUTION: Endemic to Borneo.

NOTES: Another member of section Formosae. The short mentum, warty lip with conspicuous side lobes are distinctive. Black and white photographs of this species appeared in the 1935 edition of the now discontinued Dutch Journal 'De Orchidee'. A colour photograph was published in the German Orchid Society journal 'Die Orchidee' (Wood, 1988). The flower buds are a pale jade or blue-green before opening.

DERIVATION OF NAME: The specific epithet refers to Singkawang (Singawang), the type locality in Kalimantan.

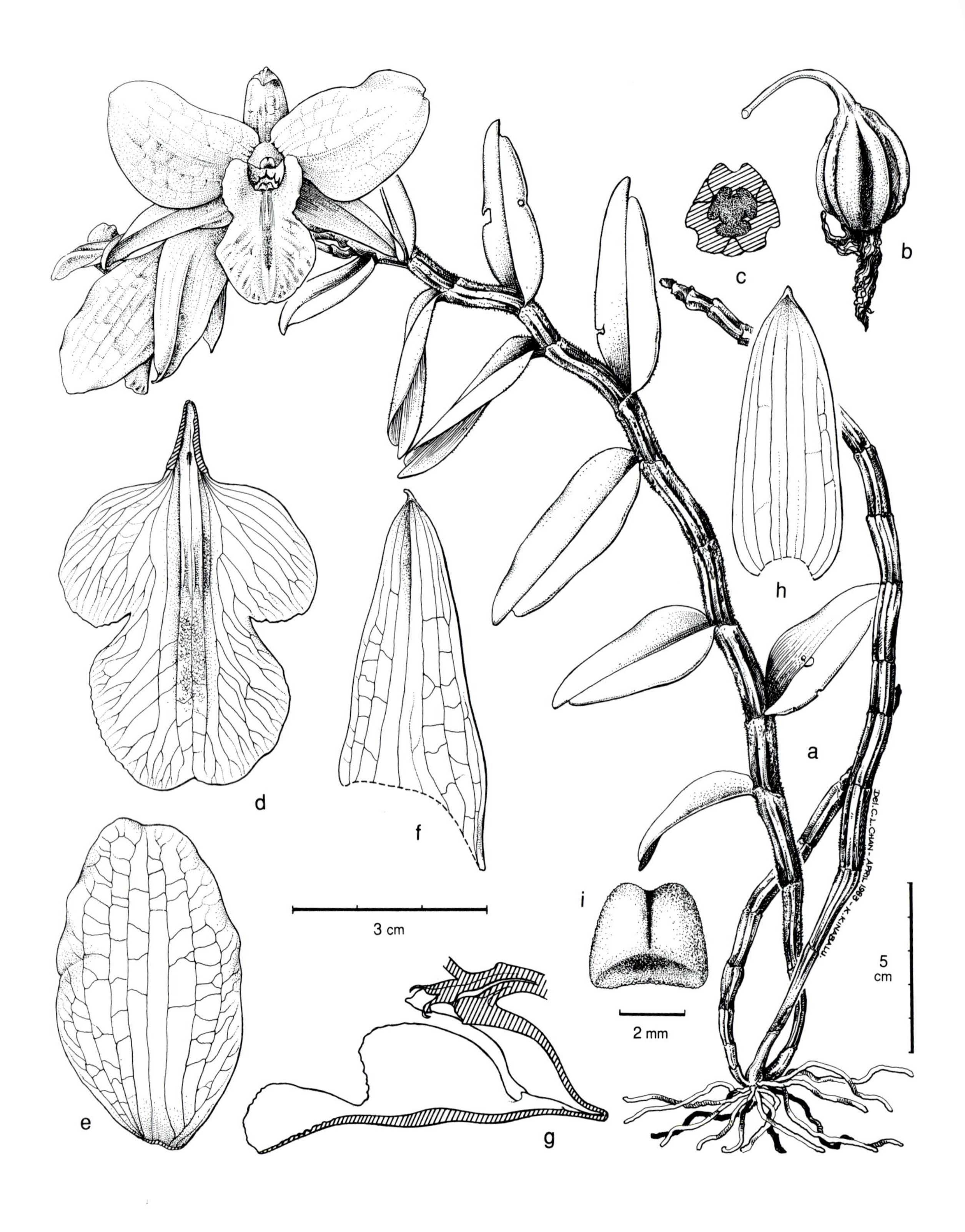

Figure 32. Dendrobium spectatissimum Rchb. f. - A: plant. - B: seed pod. - C: seed pod, transverse section. - D: lip. - E: petal. - F: lateral sepal. - G: column and lip, longitudinal section. - H: dorsal sepal. - I: anther. All drawn from *Cribb* in *Bailes* 818 by Chan Chew Lun.

32. DENDROBIUM SPECTATISSIMUM Rchb.f.

Dendrobium spectatissimum *Rchb.f.* in Linnaea 41: 41 (1877). Type: Borneo, *Lobb* s.n. (holotype W).

Dendrobium speciosissimum Rolfe in Orchid Rev. 3: 119 (1895). Type: Borneo, Sabah, Mt. Kinabalu, *Low* s.n. (holotype K).

Dendrobium reticulatum J.J. Sm. in Bull. Jard. Bot. Buitenzorg, ser.2, 13: 18 (1914). Type: Borneo, Sabah, Mt. Kinabalu, *Dumas* s.n. (holotype BO).

Epiphytic ***herb.*** ***Stems*** 30-40 cm long, 0.8 cm in diameter, internodes 2.5-2.8 cm long, erect, fractiflex, leafy. ***Leaves*** 4-4.5(-6.5) x 1.5-2.2 cm, oblong-elliptic, obtusely unequally bilobed, black-hairy, spreading, sheaths sulcate, black-hairy. ***Inflorescences*** 2-flowered, borne near stem apex; peduncle very short; floral bracts 1-2 x 0.5-0.8 cm, ovate, acute to acuminate, brownish, covered with black hairs. ***Flowers*** up to 10 cm across, long lasting, faintly fragrant, pure white, lip with a central band of colour which is mostly reddish, but clear yellow at apex, base of lip at entrance to spur deep orange edged scarlet, base of column often flushed pink to scarlet. ***Pedicel*** with ***ovary*** 2.5-6 cm long, strongly curved, narrow. ***Sepals*** and ***petals*** spreading, reticulately veined. ***Dorsal sepal*** 4-5 x 1.5-2 cm, oblong-elliptic, apex cucullate, obtuse to subacute. ***Lateral sepals*** 3.2-5.7 x 1.6-2.3 cm, obliquely oblong-triangular, carinate, apex cucullate, apiculate; mentum 0.7-1.2 cm long, narrowly conical, obtuse. ***Petals*** 3.5-5.2 x 2-4 cm, ovate or ovate-elliptic, cuneate at base, obtuse, often minutely retuse, margin slightly undulate. ***Lip*** 3-lobed, 4-6 cm long, 2.5-3.5 cm wide across side lobes, 3-3.7 cm wide across mid-lobe; side lobes c. 2 cm long, to 1.4 cm high, rounded, erect, surface minutely rugulose; mid-lobe 1.8-2.8 x 2.2-3.1 cm, transversely oval, retuse; disc with a thickened, obscurely 3-ridged central area 3 mm broad extending from the base of the lip and terminating halfway along the mid-lobe. ***Column*** 0.7-1 cm long, column-foot 1.2 cm broad at the middle, concave; anther-cap 4 x 4 mm, ovate, cucullate, retuse, papillose. Plate 6F.

HABITAT AND ECOLOGY: Epiphytic one to three metres above ground level in relatively open lower montane scrub on serpentine; most frequently found as an epiphyte on *Leptospermum*. Alt. 1600 to 1700 m.

DISTRIBUTION IN BORNEO: SABAH: Mt. Kinabalu.

GENERAL DISTRIBUTION: Endemic to Borneo (Sabah only).

NOTES: This is one of the most beautiful members of section Formosae from Borneo where it is restricted to Mt. Kinabalu. For the first week or ten days after the flower has opened the petals are curled tightly back, completely enclosing the top edges of the lateral sepals. The petals then uncurl until they form right angles to the direction of the lip. During the last two or three days of the flower's life the petals move forward and lie parallel to the lip. The flowers remain open for up to six weeks.

DERIVATION OF NAME: The specific epithet is derived from the Latin *spectatus*, beheld, seen, esteemed, in reference to the showy flowers which appear to glow in their mist shrouded habitat.

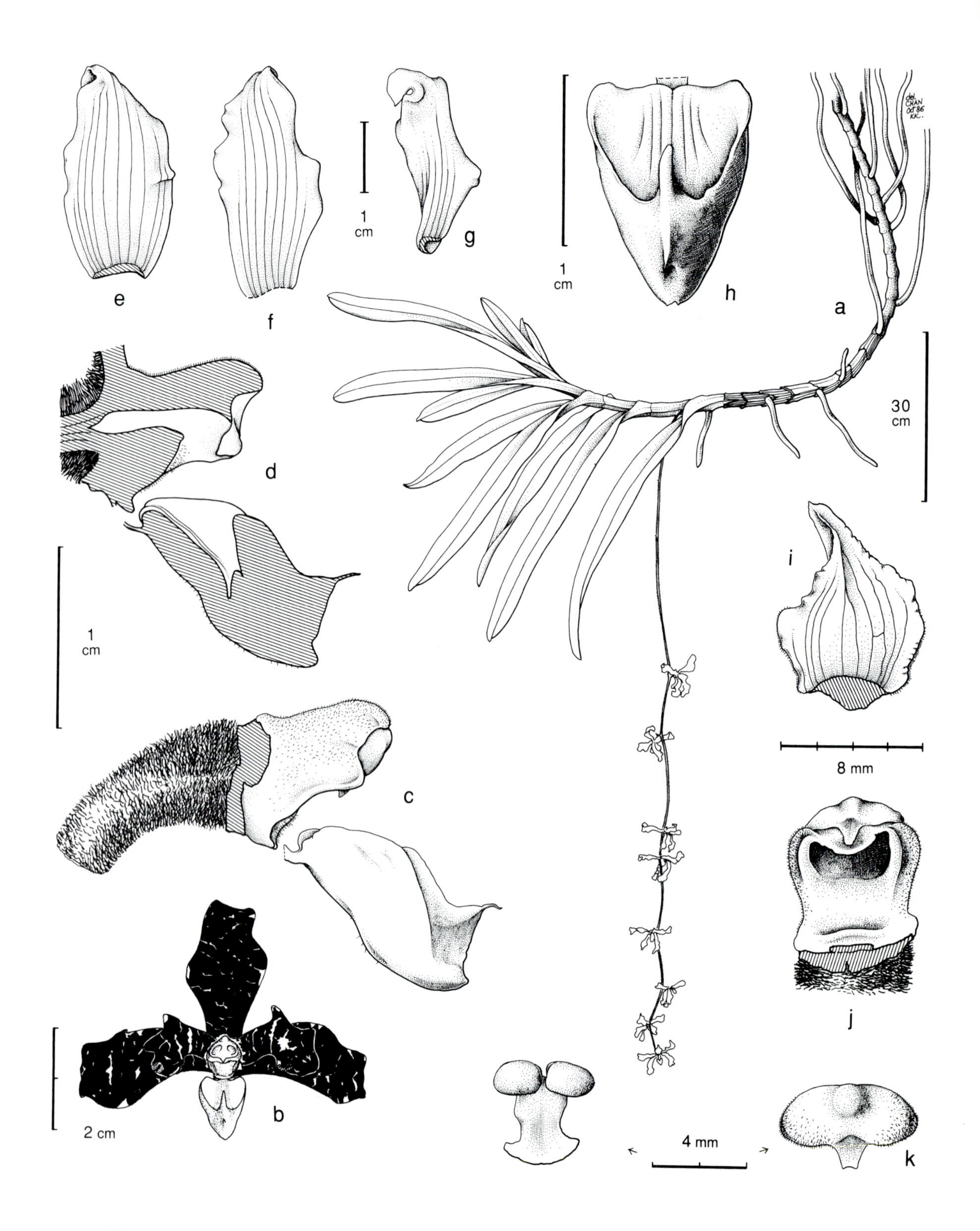

Figure 33. Dimorphorchis lowii (Lindl.) Rolfe. - A: plant. - B: upper flower. - C: flower with sepals and petals removed, lateral view. - D: column and lip, longitudinal section. - E: dorsal sepal. - F: lateral sepal. - G: petal. - H: lip. - I: floral bract. - J: column, ventral view. - K: anther. - L: pollinia. All drawn from cult. TOC by Shim Phyau Soon and Chan Chew Lun.

33. DIMORPHORCHIS LOWII (Lindl.) Rolfe

Dimorphorchis lowii (*Lindl.*) *Rolfe* in Orchid Rev. 27: 149 (1919). Type: Borneo, Sarawak, *Low* s.n. (holotype K).

Vanda lowii Lindl. in Gard. Chron. 1: 239 (1847).

Renanthera lowii (Lindl.) Rchb. f., Xenia Orch. 1: 87 (1858).

Arachnanthe lowii (Lindl.) Benth. & Hook.f., Gen. Plant. 3: 573 (1883).

Arachnis lowii (Lindl.) Rchb. f. in Beih. Bot. Centralbl. 28: 344 (1886).

Vandopsis lowii (Lindl.) Schltr. in Feddes Repert. 10: 196 (1911).

var. **lowii.**

Large epiphytic ***herb*** with long stout pendent stems up to 75 cm or more long, 1-2 cm in diameter, upturned at the apex, covered in sheathing leaf-bases, leafy in apical part. ***Leaves*** 50-70 x 4.5-5.5 cm, coriaceous, arcuate, ligulate, unequally obliquely and obtusely bilobed at the apex, articulated at base to a sheathing leaf-base 6-7 cm long. ***Inflorescences*** 1 to several, up to 300 cm long, pendent, laxly many-flowered; peduncle and rachis flexuous, tomentose, terete; floral bracts up to 3 cm long, lanceolate, acuminate. ***Flowers*** of two sorts. ***Pedicel*** with ***ovary*** c. 1 cm long, furfuraceous. ***Basal*** two to three flowers 4.5 cm high, 6 cm across, yellow spotted with purple, strongly fragrant, stellately hairy on outer surface. ***Apical*** flowers 6 cm high, 6.5 cm across, yellow to white with large irregular purple blotches all over sepals and petals, not scented, with undulate margins to the sepals and petals. ***Sepals*** 3-3.5 x 1.5 cm, spreading to recurved, ovate-lanceolate, acute. ***Petals*** 2.8-3.5 x 1.2-1.5 cm, spreading to recurved, narrowly elliptic, acute, with reflexed sides. ***Lip*** 1.2 cm long, very fleshy, 3-lobed; side lobes erect, 6 mm long, roundly triangular, with incurved margins; mid-lobe very fleshy, at an obtuse angle to the basal part of the lip, bilaterally compressed, extending into an attenuated tip; callus oblong, strongly elevated. ***Column*** 0.7 cm long, pubescent. Plate 7A, C & D.

HABITAT AND ECOLOGY: Gulley and riverine forest, often overhanging water; swamp forest; sea level to 1300 m.

DISTRIBUTION IN BORNEO: KALIMANTAN TENGAH: Bukit Raya area. SABAH: Mt. Kinabalu; Mt. Trus Madi; Crocker Range; Mt. Lotung. SARAWAK: Mt. Mulu National Park.

GENERAL DISTRIBUTION: Endemic to Borneo.

NOTES: For many years, the genus *Dimorphorchis* was considered to be monotypic, represented by a single species, *D. lowii*, with two varieties. The typical variety, which has been found in both Sabah and Sarawak, is a large plant with flowers that are markedly dimorphic. The basal flowers are bright yellow with small red spots evenly scattered over the sepals and petals while the apical flowers have slender, twisted, longitudinally rolled sepals and petals that are off-white to pale yellow and boldly blotched with reddish purple.

Var. *rohaniana* (Rchb.f.) K.W. Tan from Kalimantan and Sarawak differs from the typical variety in having narrower segments and a distinctive lip in which the fleshy mid-lobe extends into a slender fleshy point well beyond the mucronate lip apex, while the erect keel- like callus is markedly lower. Further study is needed to determine the exact status of this taxon.

DERIVATION OF NAME: The generic name is derived from the Greek *di*, two, *morphe*, form or shape, and *orchis*, an orchid, in allusion to the two sorts of flowers found in the inflorescence. The specific epithet was given in honour of Hugh Low (later Sir Hugh Low), who collected the type material in Borneo, when it first became known as Low's Necklace Vanda.

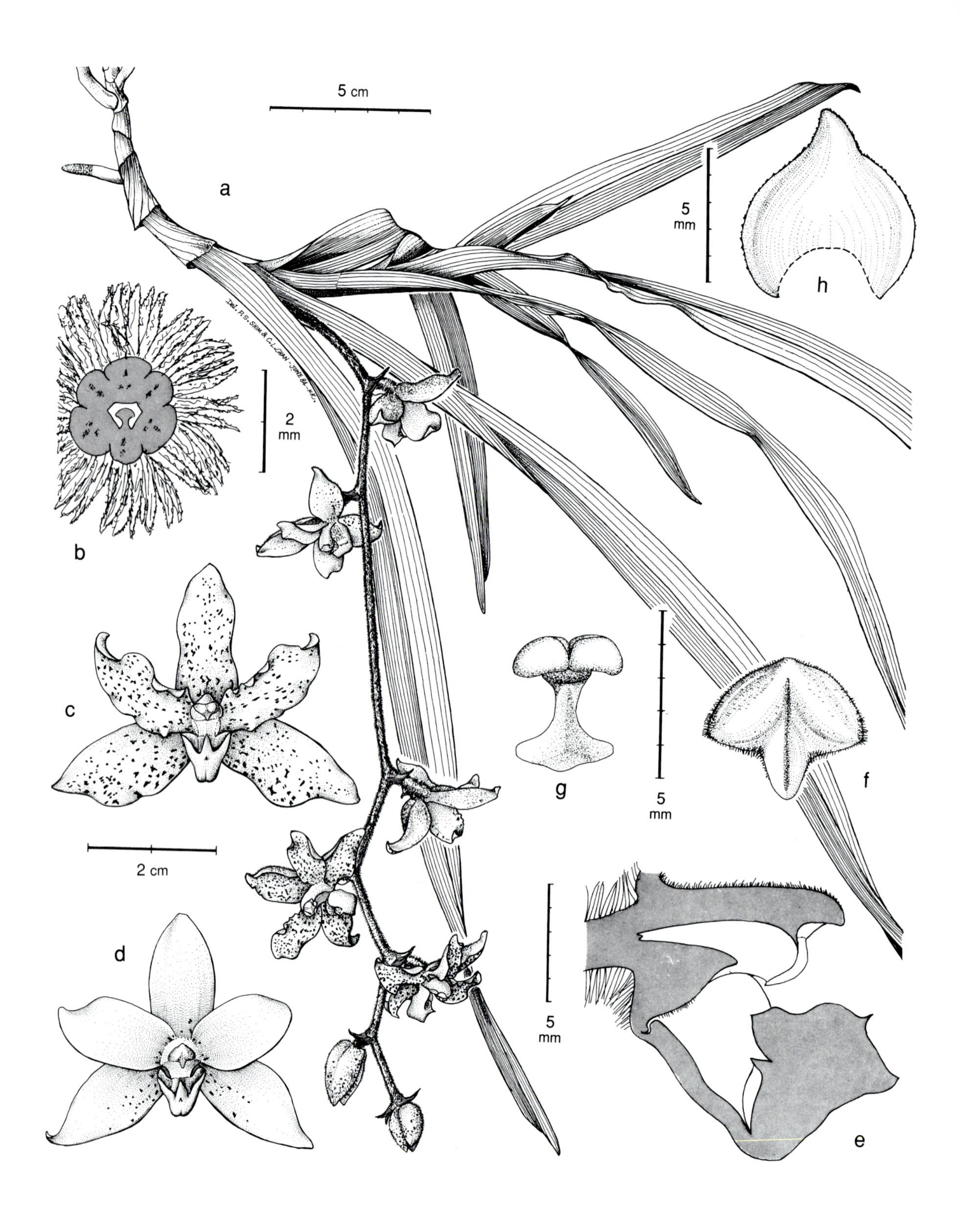

Figure 34. Dimorphorchis rossii Fowlie. - A: plant. - B: ovary, transverse section. - C: flower, upper. - D: flower, lower. - E: column and lip, longitudinal section. - F: anther. - G: pollinia. - H: floral bract. All drawn from cult. TOC by Shim Phyau Soon and Chan Chew Lun.

34. DIMORPHORCHIS ROSSII Fowlie

Dimorphorchis rossii *Fowlie* in Orchid Digest 53: 14 (1989). Type: Borneo, Sabah, Lohan River, cult. Los Angeles Arboretum, *Fowlie* & *Ross* 83P912 (holotype UCLA).

var. **rossii**

Large epiphytic ***herb*** with a spreading to pendent stem with an ascending apex, 50-200 cm long, up to 1.5 cm in diameter. ***Roots*** stout. ***Leaves*** 30-40 x 2.5-4 cm, coriaceous, arcuate, linear, unequally obliquely and acutely bilobed at apex, articulated to sheathing leaf-bases 4-6 cm long. ***Inflorescence*** 50-80 cm long, pendent, laxly few-flowered; peduncle and rachis 3-6 mm in diameter, terete, purple, tomentose; floral bracts 1.2-2.9 cm long, lanceolate or ovate-elliptic, acuminate, pubescent. ***Flowers*** of two sorts. ***Pedicel*** with ***ovary*** c. 1 cm long. ***Basal*** two flowers 3.5 cm high, 5 cm across, sweetly scented, bright yellow with a few tiny red spots at the base of the sepals and petals, and with a white lip suffused orange-yellow in the middle, and with purple-marked sides and callus. ***Apical*** flowers 3.4 cm high, 5-5.2 cm across, not scented, creamy white with small red spots all over sepals and petals, and a whitish lip with an orange-yellow centre and purple-marked sides and callus. ***Sepals*** 2.4-2.7 x 1.3-2 cm, ovate-elliptic, acute, stellately pubescent on outer surface. ***Lateral sepals*** somewhat oblique. ***Petals*** 2.1-2.7 x 1.2-1.7 cm, elliptic, acute and recurved at apex, with slightly undulate margins. ***Lip*** 0.8-0.9 cm long, fleshy, 3-lobed, minutely apiculate at apex; side lobes broadly oblong, erect; mid-lobe very fleshy, at a right angle to the basal part of the lip; callus a subquadrate erect keel with a smaller lobule behind. ***Column*** 0.7 cm long in apical flowers, 1-1.2 cm long in basal flowers. Plate 7B, 8A & B.

HABITAT AND ECOLOGY: Riverine and lower montane forest on ultramafic substrate, near or over water. Alt. 500 to 1200 m.

DISTRIBUTION IN BORNEO: SABAH: Mt. Kinabalu, etc.

GENERAL DISTRIBUTION: Endemic to Borneo (Sabah only).

NOTES: *Dimorphorchis rossii* is closely allied to *D. lowii* but differs in its slighter habit, slenderer leaves, shorter, fewer-flowered inflorescences and smaller flowers. The sepals and petals are shorter, broader and flatter in both basal and apical flowers. In the middle and lower reaches of the Labuk River and the Bidu-Bidu Hills the flowers almost lack spots but are otherwise morphologically similar (*Lamb* sight records and photographs). This species is becoming seriously endangered in the wild because of forest clearance, and fires on the ultramafics from Bukit Hempuen to Telupid.

DERIVATION OF NAME: The specific epithet refers to Earl Ross, the orchid grower at the Los Angeles Arboretum in California, who flowered the type specimen.

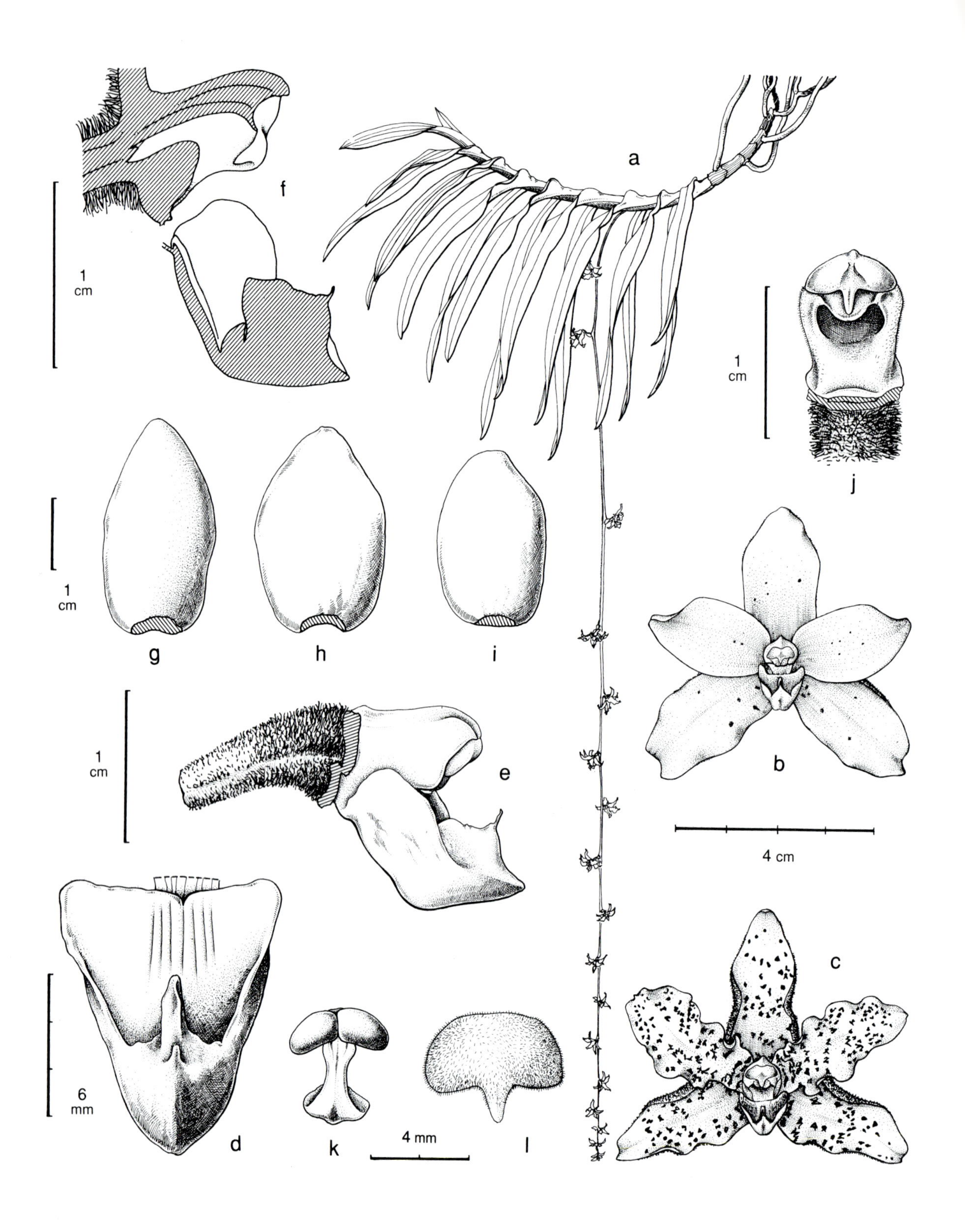

Figure 34a. Dimorphorchis rossii var. **graciliscapa** A. Lamb & Shim. - A: plant. - B: flower, lower. - C: flower, upper. - D: lip. - E: column, pedicel and lip, lateral view. - F: column and lip, longitudinal section. - G: lateral sepal. - H: dorsal sepal. - I: petal. - J: column, ventral view. - K: pollinarium. - L: anther. All from cult. *Hepburn* in *Lamb* AL 598/86, now cult. TOC. Drawn by Chan Chew Lun.

34a. DIMORPHORCHIS ROSSII Fowlie var. GRACILISCAPA A. Lamb & Shim

Dimorphorchis rossii *Fowlie* var. **graciliscapa** *A. Lamb et Shim* **var. nov.**, a varietatibus aliis habitu majore, inflorescentia longiore multiflora, floribus apicalibus punctis irregularibus parvis differt. Typus: Borneo, Sabah, Ulu Moyog, flowered in cult. 23 August 1986, *Hepburn* in *Lamb* AL 598/86 (holotypus K).

Large epiphytic ***herb*** with a stout stem to over 120 cm long, 1-1.3 cm in diameter. ***Leaves*** 47-58 x 3-6 cm, linear, unequally obliquely and obtusely bilobed at the apex, articulated to sheathing leaf-bases 4-7 cm long. ***Inflorescences*** 45-250 cm long, laxly many-flowered; peduncle and rachis terete, tomentose, flexuous; floral bracts 1.2-2 cm long, lanceolate, acuminate, pubescent at base. ***Flowers*** of two sorts, stellately pubescent on outer surface. ***Pedicel*** with ***ovary*** 0.9 cm long, densely pubescent. ***Basal*** two flowers 5.5 cm high and across, sweetly scented, orange-yellow with a few red spots on the lateral sepals. ***Apical*** flowers 5.5 cm high, 6 cm across, not scented, creamy white with small red spots all over the sepals and petals. ***Sepals*** 2.8-3.2 x 1.6-1.7 cm, oblong-ovate, obtuse. ***Lateral sepals*** of apical flowers slightly undulate and with slightly incurved margins. ***Petals*** 2.5-2.8 x 1.7-1.8 cm, elliptic, obtuse, margins in apical flowers undulate. ***Lip*** 1.2-1.3 cm long, L-shaped in side view, 3-lobed; side lobes 0.5 cm wide, erect, oblong; mid-lobe fleshy, shortly apiculate; callus with a short oblong raised keel in front and a low ridge behind. ***Column*** 0.7-0.8 cm long, fleshy pubescent. Plate 8C & D.

HABITAT AND ECOLOGY: Cool wet lower montane oak-chestnut forest. Alt. c. 300m.

DISTRIBUTION IN BORNEO: SABAH: Ulu Moyog.

GENERAL DISTRIBUTION: Endemic to Borneo (Sabah only).

NOTES: *Dimorphorchis rossii* var. *graciliscapa* is a very local plant found growing in the Crocker Range among colonies of *D. lowii*. We have treated it as a variety of *D. rossii* rather than of *D. lowii* which it resembles in its habit, because of its flower size and colouring, the shape of its petals and their attitude, the shape of the lip, the structure of the callus on the mid-lobe of the lip, and the narrow stipes of the pollinarium. Further material would undoubtedly elucidate the relationships of this taxon in what is undoubtedly a complex genus.

DERIVATION OF NAME: The varietal epithet is derived from the Latin *gracilis*, slender, and *scapus*, scape, in reference to the slender, elongate inflorescence of this variety.

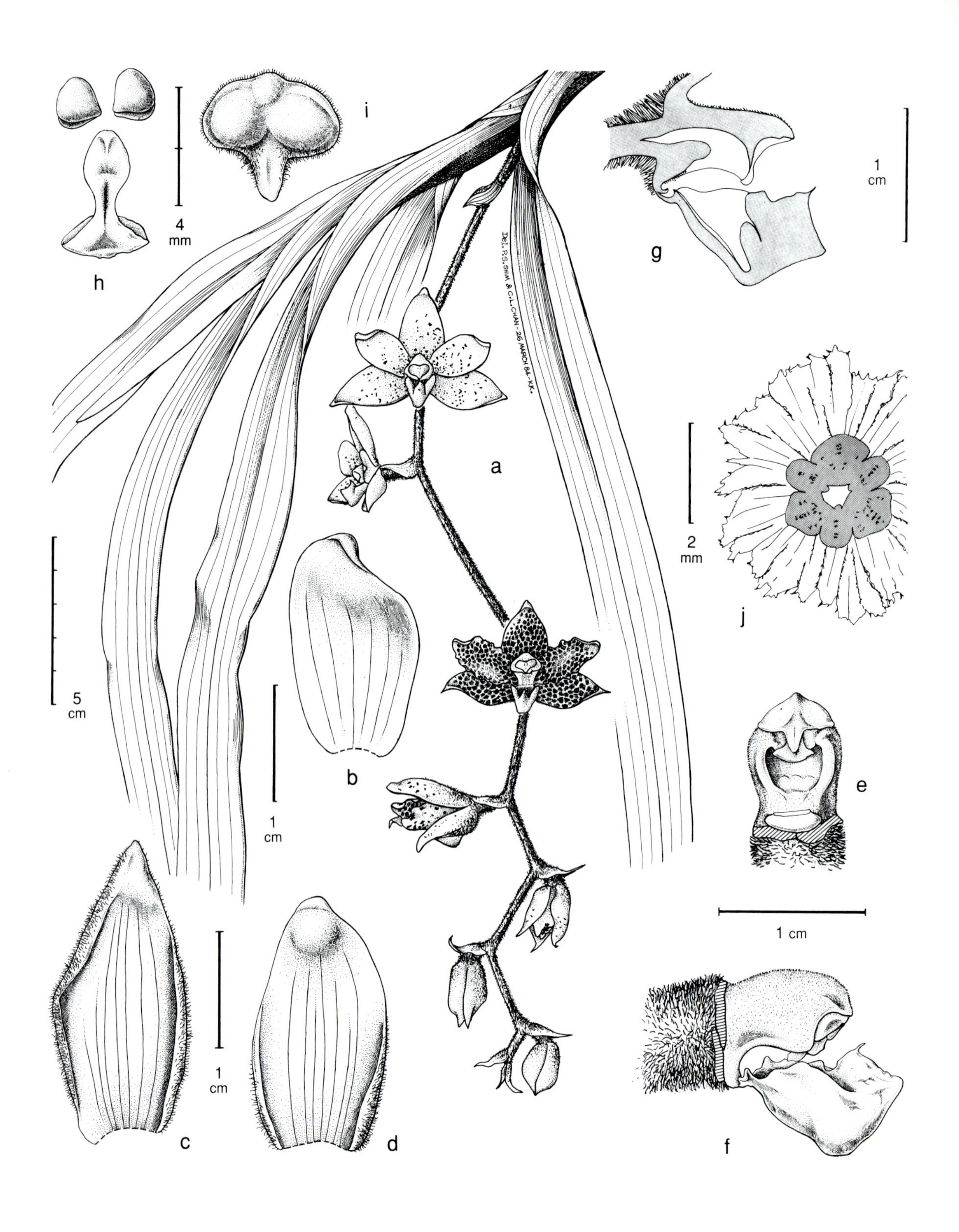

Figure 34b. Dimorphorchis rossii var. **tenomensis** A. Lamb - A: plant. - B: petal. - C: lateral sepal. - D: dorsal sepal. - E: column, ventral view. - F: column and lip, lateral view. - G: column and lip, longitudinal section. - H: pollinarium. - I: anther. - J: ovary, transverse section. All drawn from cult. TOC by Shim Phyau Soon and Chan Chew Lun.

34b. DIMORPHORCHIS ROSSII Fowlie var. TENOMENSIS A. Lamb

Dimorphorchis rossii *Fowlie* var. **tenomensis** *A. Lamb* **var. nov.** a varietate typica floribus valde purpureo-notatis et callo labelli pone lobulis ornato differt. Typus: Borneo, Sabah, Tenom District, Ulu Mentailung, above Kallang Waterfall, June 1984, *Lamb* T26 (holotypus K).

Epiphytic ***herb*** with a stout pendent or descending stem, up to 50 cm or more long, 0.6-1 cm in diameter, ascending at apex. ***Leaves*** 35-50 x 1.7-2.1 cm, arcuate, linear, obliquely and acutely bilobed at the apex, articulated to sheathing leaf-bases 4-5 cm long. ***Inflorescences*** 28-40 cm long, pendent, laxly few-flowered; peduncle and rachis terete, dark purple, tomentose; floral bracts 0.7-2 cm long, ovate, acuminate, pubescent at the base. ***Flowers*** of two sorts, showy. ***Basal*** two flowers 3.5 cm high, 5 cm across, sweetly scented, with bright yellow sepals and petals lightly spotted with red-purple all over, and with a whitish lip with a yellow centre and purple-flecked callus and sides. ***Apical*** flowers 4 cm high, 5 cm across, not scented, white heavily spotted with red purple. ***Sepals*** 2.2-2.5 x 1.2-1.3 cm, spreading, ovate, obtuse to acute. ***Petals*** 2.2-2.6 x 2-2.2 cm, elliptic, acute, with strongly undulate margins. ***Lip*** 1-1.2 cm long, fleshy, L-shaped in side view, 3-lobed; side lobes 5 mm wide, obliquely oblong, erect; mid-lobe at a right angle to the basal part, very fleshy, shortly apiculate; callus with an erect subquadrate keel in front and two smaller fleshy lobules behind. ***Column*** 0.5 cm long, papillate. Plate 8E & F.

HABITAT AND ECOLOGY: Cool, wet hill-forest and lower montane forest on sandstone ridges.Alt. 800 to 1000 m.

DISTRIBUTION IN BORNEO: SABAH: Tenom District.

GENERAL DISTRIBUTION: Endemic to Borneo (Sabah only).

NOTES: This variety is similar to the typical one in its slender habit and relatively short, few-flowered inflorescences. However it differs in its flower colour, the basal flowers being evenly spotted with red all over the sepals and petals, while the apical flowers are heavily and evenly marked with purple spots of even size that almost coalesce. The lip-callus also seems to differ slightly in having a taller keel in front and two lower ridges behind in the saccate part of the lip. So far this attractive orchid is known only from the Tenom Valley where it grows in forest on sandstone, an area some distance from the known range of the typical variety.

DERIVATION OF NAME: The specific epithet is named after Tenom District, where the type locality is situated.

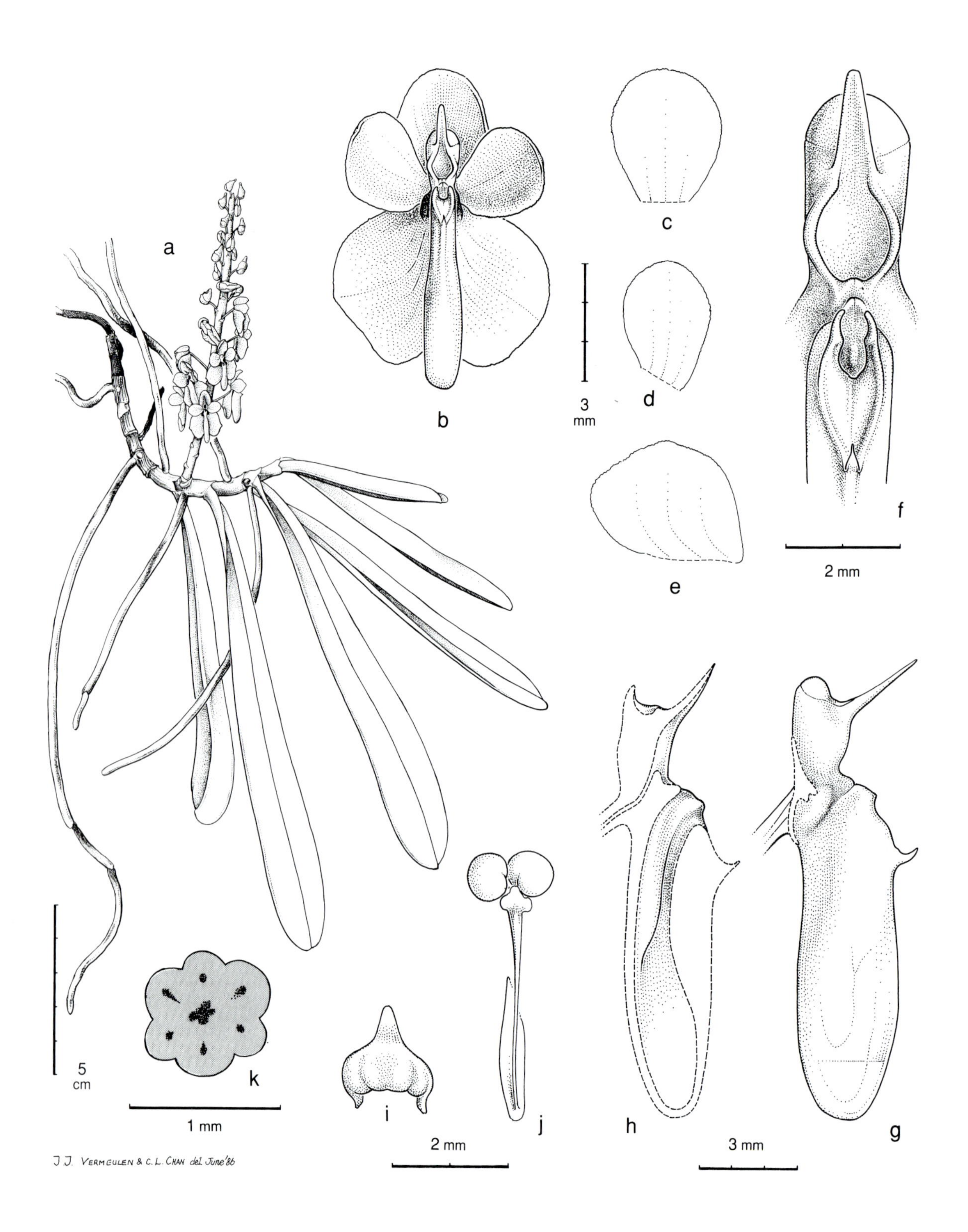

Figure 35. Dyakia hendersoniana (Rchb. f.) Christenson. - A: plant. - B: flower. - C: dorsal sepal. - D: petal. - E: lateral sepal. - F: column and entrance of spur. - G: column and lip, lateral view. - H: column and lip, longitudinal section. - I: anther. - J: pollinarium. - K: ovary, transverse section. All drawn from cult. TOC. by Jaap J. Vermeulen and Chan Chew Lun.

35. DYAKIA HENDERSONIANA (Rchb. f.) Christenson

Dyakia hendersoniana (*Rchb. f.*) *Christenson* in Orchid Digest 50: 63 (1986). Type: Borneo, cult. *Henderson & Sons*, Wellington Nursery, London (holotype W, isotype K).

Saccolabium hendersonianum Rchb.f. in Gard. Chron. n.s. 4: 356 (1875).

Ascocentrum hendersonianum (Rchb. f.) Schltr., Die Orchideen, ed. 1: 576 (1914).

Small epiphytic ***herb***. ***Stem*** 5-10 cm long. ***Leaves*** 7-15 x 1.2-3 cm, ligulate to oblanceolate, unequally obtusely bilobed at the apex. ***Inflorescence*** 5.5-24 cm long, erect or ascending, densely up to c 40-flowered; peduncle up to 3 mm in diameter; floral bracts 1-1.5 mm long, broadly ovate, obtuse. ***Flowers*** showy, deep pink to rose-purple, with a dark purple spot at the base of each lateral sepal, and with a white lip and spur. ***Pedicel*** with ***ovary*** 1-1.1 cm long, pink. ***Dorsal sepal*** 0.5-0.7 x 0.3-0.4 cm, elliptic, obtuse. ***Lateral sepals*** 0.6-0.8 x 0.5-0.6 cm, cuneate-oblong, obtuse. ***Petals*** 0.5-0.7 x 0.25-0.4 cm, obovate, obtuse. ***Lip*** 0.1-0.15 x 0.3 cm, obscurely 3-lobed, concave, ovate, acute; spur 0.9-1 cm long, pendent, cylindric-clavate. ***Column*** 0.15 cm long; pollinia 2, porate. Plate 9A.

HABITAT AND ECOLOGY: In Sarawak it is found in primary and old secondary forests which have a uniform high rainfall throughout the year on the slopes of Mt. Matang in the 1st Division. In Sabah it occurs in rather open swampy forest on ultramafic substrate; lower montane forest. Alt. sea level to 700 m. Flowering June and July.

DISTRIBUTION IN BORNEO: KALIMANTAN BARAT: Pontianak area. SABAH: Tawai Plateau. SARAWAK: Mt. Matang.

GENERAL DISTRIBUTION: Endemic to Borneo.

NOTES: Mr Au Yong Nang Yip of Kuching, who collected plants from Mt. Matang, also confirms having collected plants for cultivation from other limestone hills near Kuching. The plants often have from one to four inflorescences, and make very showy pot-plants. Flowering takes place in the rainy season. This is now a very rare species and endangered in the wild.

DERIVATION OF NAME: The generic name is derived from the Malay *dyak*, after the native people of Sarawak. The specific epithet honours Mr. E.G. Henderson who cultivated the type specimen.

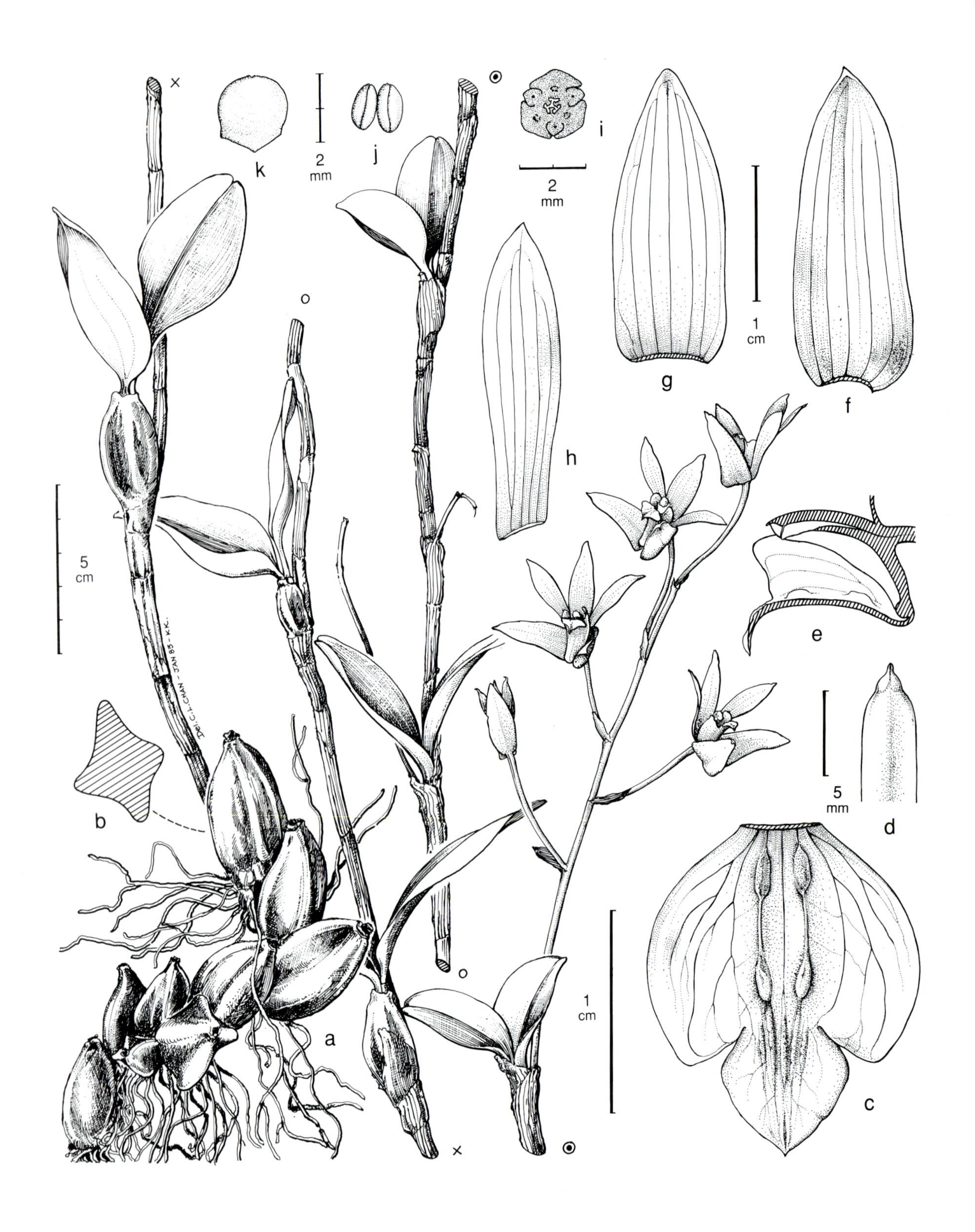

Figure 36. Epigeneium kinabaluense (Ridl.) Summerh. - A: plant. - B: pseudobulb, transverse section. - C: lip, spread out. - D: column, dorsal view. - E: column and lip, longitudinal section. - F: lateral sepal. - G: dorsal sepal. - H: petal. - I: ovary, transverse section. - J: pollinia. - K: anther. All drawn from *Lamb* AL 40/83 by Chan Chew Lun.

36. EPIGENEIUM KINABALUENSE (Ridl.) Summerh.

Epigeneium kinabaluense (*Ridl.*) *Summerh.* in Kew Bull. 12: 262 (1957). Type: Borneo, Sabah, Mt. Kinabalu, *Haviland* 1253 (holotype K).

Dendrobium kinabaluense Ridl. in Stapf in Trans. Linn. Soc. London, Bot., ser. 2. 4: 234 (1894).

Sarcopodium kinabaluense (Ridl.) Rolfe in Orchid. Rev. 18: 239 (1910).

Sarcopodium suberectum Ridl. in Kew Bull.: 211 (1914). Type: Borneo, Sarawak, Mt. Rumput, *Anderson* 172 (holotype K), **synon. nov.**

Katherinea kinabaluense (Ridl.) A.D. Hawkes in Lloydia 19: 96 (1956).

Epigeneium suberectum (Ridl.) Summerh. in Kew Bull. 12: 265 (1957), **synon. nov.**

Epiphytic, or scrambling terrestrial ***herb***. ***Rhizome*** creeping, tough, with several pseudobulbs clustered along it, producing numerous rugulose-verrucose roots. ***Stems*** erect, often clambering through undergrowth, up to 200 cm long, branching, superposed, rigid, clothed in reddish brown sheaths when young, bearing pseudobulbs at regular intervals, internodes (2-)8-10(-15) cm long. ***Pseudobulbs*** conical to ovoid, 4-angled, obtuse, 2- or rarely 3-leaved, lowermost 3-5 x 1-2(-2.5) cm, upper aerial ones generally smaller and oblong, 1-2 x 0.5-0.8 cm. ***Leaves*** (2.5-)5-6 x 1.5-3.6 cm, oblong-elliptic, equally or unequally retuse, thick, leathery, dark green, often flushed purple or with a purple margin; petiole 0.2-1 cm long. ***Inflorescences*** subapical, emerging near apex of pseudobulbs just below leaves, erect or porrect, racemose, yellowish green or crimson, (1-)8-18(-20) cm long, laxly 4- to 8(10)-flowered; floral bracts 4-8 mm long, ovate-elliptic or oblong-elliptic, obtuse to acute, golden yellow. ***Flowers*** resupinate, sweet-scented; sepals and petals pale ochre-yellow to dark yellow, pale apricot or cream, often flushed purple on reverse; lip white with purple veins on side lobes, mid-lobe yellow around basal ridges, spotted purple; column white, foot flecked purple. ***Pedicel*** with ***ovary*** 1.6-3 cm long, slender. ***Sepals*** and ***petals*** spreading. ***Dorsal sepal*** 1.7-1.8(-2) x 0.7(-0.9) cm, oblong-elliptic, acute. ***Lateral sepals*** 2(-2.3) x 0.8(-1) cm, oblong-elliptic to narrowly elliptic, acute; mentum 6 mm long, obtuse. ***Petals*** 2-2.4 x 0.6 cm, ligulate-elliptic, acute. ***Lip*** 3-lobed, 1.5 cm long, 1.5-1.6 cm wide across side lobes, 0.7 cm wide across mid-lobe; side lobes erect, oblong, rounded; mid-lobe deflexed, ovate-elliptic, apiculate, margins irregularly crenulate, surface minutely papillose, nerves somewhat raised; disc with a variable number of calli, often with 5 narrow longitudinal glabrous, fleshy calli, 3 of which are rather small and grouped at centre of lip, although the median callus is sometimes absent, the remaining 2 being larger and positioned near base of lip (as in type of *E. kinabaluense*); sometimes the 2 basal calli are enlarged and continuous with the outer central calli, the median callus again sometimes being absent (as in type of *E. suberectum*). ***Column*** 0.7 cm long, 0.4 cm wide at base, alate, column-foot 0.6 cm long; connective long; anther-cap ovate to conical; pollinia 4 in 2 groups. ***Fruit*** 6 cm, beaked. Plate 9C.

HABITAT AND ECOLOGY: Open scrubby Ericaceous-Myrtaceous vegetation; lower and upper montane mossy forest, often epiphytic on exposed *Leptospermum* branches in full sunlight; mossy rocks on exposed ridges; oak-laurel forest. Alt. 1200 to 3400 m. Flowering observed from January to March.

DISTRIBUTION IN BORNEO: SABAH: Mt. Kinabalu; Mt. Trus Madi; Crocker Range. SARAWAK: Mt. Mulu National Park; Dulit Range; Mt. Rumput.

GENERAL DISTRIBUTION: Endemic to Borneo.

NOTES: This is a common species on Mt. Kinabalu and other mountainous areas. The variation in lip morphology has given rise to the recognition of two distinct species, *E. kinabaluense* and *E. suberectum*, merely on minor details, which has caused much confusion.

DERIVATION OF NAME: The generic name is derived from the Greek *epi*, upon, and *geneion*, chin, referring to the position of the lateral sepals and petals on the column-foot. The specific epithet refers to the type locality, Mt. Kinabalu.

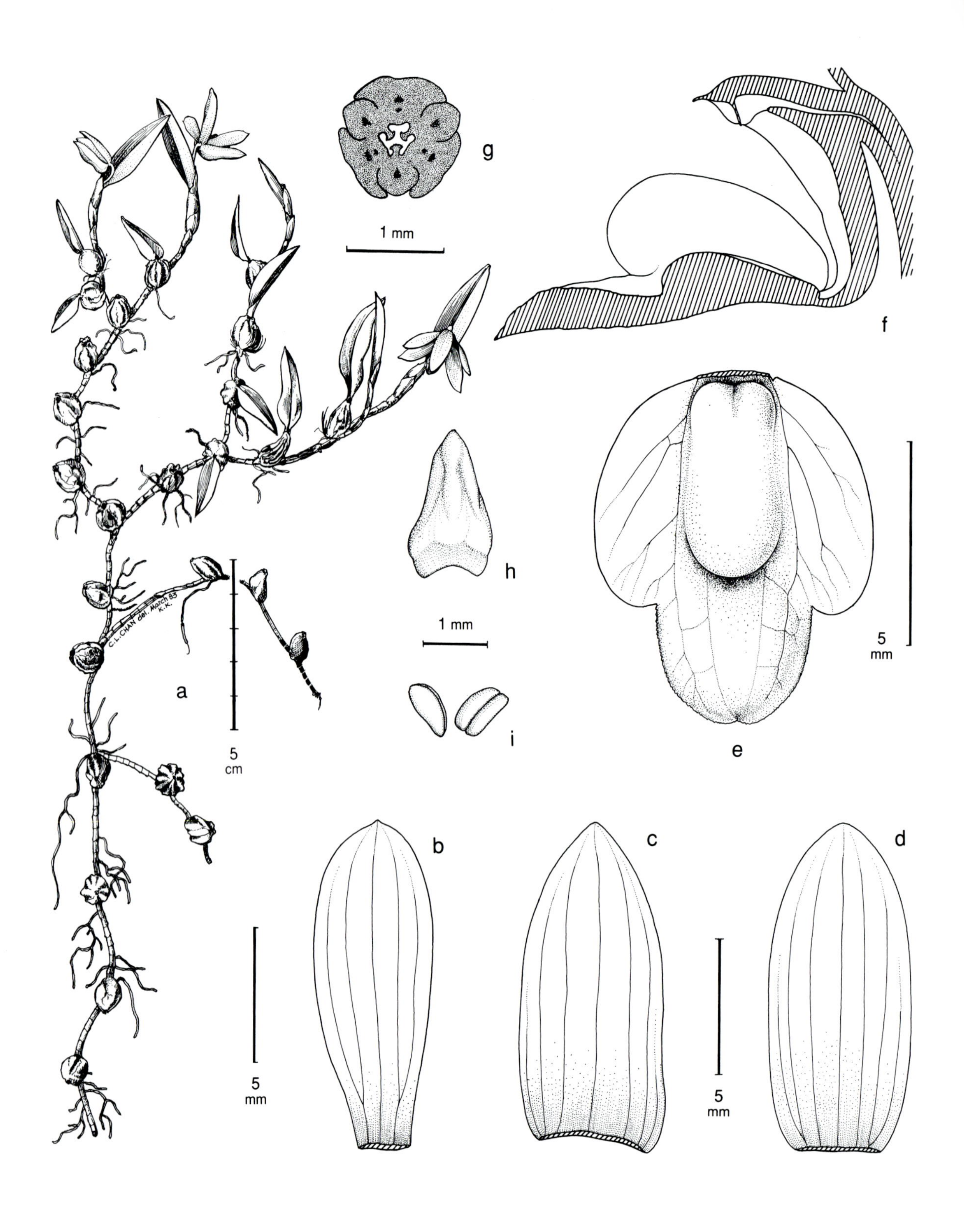

Figure 37. Epigeneium longirepens (Ames & C. Schweinf.) Seidenf. - A: plant. - B: petal. - C: lateral sepal. - D: dorsal sepal. - E: lip. - F: column and lip, longitudinal section. - G: ovary, transverse section. - H: anther. - I: pollinia. All drawn from *Lamb* AL 20/82 by Chan Chew Lun.

37. EPIGENEIUM LONGIREPENS (Ames & C. Schweinf.) Seidenf.

Epigeneium longirepens (*Ames & C. Schweinf.*) *Seidenf.* in Dansk Bot. Ark. 34(1): 18 (1980). Type: Borneo, Sabah, Mt. Kinabalu, Marai Parai Spur, *Clemens* 245 (holotype AMES, isotypes K, SING).

Dendrobium longirepens Ames & C. Schweinf. in Ames, Orch. 6: 105 (1920).

Desmotrichum longirepens (Ames & C. Schweinf.) A.D. Hawkes in Lloydia 20: 126 (1957).

Ephemerantha longirepens (Ames & C. Schweinf.) P.F. Hunt & Summerh. in Taxon 10: 105 (1961).

Flickingeria longirepens (Ames & C. Schweinf.) A.D. Hawkes in Orchid Weekly 2, 46: 456 (1961).

Epiphytic ***herb***. ***Rhizome*** long, creeping, slender, rarely branching, internodes mostly 2-3 cm long, covered in imbricate, scarious, brownish to reddish brown sheaths, rooting frequently. ***Pseudobulbs*** 1-2 x 0.8-1 cm, narrowly pyriform, sometimes fusiform, obliquely reclining, sulcate or rugose, covered by sheaths when young, 1-leafed. ***Leaves*** 1.8 x 0.5-0.8 cm, oblong-ligulate, subsessile, apex obtuse, minutely bilobed, rigid, leathery. ***Inflorescence*** 1-flowered. ***Flowers*** resupinate, ruby-red or purple-red; sepals often greenish at base. ***Pedicel*** with ***ovary*** nearly 2 cm long, slender, exserted from sheath. ***Dorsal sepal*** 0.95 x 0.45 cm, oblong-ovate, obtuse, apex somewhat mucronate. ***Lateral sepals*** 0.15 x 0.4 cm, oblong-ovate, subacute, with a dorsal mucro. ***Petals*** 1 x 0.4 cm, oblanceolate-oblong, obtuse, upper margin minutely erose. ***Lip*** 0.8 x 0.75 cm, suborbicular in outline, 3-lobed; side lobes erect, broadly rounded, free part semiorbicular; mid-lobe larger, suborbicular to oblong, shallowly retuse, minutely erose, distinctly papillose except at the centre around the base; disc with a V-shaped, shallowly triangular or oblong callus, with 1 or 2 irregular longitudinal ridges running toward the base. ***Column*** 0.4 cm long, stout, broadly winged; column-foot broad, saccate, alate; pollinia 4, in 2 groups. Plate 9B.

HABITAT AND ECOLOGY: Montane forest of various types, mostly lower. Also recorded from oak/chestnut/*Dacrydium* forest on podsolic soil and from forest on ultramafic substrate. Alt. 900 to 1700 m. Flowering observed in January, July, November, December.

DISTRIBUTION IN BORNEO: SABAH: Mt. Kinabalu.

GENERAL DISTRIBUTION: Endemic to Borneo (Sabah only).

NOTES: The creeping habit, ruby-red flowers and small purple-tinged, one-leaved pseudobulbs are distinctive.

DERIVATION OF NAME: The specific epithet is derived from the Latin *longus*, long, and *repens*, creeping, a reference to the habit.

Figure 38. Epigeneium treacherianum (Rchb. f. ex Hook. f.) Summerh. - A: plant. - B: column and lip, longitudinal section. - C: lateral sepal. - D: petal. - E: dorsal sepal. - F: column, apex, dorsal view. - G: lip, spread out. - H: anther. - I: pollinia. All drawn from plant collected from Kimanis Road, Crocker Range by Chan Chew Lun.

38. EPIGENEIUM TREACHERIANUM (Rchb.f. ex Hook.f.) Summerh.

Epigeneium treacherianum *(Rchb.f. ex Hook.f.) Summerh.* in Kew Bull. 12: 265 (1957). Type: Borneo, cult. *Low & Co.* (holotype K).

Dendrobium treacherianum Rchb.f. ex Hook.f. in Bot. Mag. 107: t. 6591 (1881).

Dendrobium lyonii Ames, Orch. 2: 177 (1908), **synon. nov**. Type: Philippines, Luzon, Bataan, *Lyon* s.n. (syntype AMES); Luzon, Bataan, Lamao, *Curran* 7153 (syntype AMES).

Sarcopodium treacherianum (Rchb.f. ex Hook.f.) Rolfe in Orchid Rev. 18: 239 (1910).

Sarcopodium lyonii (Ames) Rolfe in Orchid Rev. 18: 240 (1910), **synon. nov.**

Sarcopodium acuminatum (Rolfe) Kraenzl. var. *lyonii* (Ames) Kraenzl. in Engl., Pflanzenr. Orch. Dendrob. 1: 329 (1910), **synon. nov.**

Katherinea acuminata (Rolfe) A.D. Hawkes var. *lyonii* (Ames) A.D. Hawkes in Lloydia 19: 94 (1956), **synon. nov.**

Katherinea treacheriana (Rchb.f. ex Hook.f.) A.D. Hawkes in Lloydia 19: 97 (1956).

Epigeneium lyonii (Ames) Summerh. in Kew Bull. 12: 263 (1957), **synon. nov.**

Epiphytic ***herb***. ***Rhizome*** stout, creeping. ***Pseudobulbs*** 3-8 cm long, ovoid, curved, 4-6 angled, the angles prominent and rounded, crowded on rhizome, dull brownish green, flushed red near apex and along the angles, 2-leaved. ***Leaves*** 7-10 x 1.2-3 cm, oblong-elliptic to narrowly elliptic, retuse, carinate on reverse, subsessile, stiff and coriaceous. ***Inflorescence*** terminal, borne at apex of pseudobulb, racemose, 9-15 cm long, 2- to 7-flowered, porrect to pendent; floral bracts as long as pedicel with ovary, deciduous. ***Flowers*** resupinate, scented of dessicated coconut; sepals rose-red or lilac-pink, laterals often a deeper hue; petals pale rose-red with dark rose-red median stripe; lip side lobes dark rose-red or lilac-pink, edges of lobes white, mid-lobe white, dark rose-red or lilac pink at base. ***Pedicel*** with ***ovary*** 2-2.2 cm long, slender. ***Sepals*** and ***petals*** spreading. ***Dorsal sepal*** 2.8-3 cm long, 1 cm wide near base, narrowly ovate-elliptic, acuminate, margin somewhat revolute. ***Lateral sepals*** 3 cm long, 1 cm wide at base, narrowly triangular, acuminate, margin somewhat revolute, adnate to column-foot to form an obtuse mentum 1 cm long. ***Petals*** 2.8-3 x 0.4 cm, linear to narrowly elliptic, acute. ***Lip*** 2-2.2 cm long, 2 cm wide across side lobes, 3-lobed; side lobes oblong-triangular, obtuse, erect; mid-lobe 1.5-1.6 x 0.9-1 cm, narrowly elliptic to rhomboid, acute to acuminate; disc with a short, low, narrow central ridge with 2 small keels at its base and a group of small papillae each side, sometimes also with an odd small keel near apex of central ridge. ***Column*** 0.7-0.8 x 0.6 cm, column-foot 1 cm long, curved, anther connective uncinate; anther-cap 4 x 2 mm, oblong, sulcate; pollinia 4, in 2 groups. Plate 9D.

HABITAT AND ECOLOGY: On large branches in the crowns of forest trees that have a well-fissured bark, particularly dipterocarps. It prefers the crowns of the largest trees in primary mixed hill-dipterocarp forest, often in valleys where thick early morning mists collect, providing a cool moist environment, before exposure to bright sunshine later in the day. Alt. 100 to 400 m. Flowering June and July.

DISTRIBUTION IN BORNEO: Philippines and Borneo.

NOTES: In Sabah, plants appear to be triggered into flowering after a short dry season or after the onset of the rainy season. In June and July colonies of plants along the branches of large dipterocarp trees are cloaked in masses of pink flowers, making a wonderful sight. The flowers smell of dessicated coconut and are visited by the bee *Apis dorsata*. The habit of this orchid of growing in the crowns of the largest trees has no doubt been the reason for the scarcity of herbarium specimens, as plants can only be collected with ease from fallen branches or during felling.

DERIVATION OF NAME: The specific epithet honours W.H. Treacher, a Colonial Secretary in Labuan.

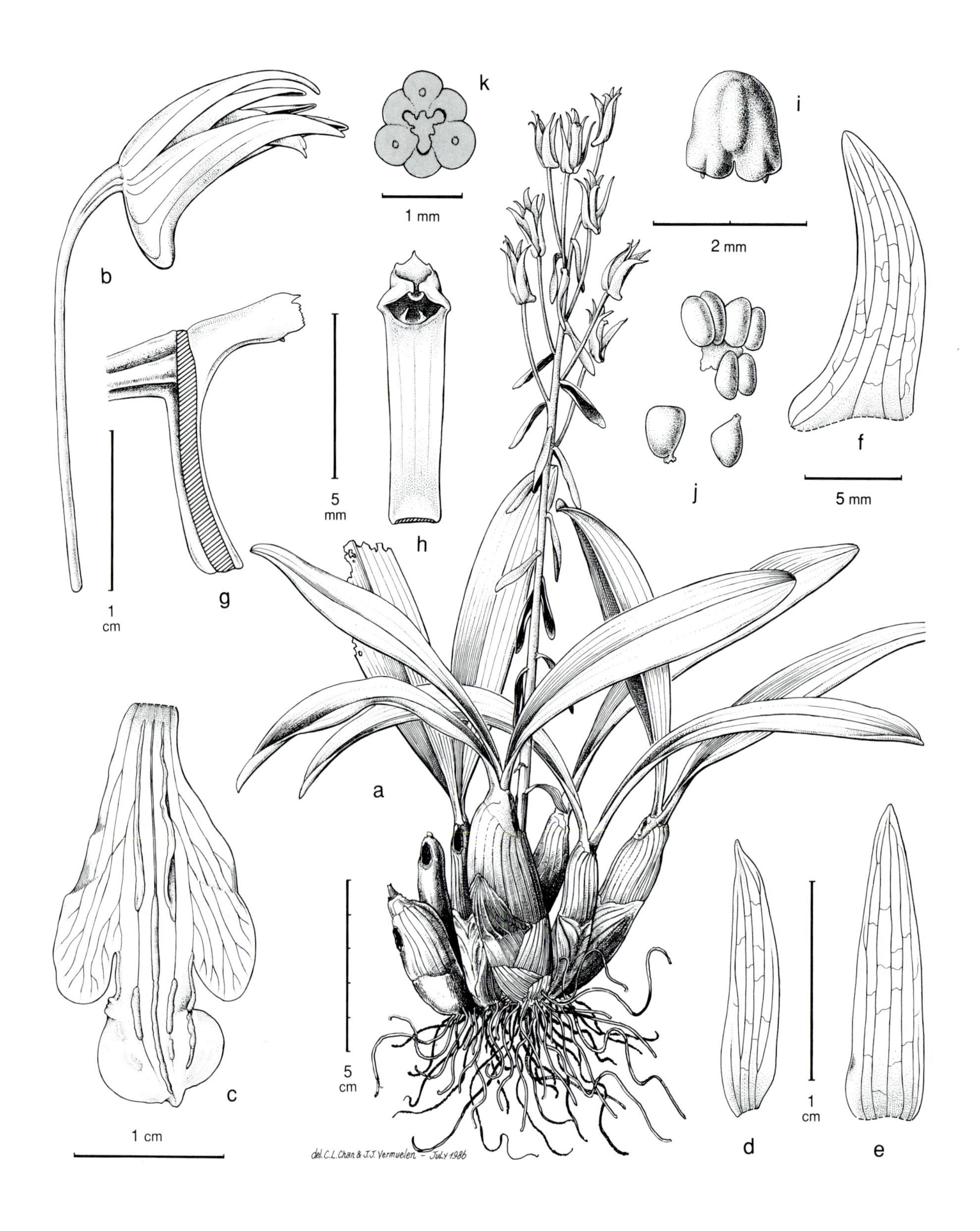

Figure 39. Eria ignea Rchb. f. - A: plant. - B: flower, lateral view. - C: lip, spread out. - D: petal. - E: dorsal sepal. - F: lateral sepal. - G: column and ovary, lateral view. - H: column, ventral view. - I: anther. - J: pollinia. - K: ovary, transverse section. All drawn from *Chan* 95/84 by Chan Chew Lun and Jaap J. Vermeulen.

39. ERIA IGNEA Rchb. f.

Eria ignea *Rchb. f.* in Gard. Chron. ser. 2, 15: 782 (1881). Type: Borneo, imported by *Veitch* (holotype W).

Eria cinnabarina Rolfe in Kew Bull.: 183 (1894). Type: Borneo, introduced by *Linden* (holotype K).

Epiphytic ***herb***. ***Pseudobulbs*** 2.5-7 x 3 cm, ovoid to oblong, clustered, 2- to 3-leaved, enclosed by several imbricate sheaths at base. ***Leaves*** 10-16.5 x 1-2.6 cm, oblong-ligulate, obtuse to subacute, subsessile. ***Inflorescence*** subapical, porrect to erect, 6- to 8-flowered, lax; peduncle, rachis and floral bracts deep cinnabar-orange; peduncle 2-4 cm long, with several basal bracts; rachis 4-8 cm long; floral bracts 1-1.7 x 0.3-0.4 cm, oblong-elliptic, obtuse, deflexed. ***Flowers*** deep cinnabar-orange. ***Pedicel*** with ***ovary*** 1.8-2.5 cm long, slender. ***Dorsal sepal*** 1.4 x 0.25-0.3 cm, oblong, acute. ***Lateral sepals*** 1-1.2 cm long, 0.6 cm wide at base, narrowly triangular-ovate, somewhat falcate, acute; mentum 0.6 cm long. ***Petals*** 1.1 x 0.2 cm, linear, acute. ***Lip*** 3-lobed, 1.3 cm long, 0.45 cm wide across side lobes, c 1.8 mm wide across mid-lobe, cuneate at base; side lobes oblong, rounded; mid-lobe obovate- spathulate, obtuse; disc with a narrow central keel which is thickened from about centre of lip to apex of mid-lobe, and 2 lateral keels extending from near base and terminating at about centre of lip, mid-lobe with 3-4 irregular lateral keels. ***Column*** 0.35-0.4 cm long, apical wings truncate, erose, column-foot 0.6 cm long; anther-cap cucullate; pollinia 8. Plate 9E.

HABITAT AND ECOLOGY: Epiphytic low down on small trees in podsolic heath forest with *Dacrydium pectinatum* and *Rhododendron malayanum*, and in the light open crowns of trees on ultramafic substrate. Alt. 200 to 500 m.

DISTRIBUTION IN BORNEO: SABAH: Nabawan area; Telupid area.

GENERAL DISTRIBUTION: Endemic to Borneo (Sabah only).

NOTES: This very brightly coloured orchid with orange bracts and flowers is similar to *Eria atrovinosa* Ridl. which grows in the same habitat, but has white bracts and flowers. Lamb has found that many species of podsolic heath forest also occur on ultramafic substrates.

DERIVATION OF NAME: The generic name is derived from the Greek *erion*, wool, referring to the woolly indumentum usually covering the inflorescence and parts of the flower. The specific epithet is derived from the Latin *igneus*, fire-red, flame-coloured, in reference to the inflorescence and flower colour.

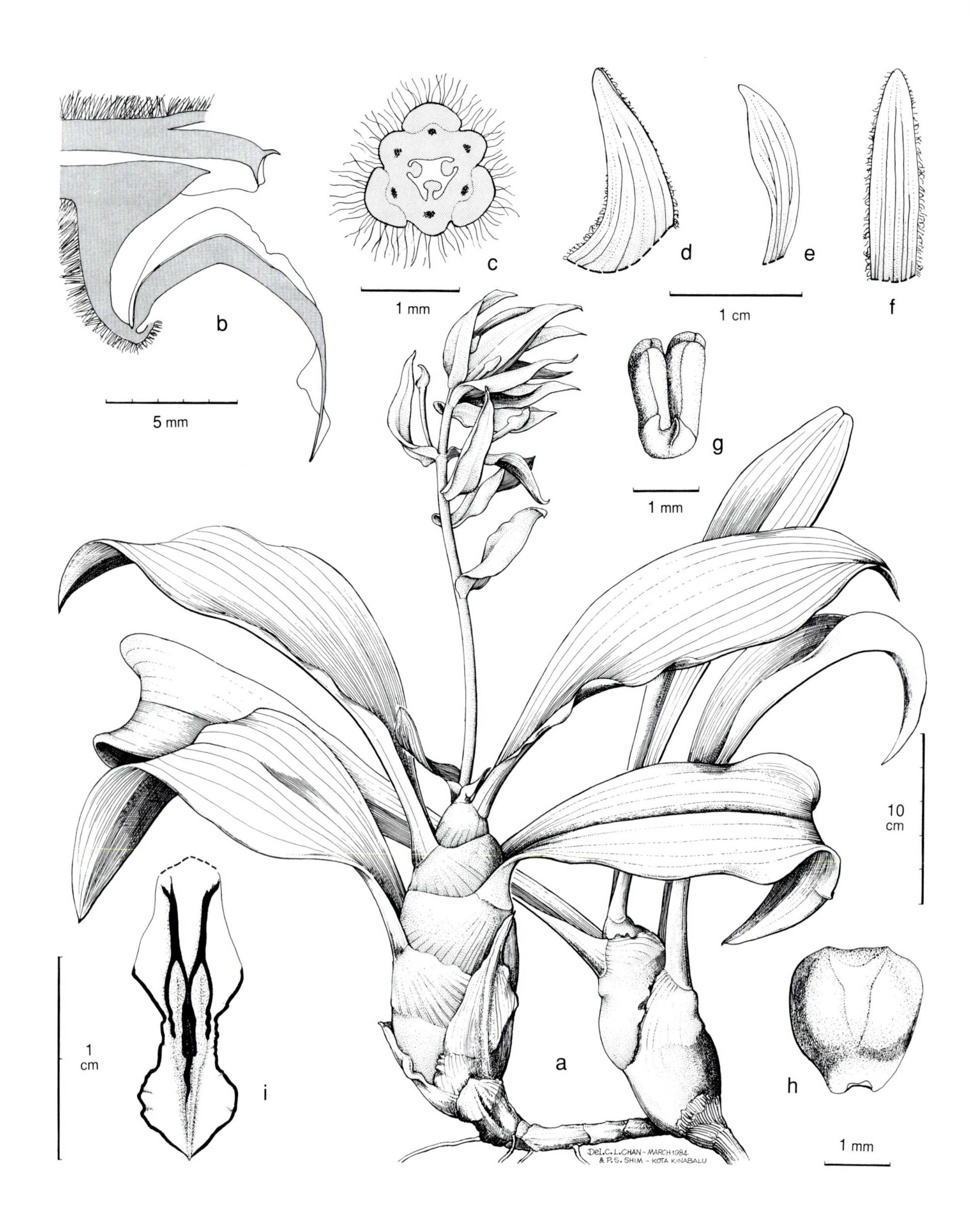

Figure 40. Eria ornata (Blume) Lindl. - A: plant. - B: column and lip, longitudinal section. - C: ovary, transverse section. - D: lateral sepal. - E: petal. - F: dorsal sepal. - G: pollinia. - H: anther. - I: lip, spread out. All drawn from cult. TOC by Chan Chew Lun and Shim Phyau Soon.

40. ERIA ORNATA (Blume) Lindl.

Eria ornata *(Blume) Lindl.*, Gen. Sp. Orch. Pl.: 66 (1830). Types: Java, Buitenzorg & Bantam Provinces, *Blume* s.n. (syntypes BO).

Dendrolirium ornatum Blume, Bijdr.: 345 (1825).

Eria armeniaca Lindl. in Bot. Reg. 27: misc. 38, t. 42 (1841). Type: Philippines, *Cuming* s.n. (holotype K).

Pinalia ornata (Blume) Kuntze, Rev. Gen. Pl. 2: 678 (1891).

Robust epiphytic ***herb*** to 50 cm high. ***Rhizomes*** long creeping, to 0.8 cm in diameter. ***Pseudobulbs*** 5-11 x 3-4 cm, ovoid or oblong-elliptic, flattened, bearing 3 to 5 leaves. ***Leaves*** 11-22 x 2.8-7 cm, elliptic or oblong-elliptic, acute, yellowish green, narrowed to a sheathing petiole 2.5-4.5 cm long, sheaths with yellow veins. ***Inflorescences*** 26-45 cm long, covered with reddish brown hairs, densely many-flowered; peduncle enclosed by several imbricate, acute sheaths below; floral bracts 2.5-8 x 0.8-1.5(-2.5) cm, ovate-elliptic or narrowly elliptic, acute to acuminate, bright orange-red. ***Flowers*** not widely opening, greenish yellow, the petals with a red median streak, mid-lobe of lip reddish or dark brown with a darker red margin. ***Pedicel*** with ***ovary*** 2.5-4 cm long, narrowly clavate, densely reddish brown pubescent. ***Sepals*** densely pubescent. ***Dorsal sepal*** 1.3-1.5 x 0.4 cm, oblong-elliptic, acute. ***Lateral sepals*** 1.5 cm long, 0.8 cm wide at base, triangular-ovate, acute; mentum 0.6 cm long. ***Petals*** 1.3-1.4 x 0.3 cm, narrowly elliptic, acute. ***Lip*** 1.3-1.4 cm long, 0.5 cm wide across side lobes, 0.4 cm wide across mid-lobe, obscurely 3-lobed, curved, glabrous; side lobes erect, rounded; mid-lobe oblong to oblong-ovate, apiculate, margin strongly undulate; lower half of lip with a central raised band extending from base of lip to base of mid-lobe, producing 2 short low lateral keels near the middle which run at an angle and terminate at base of mid-lobe, mid-lobe with a sulcate, elliptic central callus. ***Column*** 0.5 cm long, column-foot 0.6 cm long, curved; pollinia 8, in 4 groups. Plate 9F.

HABITAT AND ECOLOGY: Hill-forest and lower montane forest on limestone. Alt. 300-1500 m. Flowering recorded in April, May, June, July and September.

DISTRIBUTION IN BORNEO: KALIMANTAN SELATAN: Banjarmasin area; Martapura area. SABAH: Mt. Kinabalu; Mt. Alab. SARAWAK: Mt. Bidi; Mt. Meraja; Bukit Krian, etc.

GENERAL DISTRIBUTION: Thailand, Peninsular Malaysia, Sumatra, Java, Sulawesi and the Philippines.

NOTES: This species seems to adapt readily to colonising trees in secondary forest and is widespread in Sabah and Sarawak. It is common on trees along roads around Ranau and Poring near Mt. Kinabalu.

DERIVATION OF NAME: The specific epithet is derived from the Latin *ornatus*, adorned, embellished, in reference to the showy floral bracts.

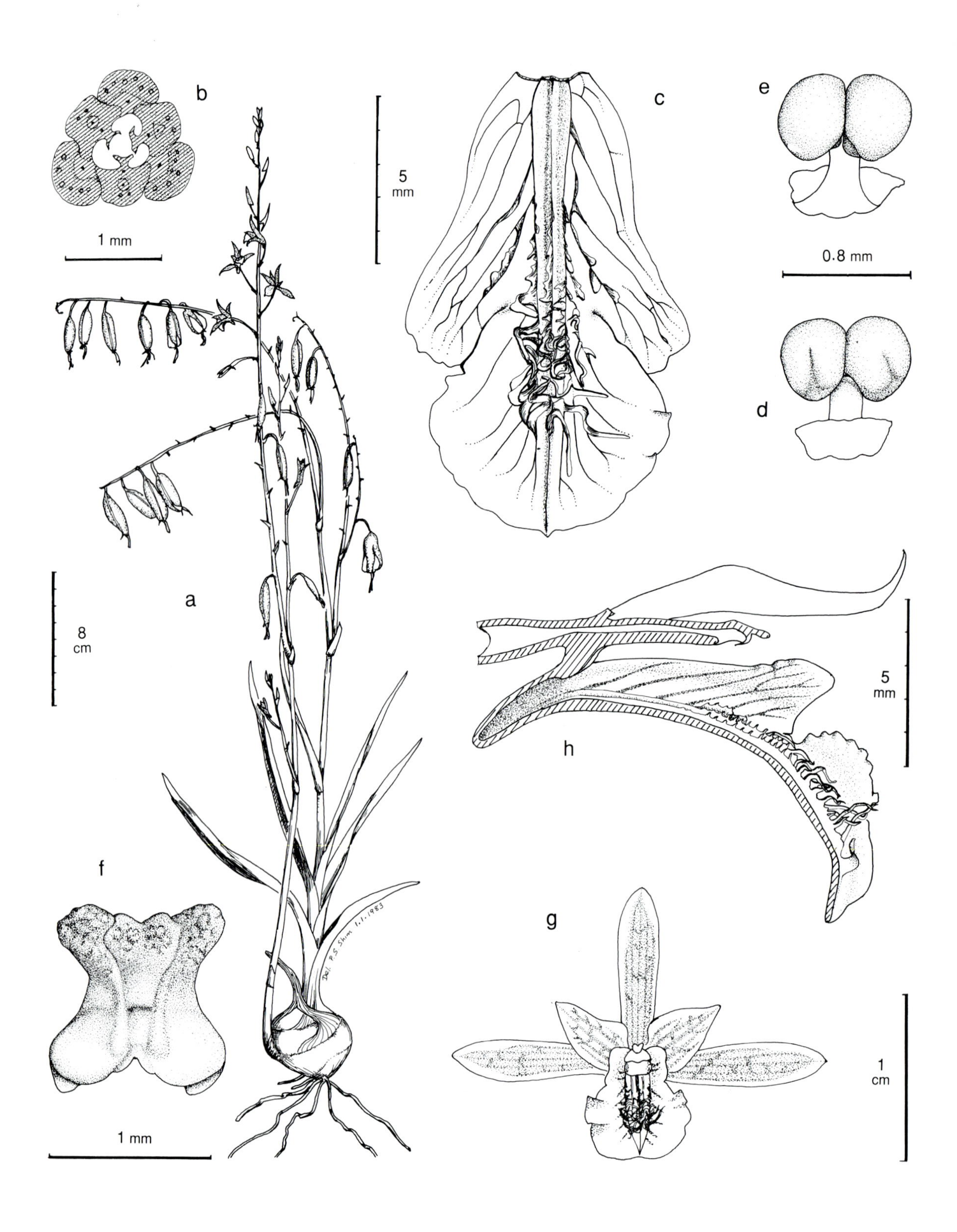

Figure 41. Eulophia graminea Lindl. - A: plant. - B: ovary, transverse section. - C: lip, spread out. - D: pollinarium, abaxially. - E: pollinarium, adaxially. - F: anther. - G: flower, front view. - H: column and lip, longitudinal section. All drawn from *Shim* s.n. by Shim Phyau Soon.

41. EULOPHIA GRAMINEA Lindl.

Eulophia graminea *Lindl.*, Gen. Sp. Orch. Pl.: 182 (1833). Type: Singapore, *Wallich* 7372 (holotype K).

Eulophia sinensis Miq. in J. Bot. Néerl. 1: 91 (1861). Type: China, *Krone* s.n. (holotype U).

Eulophia decipiens Kurz in J. As. Soc. Bengal 75: 155, t.13 (1876). Type: Nicobar Islands, *Kurz* s.n. (holotype CAL, isotype K).

Graphorkis graminea (Lindl.) Kuntze, Rev. Gen. Pl. 2: 662 (1891).

Eulophia ramosa Hayata in J. Coll. Sci. Imp. Univ. Tokyo 30: 332 (1911). Type: Taiwan, *Kawakami & Mori* 6281 (holotype TAI).

Eulophia gusukumai Masamune in Trans. Nat. Hist. Soc. Formosa 24: 208 (1934). Type: Taiwan, *Masamune* s.n. (holotype ?TAI).

For full synonymy see Seidenfaden in Opera Botanica 72: 29 (1983).

Large terrestrial ***herb*** to 70 cm high. ***Pseudobulbs*** large, 5-15 cm, subterranean, rhizomatous, several-noded. ***Roots*** white, elongate, 1-2.5 mm in diameter. ***Leaves*** c. 5, 10-20 x 0.8-1.5 cm, coriaceous, basal, linear-lanceolate, acute, arcuate. ***Inflorescences*** 1 or 2, erect, branched, 65-80 cm high, laxly many-flowered; peduncle stout, often bearing 1-3 leafy sterile bracts; floral bracts small, triangular to lanceolate, 2-8 mm long. ***Flowers*** spreading, about 3 cm across; sepals pale green with purple venation; petals pale green; lip white with a greenish base and purple-veined side lobes; column green with purple on edges and at base. ***Pedicel*** with ***ovary*** 0.8-1.4 cm long. ***Dorsal sepal*** 1-1.5 x 0.2-0.3 cm, erect to reflexed, linear-oblanceolate, acute. ***Lateral sepals*** 1-1.2 x 0.2 cm, spreading to reflexed, linear-oblanceolate, acute. ***Petals*** 0.8-1 x 0.25-0.35 cm, spreading to subporrect, obliquely lanceolate to oblong, acuminate or acute. ***Lip*** 3-lobed in basal half, 1-1.2 x 0.7-0.9 cm; side lobes obliquely oblong, rounded; mid-lobe subcircular, longly papillose in centre; spur 2 mm long, cylindrical. ***Column*** 0.6 cm long, straight, acute at apex, subclavate; anther-cap 1 mm long; pollinia 2. Plate 10A.

HABITAT AND ECOLOGY: In Borneo often growing in open grassy areas in poor sandy soils near the coast, or inland on sandy river banks that are open and exposed to plenty of light. Alt. sea level to 500 m. Flowering January & February.

DISTRIBUTION IN BORNEO: BRUNEI: Tutong District, Bukit Beruang. SABAH: Kota Kinabalu area; Likas Bay; Mt. Kinabalu. SARAWAK: Simangang.

GENERAL DISTRIBUTION: India, Sri Lanka, Peninsular Malaysia, Singapore, Sumatra, Java, Natuna, Borneo, China, Taiwan, Hong Kong, Ryukyu Islands.

DERIVATION OF NAME: The generic name is derived from the Greek *eu*, well, true, and *lophos*, plume, in reference to the crests on the lip. The specific epithet is derived from the Latin *gramineus*, grassy or grass-like, and refers to the habit and the leaves.

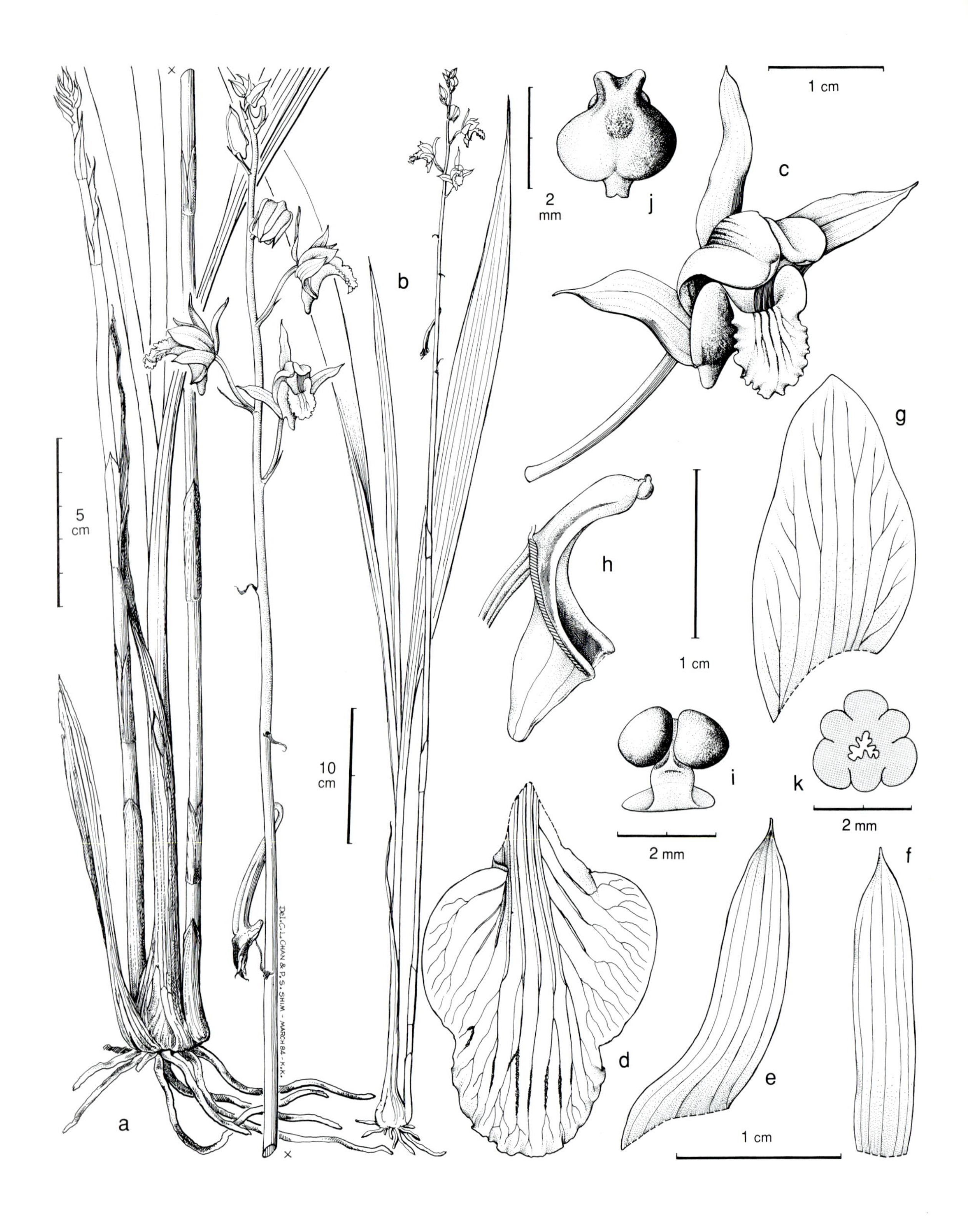

Figure 42. Eulophia spectabilis (Dennst.) Suresh - A & B: plant. - C: flower, front view. - D: lip, spread out. - E: lateral sepal. - F: dorsal sepal. - G: petal. - H: column, lateral view. - I: pollinarium. - J: anther. - K: ovary, transverse section. All drawn from cult. TOC by Chan Chew Lun.

42. EULOPHIA SPECTABILIS (Dennst.) Suresh

Eulophia spectabilis (*Dennst.*) *Suresh* in Regnum Veg. 119: 300 (1988). Type: India, Malabar coast, *Rheede* s.n. (lectotype Rheede's t. 36 in Hort. Malab. 11 (1692)).

Wolfia spectabilis Dennst., Schlüssel Hortus Malab., 11, 25, 38 (1818).

Eulophia nuda Lindl., Gen. Sp. Orch. Pl.: 180 (1833). Type: Nepal, Morang, *Hamilton* s.n., *Wallich* no. 7371 (holotype K).

E. squalida Lindl. in Bot. Reg. 27: misc. 77 (1841). Type: Philippines, *Cuming* s.n. (holotype K).

For full synonymy see Seidenfaden in Opera Bot. 72: 40 (1983).

Erect terrestrial ***herb***. ***Pseudobulbs*** subterranean, 1.5-2 x 3 cm, 3- to 4-leaved, enclosed by several lanceolate sheaths. ***Leaves*** lanceolate, acuminate, plicate; blade 35-50 x 2-5(-6.5) cm; petiole 15-30 cm long, sheathing, sulcate. ***Inflorescence*** laxly 12-20 (or more)-flowered; peduncle 30-70 cm long, fleshy, bearing several 2-6 cm long ovate-elliptic, acute to acuminate sterile sheaths; rachis (15-)20-30 cm long; floral bracts 1-1.5 cm long, linear-lanceolate, acuminate. ***Flowers*** 2.5-3 cm across; sepals olive-green with darker green or maroon nerves on exterior; petals creamy or white with pink nerves at the base; lip pale mauve-pink or white with pale pink margin, raised nerves on disc reddish brown; spur pale green; column pink; anther cream. ***Pedicel*** with ***ovary*** 2-2.5 cm long, slender. ***Sepals*** spreading, acute to acuminate. ***Dorsal sepal*** 2 x 0.4-0.5 cm, narrowly oblong-elliptic. ***Lateral sepals*** 2.3-2.4 cm long, 0.6-0.7 cm wide at base, 0.4-0.5 cm wide above, obliquely oblong-elliptic, somewhat falcate, adnate to column-foot. ***Petals*** 1.7-2 x 0.8 cm, ovate-elliptic or oblong-elliptic, obtuse or ± acute, adnate to column-foot, often reflexed at apex. ***Lip*** ± 3-lobed, 1.8 cm long, 1.4 cm wide across side lobes, ovate-oblong, strongly curved at right angles to spur; side lobes obscure, rounded; mid-lobe 0.8-1 cm wide, obtuse, margin crenulate, nerves minutely papillose; disc with somewhat raised nerves; spur 0.6-0.8 x 0.5-0.6 cm, conical, obtuse, flattened, pointing down between the divergent halves of the column-foot. ***Column*** 0.9-1 x 0.4-0.5 cm, narrowly winged below, column-foot 0.8-0.9 cm long; anther-cap ovate, with a small retuse flange at base, apex truncate. Plate 10B.

HABITAT AND ECOLOGY: Open grassy places; secondary regrowth; roadsides; disturbed lowland dipterocarp forest; rubber plantations; in light shade or full sunlight. Alt. sea level to 900 m.

DISTRIBUTION IN BORNEO: KALIMANTAN TIMUR: Tabang. SABAH: Labuan; Sandakan; Tawau; Mt. Kinabalu; Maliau River area; Sipitang District. SARAWAK: locality unknown.

GENERAL DISTRIBUTION: Widespread from India, Sri Lanka and the Himalayan region to China and Indochina through Malaysia and Indonesia, the Philippines, New Guinea east to Fiji and Tonga.

DERIVATION OF NAME: The specific epithet is derived from the Latin *spectabilis*, notable, remarkable, probably in reference to the flowers.

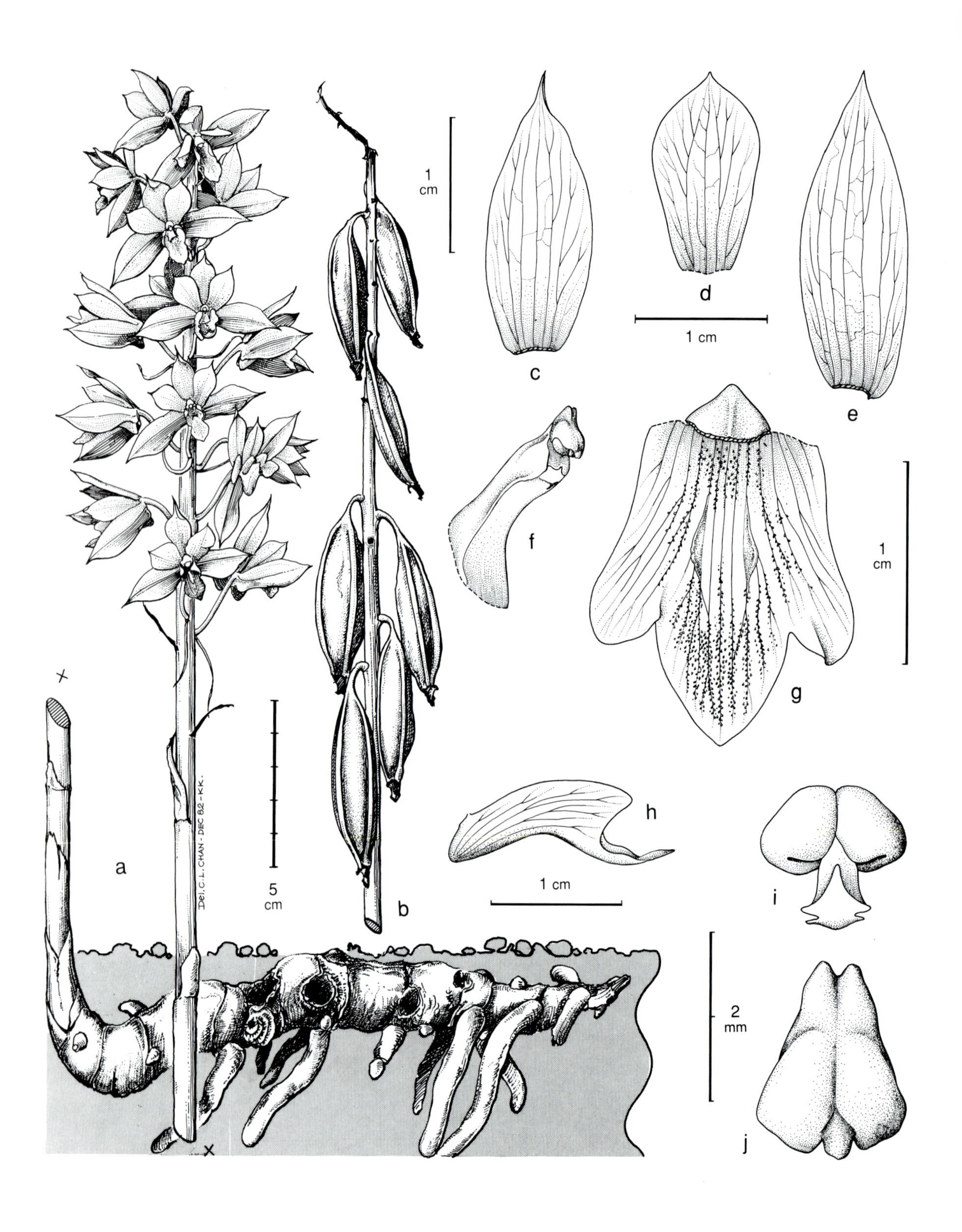

Figure 43. Eulophia zollingeri (Rchb. f.) J.J. Sm. - A: plant. - B: seed pods. - C: dorsal sepal. - D: petal. - E: lateral sepal. - F: column, lateral view. - G: lip, spread out. - H: lip, lateral view. - I: pollinarium. - J: anther. All drawn from *Lamb* AL 13/82 by Chan Chew Lun.

43. EULOPHIA ZOLLINGERI (Rchb.f.) J.J.Sm.

Eulophia zollingeri *(Rchb.f.) J.J.Sm.*, Orch. Java: 228 (1905). Type: Java, Lampong Province & Gebbok Klakka, *Zollinger* 585 (holotype W).

Cyrtopera zollingeri Rchb.f. in Bonplandia 5: 38 (1857).

C. sanguinea Lindl. in J. Linn. Soc., Bot. 3: 32 (1858). Types: India, Sikkim, *Cathcart* s.n., *Hooker.f.* 223 & 361 (syntypes K).

Eulophia macrorhiza Blume, Fl. Javae Orch.: 155, t. 63, f. 2 (1859). Type: Java, *Blume* s.n. (holotype BO).

Cyrtopera papuana Ridl. in J. Bot. (Schrader) 24: 354 (1886) non Kraenzl. (1898). Type: New Guinea, *Forbes* 391 (holotype BM).

Eulophia sanguinea (Lindl.) Hook.f., Fl. Brit. Ind. 6: 8 (1890).

E. papuana (Ridl.) J.J. Smith in Nova Guinea 8: 26 (1909).

E. macrorhiza Blume var. *papuana* (Ridl.) Schltr. in Feddes Repert. Beih. 1: 417 (1912).

E. carrii C.T.White in Proc. Roy. Soc. Queensl. 47: 82 (1936). Type: Australia, N. Queensland, *Carr* s.n. (holotype BRI).

Erect, leafless saprophytic ***herb***. ***Pseudobulbs*** subterranean, 3-12 x 1.5-3 cm, horizontal. ***Roots*** thick. ***Inflorescence*** 8- to 20 (or more) -flowered, lax to rather dense; peduncle 24-30(-35) cm long, thick and fleshy, with several loose ovate-elliptic, obtuse to acute sheaths 2-5 cm long; rachis (10-)20-30 cm long; floral bracts 1.5-2.5 cm long, linear, acuminate. ***Flowers*** with reddish brown, purplish or greenish brown sepals and petals; lip reddish brown, sometimes yellow, flushed crimson on side lobes; column creamy yellow. ***Pedicel*** with ***ovary*** 1.5 cm long, slender. ***Dorsal sepal*** 1.6-2 x 0.5-0.8 cm, elliptic or oblong-elliptic, acuminate. ***Lateral sepals*** 2-2.3 x 0.8-0.9 cm, elliptic, acute, slightly oblique, mid-nerve raised on exterior. ***Petals*** 1.4-1.5 x 0.9 cm, obovate, mucronate. ***Lip*** distinctly 3-lobed, not spurred, 1.6-1.7 cm long, 1.4 cm wide across side lobes, borne at an acute angle to the column-foot, shortly saccate at base; side lobes 4 mm long, obtuse, erect or incurved, nerves slightly papillose; mid-lobe 6-7 x 6 mm, triangular-ovate, subacute, slightly deflexed, concave, papillose, central 3 nerves slightly raised, disc with 2 short, entire central keels. ***Column*** 0.6-0.7 cm long, column-foot 2.5-3 mm long; anther-cap ovate, retuse. Plate 10C.

HABITAT AND ECOLOGY: Hill-forest and lower montane forest; secondary forest; in leaf litter in shade, often forming colonies. Alt. 500 to 1500 m, probably also occurring at much lower altitudes.

DISTRIBUTION IN BORNEO: SABAH: Mt. Kinabalu.

GENERAL DISTRIBUTION: Widespread from India, Malaysia, Indonesia and the Philippines, north to Taiwan and Japan, east to New Guinea and Australia (Queensland).

DERIVATION OF NAME: The specific epithet honours the botanist Heinrich Zollinger (1818-1859) who collected the type specimen.

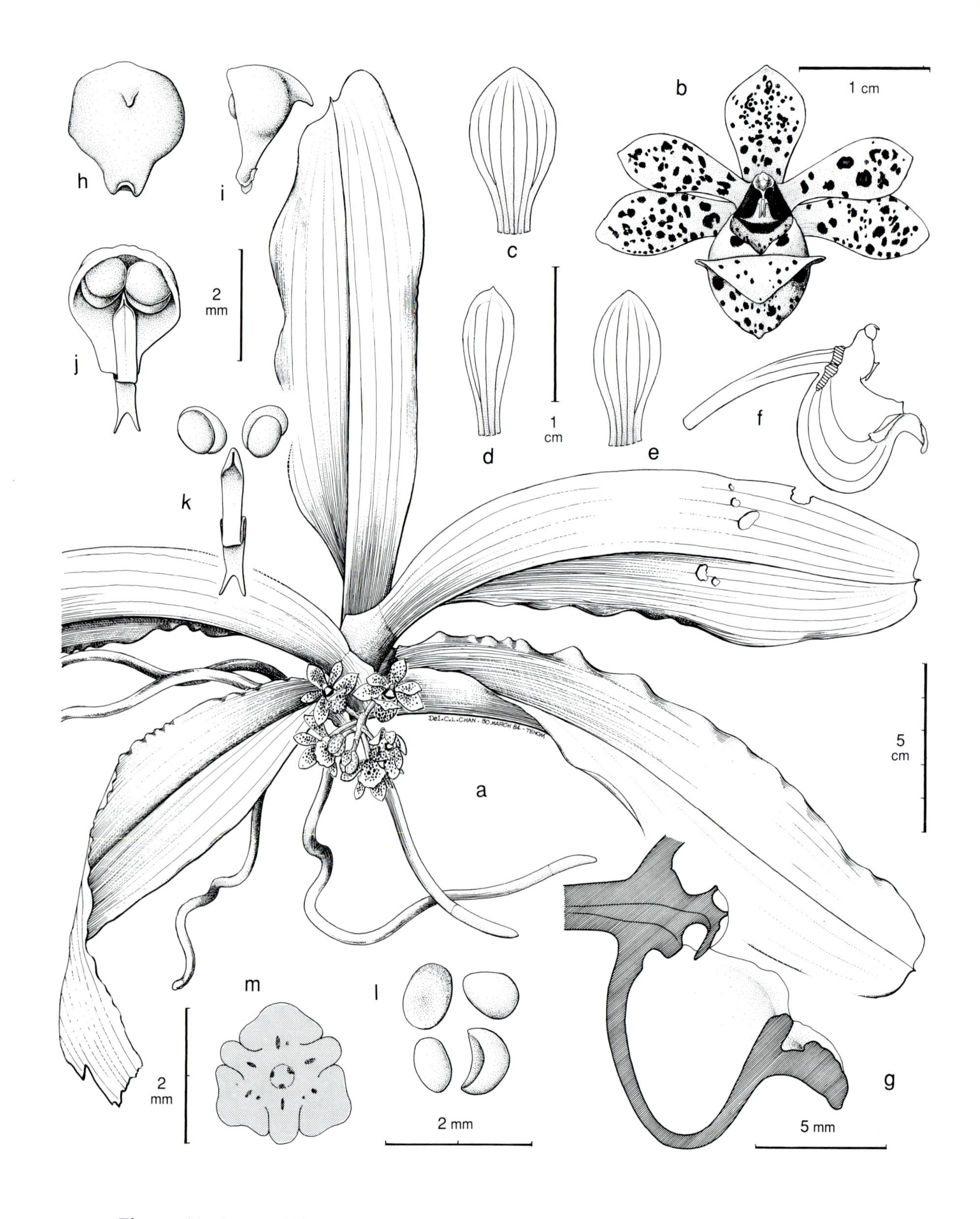

Figure 44. Gastrochilus patinatus (Ridl.) Schltr. - A: plant. - B: flower, front view. - C: dorsal sepal. - D: petal. - E: lateral sepal. - F: column, pedicel and lip, lateral view. - G: column and lip, longitudinal section. - H: anther, front view. - I: anther, lateral view. - J: anther, back view with pollinarium. - K: pollinarium. - L: pollinia. - M: ovary, tranverse section. All drawn from cult. TOC by Chan Chew Lun.

44. GASTROCHILUS PATINATUS (Ridl.) Schltr.

Gastrochilus patinatus (*Ridl.*) *Schltr.* in Feddes Repert. 12: 314 (1913). Type: Peninsular Malaysia, Pahang, Kota Glanggi, *Ridley* s.n. (holotype SING).

Saccolabium patinatum Ridl. in J. Straits Branch Roy. Asiat. Soc. 39: 84 (1903).

Epiphytic ***herb***. ***Stem*** 5 cm long. ***Leaves*** 5-7, blade 17-26 x 3.5-5.7 cm, oblong to oblong-elliptic, unequally bilobed, margins undulate, leathery, dark green; sheaths 1.5-3 cm long. ***Inflorescence*** emerging below leaves, 5- to 6-flowered, 5 cm across; peduncle 1.5 cm long; rachis up to 2 cm long, 0.3 cm thick; floral bracts 0.2-0.3 cm long. ***Flowers*** 2-2.7 cm across, fleshy, sweetly scented; sepals and petals pale cream turning pale yellow, with fine red spots, or yellow with larger red spots; lip with a white sac flushed yellow, epichile white with purple spots; column reddish brown. ***Sepals*** and ***petals*** spathulate. ***Dorsal sepal*** 0.9-1.2 x 0.5-0.7 cm. ***Lateral sepals*** 0.9-1.2 x 0.4-0.6 cm. ***Petals*** 0.8-1.1(-1.2) x 0.3-0.4(-0.5) cm. ***Lip*** sac 0.6-0.9 cm long, mouth 0.5-0.6 cm across, median line raised; epichile 0.3-0.4 x 0.8-1 cm, with small lateral wings, broadly triangular-ovate, obtuse, entire, without a central fleshy cushion, glabrous, sometimes recurved. ***Column*** 0.2 cm long; anther-cap 2.5 x 2 mm, with a dorsal keel, papillose; pollinia 4, unequal. Plate 10D.

HABITAT AND ECOLOGY: Mixed lowland dipterocarp forest on limestone, sandstone and shales, often beside rivers. Alt. sea level to 300 m. Flowering seems to occur throughout the year except during the latter part of the dry season or during exceptionally dry periods.

DISTRIBUTION IN BORNEO: SABAH: Lahad Datu District, Danum Valley; Ulu Tomani.

GENERAL DISTRIBUTION: Peninsular Malaysia, Sumatra and Borneo.

NOTES: *G. patinatus* differs from all other species of the genus in having four unequal instead of two porate pollinia. Plants from Tenom District have smaller flowers than typical, about 2 cm across with yellow sepals and petals usually under 1 cm long and conspicuously spotted red. Larger pale cream to pale yellow-flowered plants occur in the Danum Valley. These have finer red spots on the sepals and petals.

DERIVATION OF NAME: The generic name is derived from the Greek *gaster*, belly, and *glottis*, tongue, referring to the swollen lip sac. The specific epithet is derived from the Latin *patina*, a broad, shallow dish, referring to the shape of the blade of the lip.

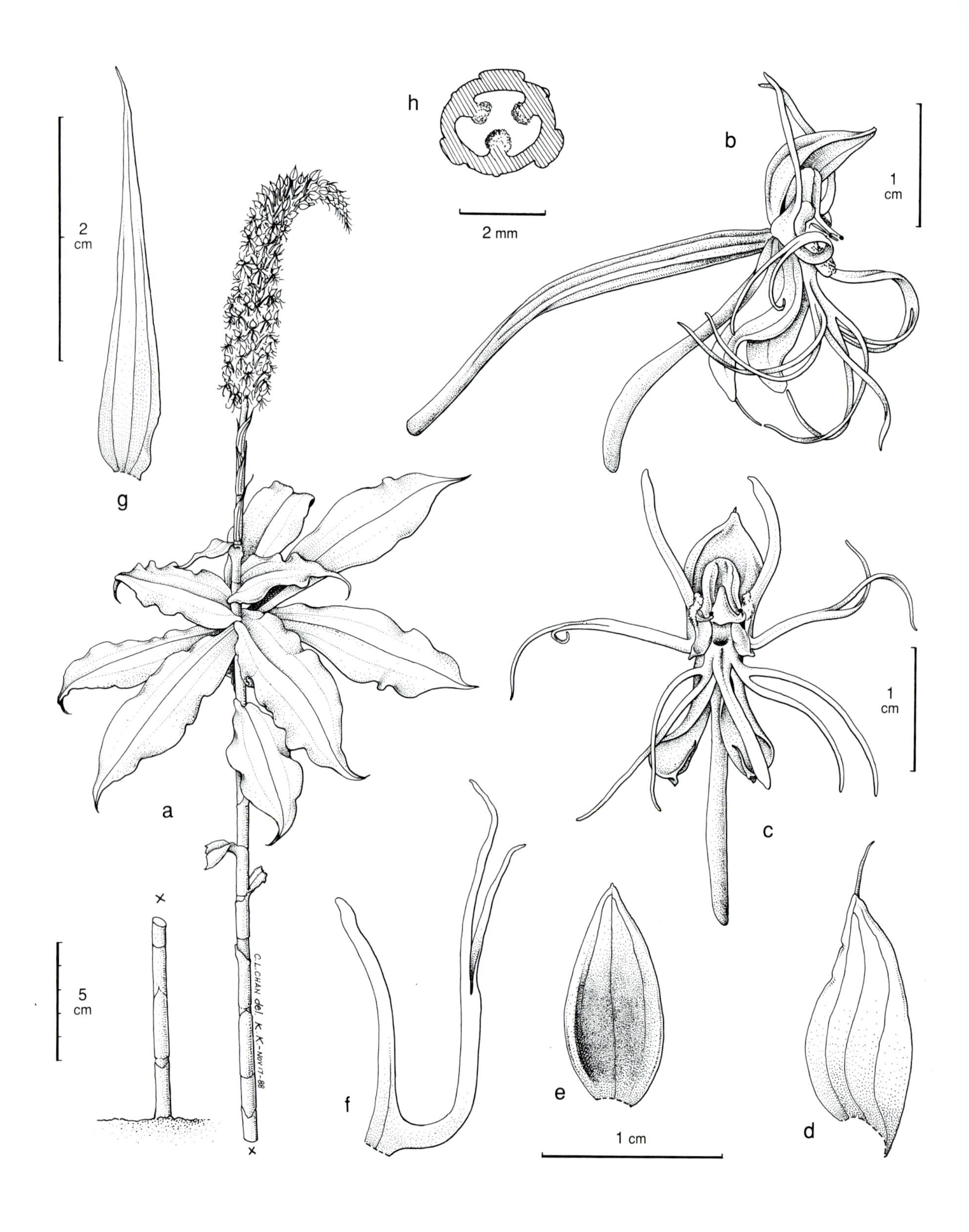

Figure 45. Habenaria setifolia Carr - A: plant. - B: flower, lateral view. - C: flower, front view. - D: lateral sepal. - E: dorsal sepal. - F: petal. - G: floral bract. - H: ovary, tranverse section. All drawn from *Phillipps, Lamb & Bacon* in *Lamb* AL 746/87 by Chan Chew Lun.

45. HABENARIA SETIFOLIA Carr

Habenaria setifolia *Carr* in Gard. Bull. Straits Settlem. 8: 171 (1935). Type: Borneo, Sabah, Mt. Kinabalu, Tenompok, *Clemens* 28323 (holotype BM, isotype AMES).

Terrestrial ***herb***, arising from tubers, each 2-9 x 1-1.5 cm. ***Stem*** 34-90(-180) cm tall, erect, stout, to 1.5 cm thick, sheathed below; internodes to 9 cm; bracts 2-4 cm, with 5 or 6 leaves above. ***Leaves*** 15-25 x 5-7 cm, obovate, oblong-obovate to obovate-elliptic, shortly acuminate, with an apical seta up to 2 mm long. ***Inflorescence*** laxly many-flowered; peduncle 12.5-20 cm long, with about 3 narrowly elliptic, acuminate sheaths up to 3.3 cm long; rachis up to 12 cm long; floral bracts 2.3-3 x 0.4-0.5 cm, narrowly elliptic, long-acuminate, spreading. ***Flowers*** green with white stigmas and cream pollinia. ***Pedicel*** with ***ovary*** 2.5 cm long. ***Dorsal sepal*** 1 x 0.6 cm, ovate, acuminate, deltoid in side view, cucullate-convex, nerves 3, elevated on exterior. ***Lateral sepals*** 1.2-1.3 x 0.43-0.5 cm, narrowly ovate, acuminate, the length measurement including a subapical 2-3 mm long seta. ***Petals*** free, bilobed about 0.17 cm above base; posterior lobule 1 cm long, subacute, 3-nerved; anterior lobule 1.45 cm long, sigmoid, narrowly oblong in the lower half and bilobed from about the middle, lobes subulate, the anterior slightly shorter. ***Lip*** about 1.6 cm long, 3-lobed, spurred; side lobes bilobed about 0.17 cm above base, lobules narrowly subulate, the anterior lobule longer, 1.7 cm long; mid-lobe 1.25 cm long, broader, subacute; spur 1.8-2 cm long, cylindric, somewhat dilated at apex. ***Column*** 3.5 mm long; anther-canals 2.5 mm long; rostellar arms 3.5 mm long; stigmas 2, separate; pollinia 2. Plate 10E.

HABITAT AND ECOLOGY: Hill-forest on colluvial or alluvial soils, usually beside streams; in very damp, lower montane forest. Alt. 1000 to 1500 m. Flowering January - March.

DISTRIBUTION IN BORNEO: SABAH: Mt. Kinabalu.

GENERAL DISTRIBUTION: Endemic to Borneo (Sabah only).

NOTES: A plant collected by Phillipps and Lamb stood nearly 2 m tall, with leaves to 32 x 8 cm, and with a scape to nearly 35 cm, with over 100 flowers.

DERIVATION OF NAME: The generic name is derived from the Latin *habena*, reins, from the long, strap-like divisions of the petals and the lip. The specific epithet is derived from the Latin *seta*, bristle, and *folium*, leaf, referring to the seta at the leaf apex.

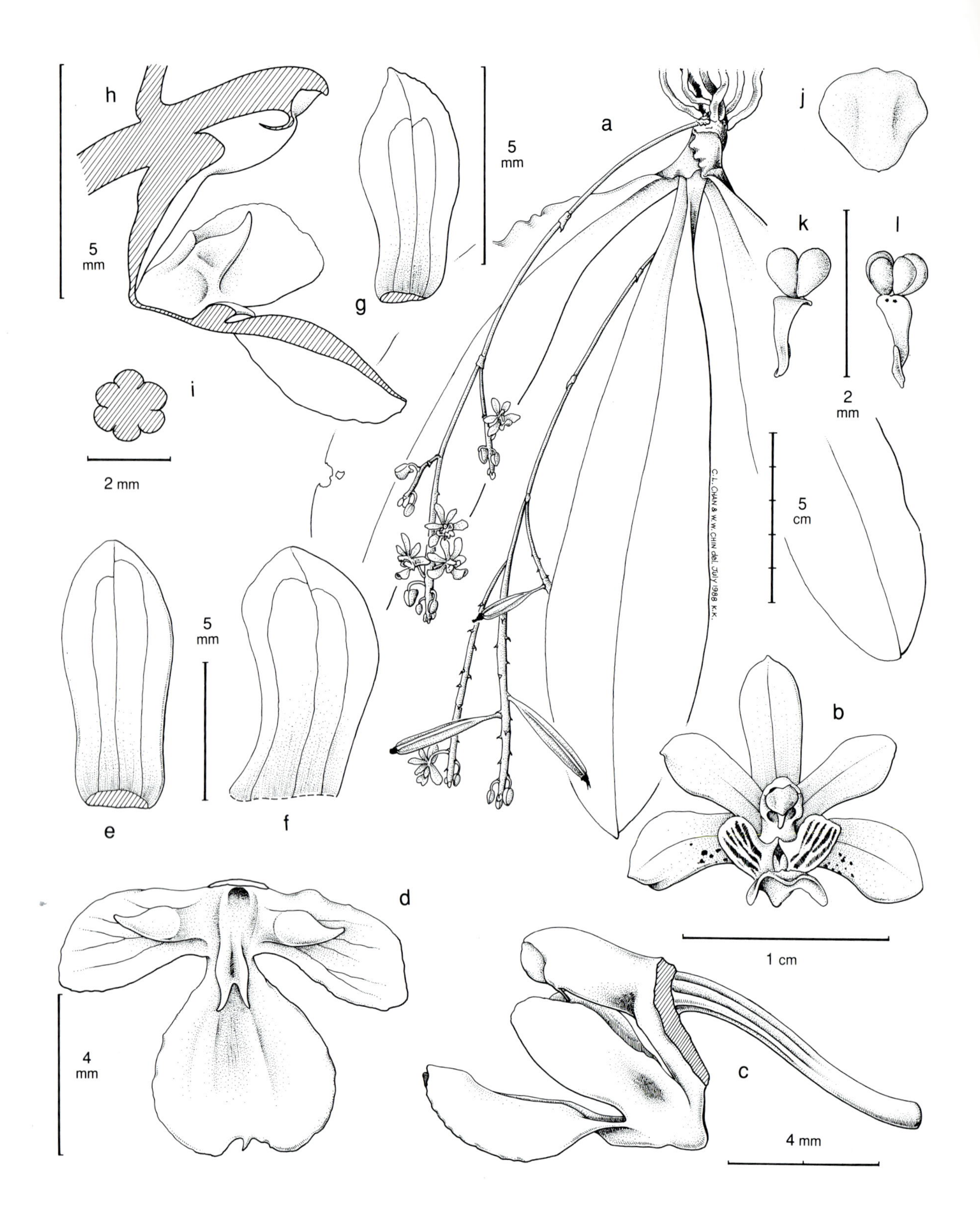

Figure 46. Kingidium deliciosum (Rchb. f.) Sweet - A: plant. - B: flower, front view. - C: column, pedicel and lip, lateral view. - D: lip, spread out. - E: dorsal sepal. - F: lateral sepal. - G: petal. - H: column and lip, longitudinal section. - I: ovary, transverse section. - J: anther. - K: pollinarium, abaxially. - L: pollinarium, adaxially. All drawn from cult. TOC by Chan Chew Lun and Chin Wan Wai.

46. KINGIDIUM DELICIOSUM (Rchb.f.) Sweet

Kingidium deliciosum (*Rchb.f.*) *Sweet* in Amer. Orchid Soc. Bull. 39: 1095 (1970). Type: origin unknown (holotype W).

Phalaenopsis deliciosa Rchb.f. in Bonplandia 2: 93 (1854).

Phalaenopsis bella Teijsm. & Binn. in Natuurk. Tijdschr. Ned. - Indië 24: 321 (1862). Type: Java, Salak, *Teijsmann* s.n. (syntype BO); Sumatra, Radja Basa, Lampongs, *Teijsmann* s.n. (syntype BO).

Phalaenopsis wightii Rchb.f. in Bot. Zeitung (Berlin) 20: 214 (1862).

Kingiella decumbens (Griff.) Rolfe in Orchid Rev. 25: 197 (1917).

Phalaenopsis decumbens (Griff.) Holttum in Gard. Bull. Singapore 11: 286 (1947).

Kingidium decumbens (Griff.) P.F.Hunt in Kew Bull. 24, 1: 97 (1970).

Kingidium deliciosum (Rchb.f.) Sweet var. *bellum* (Teijsm. & Binn.) O. Gruss & Röllke in Orchidee (Hamburg) 44: 225 (1993).

For full synonymy see Seidenfaden in Opera Bot. 95: 183 (1988).

Epiphytic ***herb***. ***Stems*** 0.3-1 cm long, forming large tortuous tufts. ***Leaves*** 7.5-15.5 x 3.5 cm, few, obovate-oblong, obtuse or subacute, sessile, rather thin-textured. ***Inflorescence*** 1 to several, simple or branched; peduncle 6-12 cm long, slender; rachis hardly thickened; floral bracts small, triangular. ***Flowers*** small, resupinate, facing all ways, with white sepals and petals; lateral sepals spotted purple at base; lip purple. ***Dorsal sepal*** 0.6-0.9 x 0.3-0.4 cm, narrowly elliptic, subobtuse. ***Lateral sepals*** 0.6 x 0.3-0.4 cm, oblong-falcate, obtuse. ***Petals*** 0.7-0.9 x 0.4 cm, broadly clawed, elliptic, rounded. ***Lip*** 3-lobed, up to 1.3 x 0.8 cm, spreading, broader above, cuneate-obovate, each with a tooth-like appendage near the back edge; mid-lobe 0.6 x 0.6 cm, obcordate, apex broad, deeply cleft; callus flattened, divided into 2 spreading teeth. ***Column*** somewhat elliptic in front view; pollinia 4, unequal. Plate 10F.

HABITAT AND ECOLOGY: Riverine forest. Alt. lowlands to 300 m. Flowering recorded in April and November.

DISTRIBUTION IN BORNEO: KALIMANTAN SELATAN: Banjarmasin area. SABAH: Tenom District; Labuk Valley.

GENERAL DISTRIBUTION: Widespread in China, India, Sri Lanka, Nepal, Bhutan, Burma, Thailand, Indochina, Peninsular Malaysia, Sumatra, Java, Borneo, Sulawesi, Maluku and the Philippines.

NOTES: In Sabah, this species is found mostly to the south on the drier eastern foothills of the Crocker Range which are in more of a rain-shadow. This perhaps reflects its wider distribution in drier countries in the northern latitudes. The west slopes of the Crocker Range are probably too wet. However, a favoured habitat is on the branches of trees overhanging streams where there is probably more light.

DERIVATION OF NAME: The generic name honours Sir George King who jointly authored "The Orchids of Sikkim-Himalaya" with Robert Pantling. The specific epithet is derived from the Latin *deliciosus*, delicious or delicate, in reference to its delicate appearance.

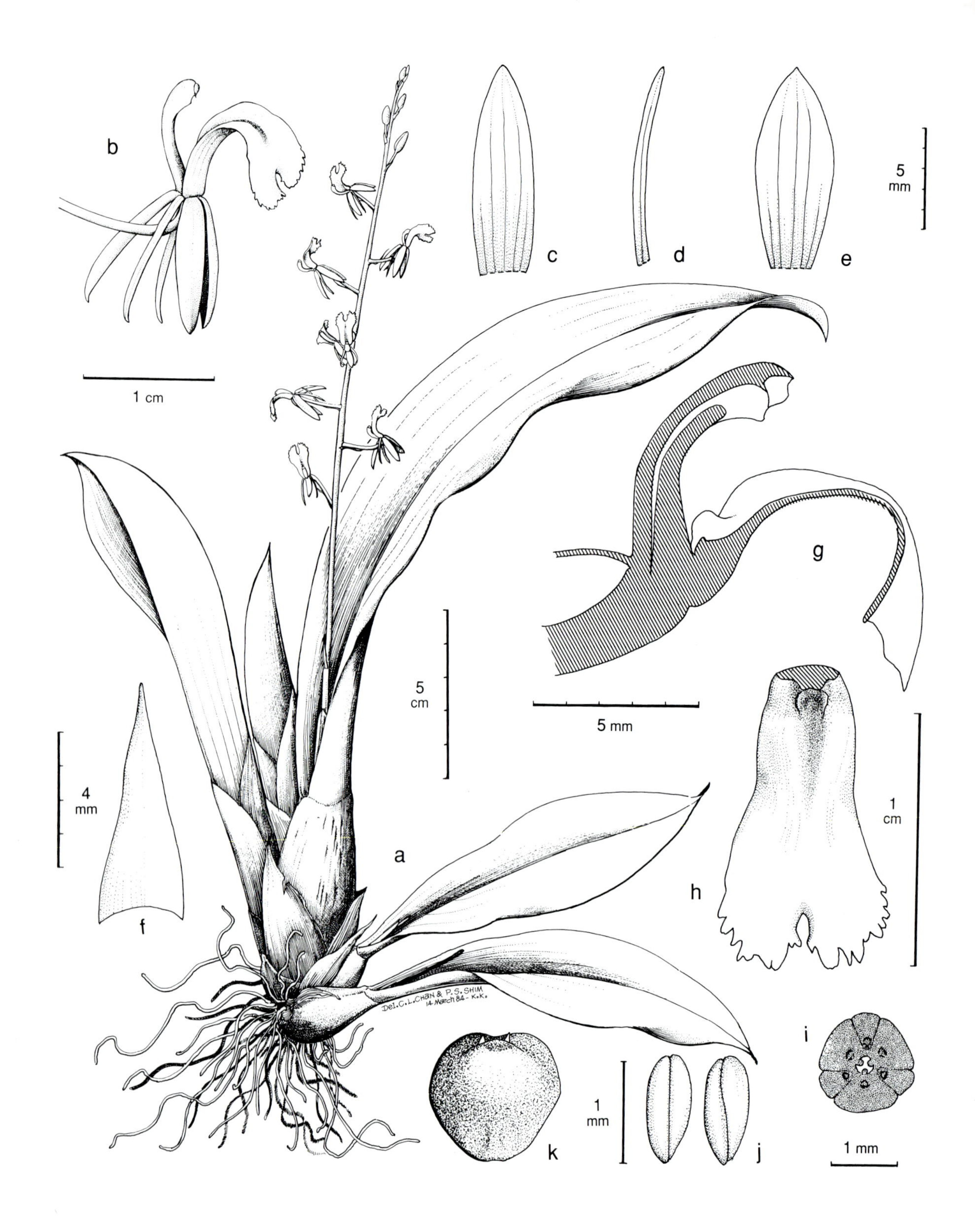

Figure 47. Liparis latifolia (Blume) Lindl. - A: plant. - B: flower, lateral view. - C: dorsal sepal. - D: petal. - E: lateral sepal. - F: floral bract. - G: column and lip, longitudinal section. - H: lip. - I: ovary, transverse section. - J: pollinia. - K: anther. All drawn from cult. TOC by Chan Chew Lun and Shim Phyau Soon.

47. LIPARIS LATIFOLIA (Blume) Lindl.

Liparis latifolia (*Blume*) *Lindl.*, Gen. Spec. Orch. Pl.: 30 (1830). Type: Java, *Blume* s.n. (holotype BO).

Malaxis latifolia Blume, Bijdr.: 393 (1825).

Liparis scortechinii Hook.f., Icon. Plant.: t. 2009 (1890). Type: Malaya, Perak, *Scortechini* s.n. (holotype K).

Liparis robusta Hook.f. in Icon. Plant.: t. 2012 (1890). Type: Malaya, Perak, Maxwell Hill, *Wray* 2808 (holotype K).

Epiphytic ***herb*** up to 40 cm high. ***Roots*** wiry, dense, slender, elongate. ***Pseudobulbs*** 4-11 cm long, 0.7-3 cm in diameter, clustered, fleshy, ovoid to cylindrical, bilaterally somewhat compressed, pale green drying pale yellow, 1-2-leaved at apex, subtended by 2 to 3 lanceolate sheaths. ***Leaves*** 15-30 x 2-7 cm, suberect, oblanceolate to elliptic, acute or subacute, fleshy. ***Inflorescence*** erect to arching, subdensely many-flowered; peduncle fleshy, terete, bearing several lanceolate, sterile bracts; floral bracts spreading, lanceolate, acuminate, 0.6-6 cm long, flesh-coloured. ***Flowers*** suberect; sepals and petals pale buff-pink or cream; lip orange to brick-red; column cream or yellow. ***Pedicel*** with ***ovary*** 1.5-2.2 cm long, flesh-coloured. ***Dorsal sepal*** 0.9-1.1 x 0.3 cm, strongly reflexed, oblong, obtuse, side margins strongly recurved. ***Lateral sepals*** 0.9-1.1 x 0.3 cm, strongly reflexed, oblong-elliptic, obtuse, with side margins strongly recurved. ***Petals*** 0.9-1 x 0.5 cm, strongly reflexed, linear, acute or obtuse. ***Lip*** 1 x 0.9 cm, narrowly clawed below, abruptly recurved in middle, obovate and deeply emarginate above; basal half of the lip lying parallel to the column and the apical half reflexed down through 180°. ***Column*** 0.4-0.6 cm long, subsigmoid, erect, cylindrical, with two small subtriangular flaps near apex; pollinia 4 in 2 groups. Plate 11A.

HABITAT AND ECOLOGY: Epiphytic relatively close to the forest floor, in shady moist habitats, particularly on trees besides rivers and streams; on cultivated *Citrus*. Alt. 200 to 1500 m. Flowering occurs twice a year.

DISTRIBUTION IN BORNEO: KALIMANTAN: Mt. Pamattin; SABAH: Mt. Kinabalu; Tenom District; Mt. Alab.

GENERAL DISTRIBUTION: China (Hainan), Peninsular Malaysia, Thailand, Sumatra, Java, Borneo, Timor and New Guinea.

NOTES: The way the flower is held is reminiscent of a small praying mantis. The cream column and contrasting orange-red sepals and petals and deep red lip produce a showy display.

DERIVATION OF NAME: The generic name is derived from the Greek *liparos*, fat, greasy or shining, referring to the smooth shiny leaves of many species. The specific epithet is derived from the Latin *lati*, broad or wide, and *folius*, leaf, referring to the broad leaves of this species.

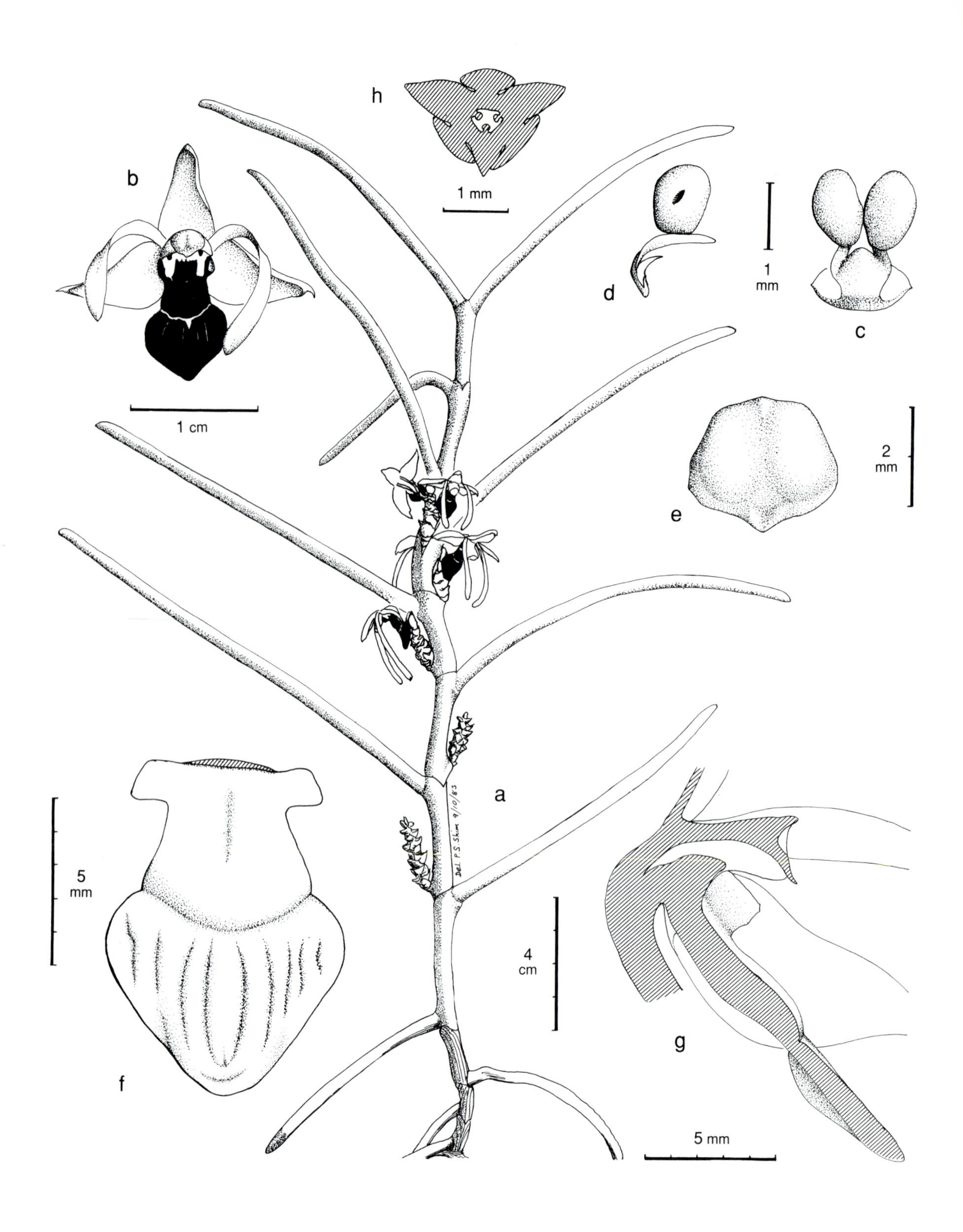

Figure 48. Luisia curtisii Seidenf. - A: plant. - B: flower, front view. - C: pollinarium, front view. - D: pollinarium, lateral view. - E: anther. - F: lip. - G: column and lip, longitudinal section. - H: ovary, transverse section. All from living plant collected from Lohan River, Mt. Kinabalu. Drawn by Shim Phyau Soon.

48. LUISIA CURTISII Seidenf.

Luisia curtisii *Seidenf.* in Bot. Tidsskr. 68: 83 (1973). Type: Peninsular Malaysia, Penang, Bukit Penara, *Curtis* 1176 (holotype K).

Luisia tristis sensu Hook.f., Fl. Brit. Ind. 6: 25 (1890).

Epiphytic ***herb***. ***Stem*** up to 50 cm long. ***Leaves*** (5-)15-18 x 0.5 cm, terete, porrect, dark green; sheaths 1.5-4 cm long, blackish green. ***Inflorescence*** 1.5-2 cm long, erect, close to stem, with several flowers, only 1 or 2 open at a time; floral bracts 2-3 x 3-3.5 mm, ovate, obtuse, concave. ***Flowers*** yellowish white, flushed purple, the long petals with distinct cream to yellow tips; lip blackish purple; column white with maroon spots. ***Pedicel*** with ***ovary*** 1.2-1.8 cm long, twisted. ***Dorsal sepal*** (0.6-)0.8-1 x 0.4 cm, oblong, apex slightly cucullate, obtuse, erect. ***Lateral sepals*** 1.2 x 0.3-0.4 cm, ovate-oblong, strongly dorsally carinate, somewhat concave, spreading. ***Petals*** (1.7-)2-2.5 x 0.2-0.35 cm, linear, obtuse. ***Lip*** 1-1.1 cm long, about equally divided by a curved line into hypochile and epichile; hypochile 0.3-0.4 x 0.3-0.4 cm, with very small erect rounded side lobes to 1-1.5 x 1-1.5 mm; epichile 0.4 x 0.6-0.7 cm, cordate, obtuse, upper surface slightly convex, with obscure longitudinal furrows. ***Column*** 0.3-0.4 mm long. Plate 11B.

HABITAT AND ECOLOGY: Lowland dipterocarp and lower montane forest on granitic, sandstone and ultramafic substrates. Alt. 900 to 1500 m. Flowering observed in May and December.

DISTRIBUTION IN BORNEO: SABAH: Mt. Kinabalu; Ulu Dusun; Sinsuron Road.

GENERAL DISTRIBUTION: Peninsular Malaysia, Thailand, Vietnam and Borneo.

NOTES: Populations from Peninsular Malaysia have larger flowers with a lip to 1.8 cm long and 1.3 cm wide, when flattened.

DERIVATION OF NAME: The generic name is dedicated to Don Luis de Torres, a nineteenth century Spanish botanist. The specific epithet honours Charles Curtis (1852-1928), Superintendent of Gardens and Forests, Penang, who collected the type.

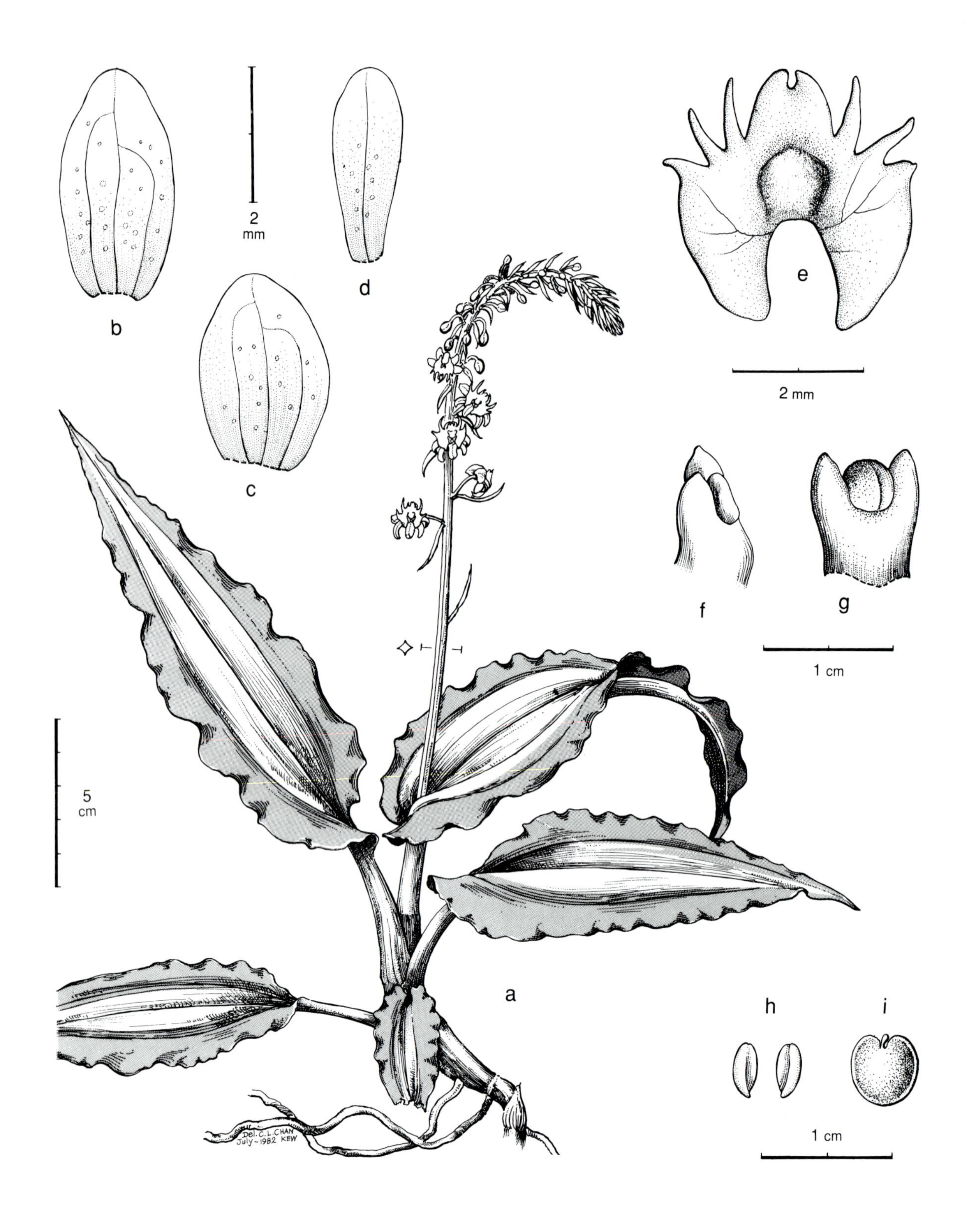

Figure 49. Malaxis lowii (E. Morren) Ames - A: plant. - B: dorsal sepal. - C: lateral sepal. - D: petal. - E: lip. - F: column, lateral view. - G: column, dorsal view. - H: pollinia. - I: anther. All from *Lamb* K14, Hort. Bot. Kew. Drawn by Chan Chew Lun and Philip Cribb.

49. MALAXIS LOWII (E.Morren) Ames

Malaxis lowii (*E.Morren*) *Ames* in Merr., Enum. Born. Pl.: 151 (1921). Type: Borneo, Cult. *Jacob-Mahoy & Co.*, *Low* s.n. (holotype not located).

Microstylis lowii E.Morren in Belgique Hort. 34: 281, t. 14, fig. 2 (1884).

Terrestrial ***herb***. ***Rhizome*** creeping, fleshy. ***Stems*** to 10 cm high, fleshy and swollen below. ***Leaves*** 4-10; blade 5-8(-10) x 1.5-2.5 cm, narrowly elliptic, acuminate, asymmetric at base, margin undulate, dull brownish purple to golden brown with a broad greenish white or pale green central band above, pale green with purple veins below; petiole 1-2(-2.5) cm long, sheathing at base, purple. ***Inflorescence*** many-flowered, lax or somewhat dense; peduncle 5-9 cm long, bright purple; rachis 4-15 cm long, bright purple; floral bracts 3-5 mm long, linear, acuminate, reflexed, dark purple. ***Flowers*** greenish yellow, ochre or dark purple; lip often flushed green. ***Pedicel*** with ***ovary*** 0.3-0.5 cm long, curved. ***Sepals*** and ***petals*** reflexed. ***Dorsal sepal*** 0.2-0.3 x 0.1-0.15 cm, oblong, obtuse. ***Lateral sepals*** 0.2-0.3 x 0.15-0.2 cm, ovate to ovate-oblong, obtuse. ***Petals*** 0.2-0.25 cm x 0.3-1 mm, ligulate, obtuse. ***Lip*** 0.35-0.4 cm wide across auricles, sagittate, auriculate, pectinate; mid-lobe 1-2 mm long, bifid, with 2 teeth each side; auricles 1-2 mm long, acute; disc with an obscure or distinct horseshoe-shaped thickening surrounding the fovea. ***Column*** 0.8-1 mm long. Plate 11C.

HABITAT AND ECOLOGY: In leaf-litter or amongst rocks, sometimes in swampy alluvial soils; mixed lowland and hill-dipterocarp forest to lower montane forest, on sandstone, with scattered limestone rocks; limestone cliffs and ultramafic substrates. Alt. sea level to 1500(2100) m. Flowering observed in March, August, September and November.

DISTRIBUTION IN BORNEO: SABAH: Mt. Kinabalu; Lahad Datu District, Ulu Sungai Danum. SARAWAK: Mt. Mulu National Park.

GENERAL DISTRIBUTION: Endemic to Borneo.

NOTES: Widely scattered particularly in Sabah but nowhere common. In Mulu National Park, Sarawak, plants have been seen in a seasonally swampy alluvial valley, with very moist, humid conditions. Normally a species of low to middle elevations, it ascends to as high as 2100 m in the Mesilau Valley on Mt. Kinabalu. Here tall, humid forest thrives in an unusually deep, protected valley.

DERIVATION OF NAME: The generic name is derived from the Greek *malaxis*, softening, referring to the soft and tender texture of the leaves. The specific epithet honours Sir Hugh Low who collected the type.

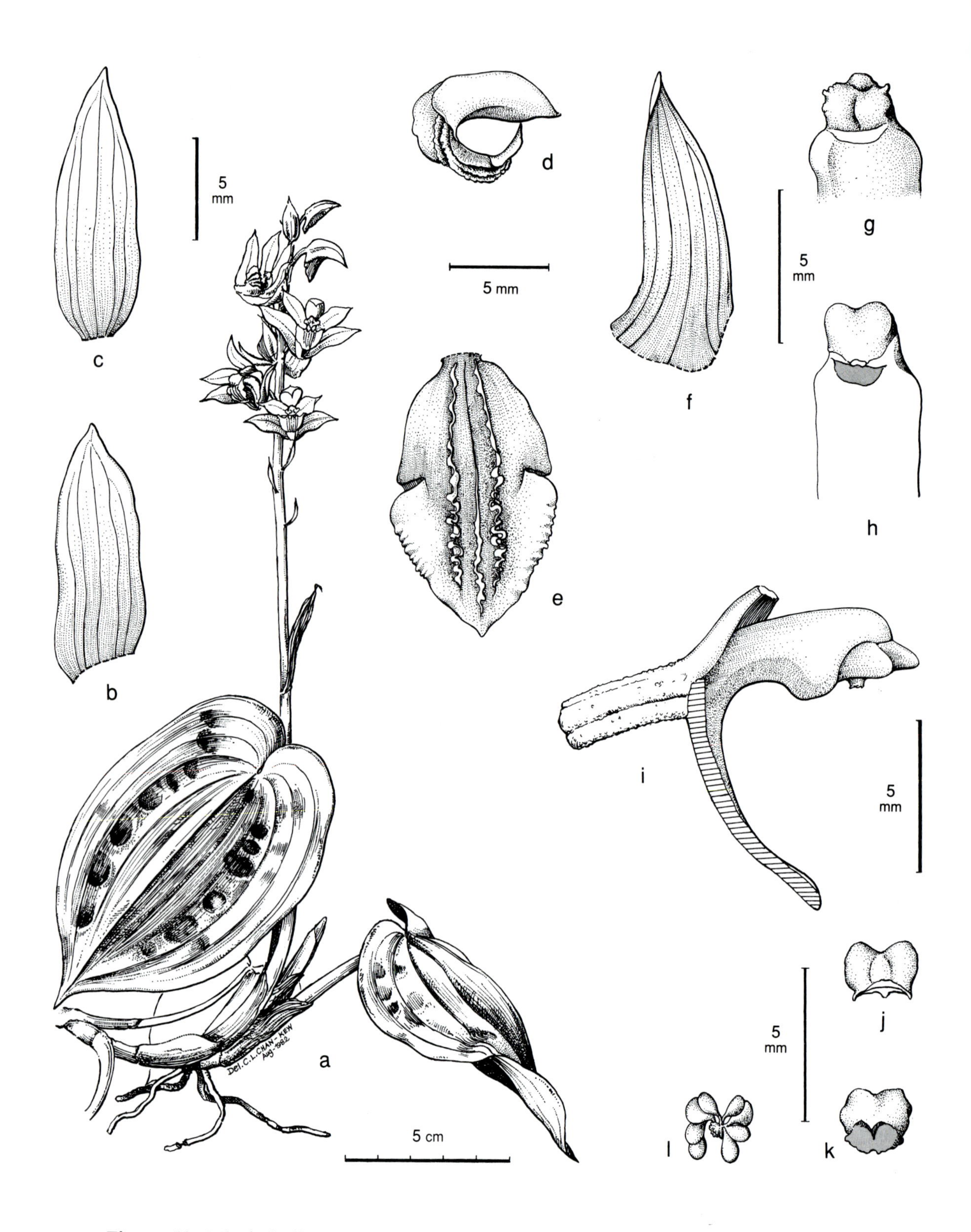

Figure 50. Mischobulbum scapigerum (Hook. f.) Schltr. - A: plant. - B: dorsal sepal. - C: petal. - D: lip, lateral view. - E: lip. - F: lateral sepal. - G: column, apex, with anther and pollinia removed. - H: column, apex, with anther attached. - I: column and pedicel, lateral view. -J: anther, abaxially. - K: anther, adaxially. -L: pollinia. All drawn from Hort. Bot. Kew by Chan Chew Lun and Philip J. Cribb.

50. MISCHOBULBUM SCAPIGERUM (Hook.f.) Schltr.

Mischobulbum scapigerum (*Hook.f.*) *Schltr.* in Feddes Repert. Beih. 1: 98 (1911). Type: Borneo, cult. *Low* (holotype K).

Nephelaphyllum scapigerum Hook.f. in Bot. Mag. ser. 3, 19: t. 5390 (1863).

Tainia scapigera (Hook.f.) J.J.Smith in Bull. Jard. Bot. Buitenzorg, ser. 2, 8: 6 (1912).

See H. Turner. A Revision of the Orchid Genera *Ania*, *Hancockia*, *Mischobulbum* and *Tainia*. Orchid Monographs 6: 64-73 (1992).

Glabrous terrestrial ***herb*** with a creeping rhizome. ***Roots*** 1-2 mm in diameter. ***Pseudobulbs*** petiole-like, fusiform, curved, olive-green to dull purple, 6-9 x 0.2-0.8 cm, lower half enclosed in the reticulate remains of two sheaths. ***Leaves*** sessile, cordate-ovate, acute to acuminate, 7-13 x 5-9.5 cm, dull dark green, surface uneven or rather undulate, the mature leaves becoming puckered with rows of darker green concavities. ***Inflorescence*** with an erect scape 10 cm long, bearing 3 membranous, reticulate-nerved, acute brown sheaths up to 3.5 cm long; rachis up to 7 cm long, minutely ramentaceous, 4- to 20-flowered; floral bracts linear, acute, 0.6 cm long. ***Flowers*** faintly sweet-scented; sepals and petals greenish mustard-yellow to light golden to orange-brown, speckled and veined purple-red; lip white with fine purple spots, crests purple, apex orange-yellow to yellow with reddish purple keels, base of lip purple-red, sometimes with yellow; column white with purple spots or plain purple, anther-cap pale creamy yellow, column-foot purple, yellow at base; the plant illustrated in Bot. Mag. t. 5390 has pale yellow-green sepals and petals with purple veins, the lower half of the lip white, blotched and spotted purple and the upper half golden yellow. ***Pedicel*** with ***ovary*** 1-1.4 cm long, 6-ribbed, papillose and minutely ramentaceous. ***Dorsal sepal*** 1-1.5 x 0.3-0.4 cm, ovate to elliptic, acute. ***Lateral sepals*** 1-1.6 x 0.5-0.7 cm, obliquely triangular, elliptic to acuminate, acute; mentum 0.6-0.7 cm long. ***Petals*** 0.8-1.3 x 0.3-0.55 cm, obliquely elliptic to ovate, acute. ***Lip*** blade 1.1-1.5 x 0.5-0.7 cm, entire, strongly recurved, elliptic, acute, gently undulate distally, disc with three undulate keels, 1-1.1 cm long, apex rounded to slightly acuminate. ***Column*** 0.5-0.8 cm; pollinia 8 in 4 pairs, subequal. Plate 11D.

HABITAT AND ECOLOGY: Among rocks or in stony colluvial soil, in deep shade under tall mixed hill-dipterocarp forest, usually on sandstone ridges. Alt. 250 to 900 m. Flowering observed May to July, September to January.

DISTRIBUTION IN BORNEO: SABAH: Mt. Kinabalu; Tenom District; Mt. Lotung. SARAWAK: Mt. Matang.

GENERAL DISTRIBUTION: Endemic to Borneo.

NOTES: Plants seen and photographed in 1976 in Sabah on the lower slopes of Mt. Lotung, above Lake Linumunsut, had a taller erect scape and much larger flowers with bright yellow sepals and petals and a bright yellow lip apex.

DERIVATION OF NAME: The generic name is derived from the Greek *mischos*, stalk, and *bolbos*, bulb, describing the shape of the pseudobulbs. The specific epithet is derived from the Latin *scapiger*, bearing a scape, ie. a leafless or almost leafless stem.

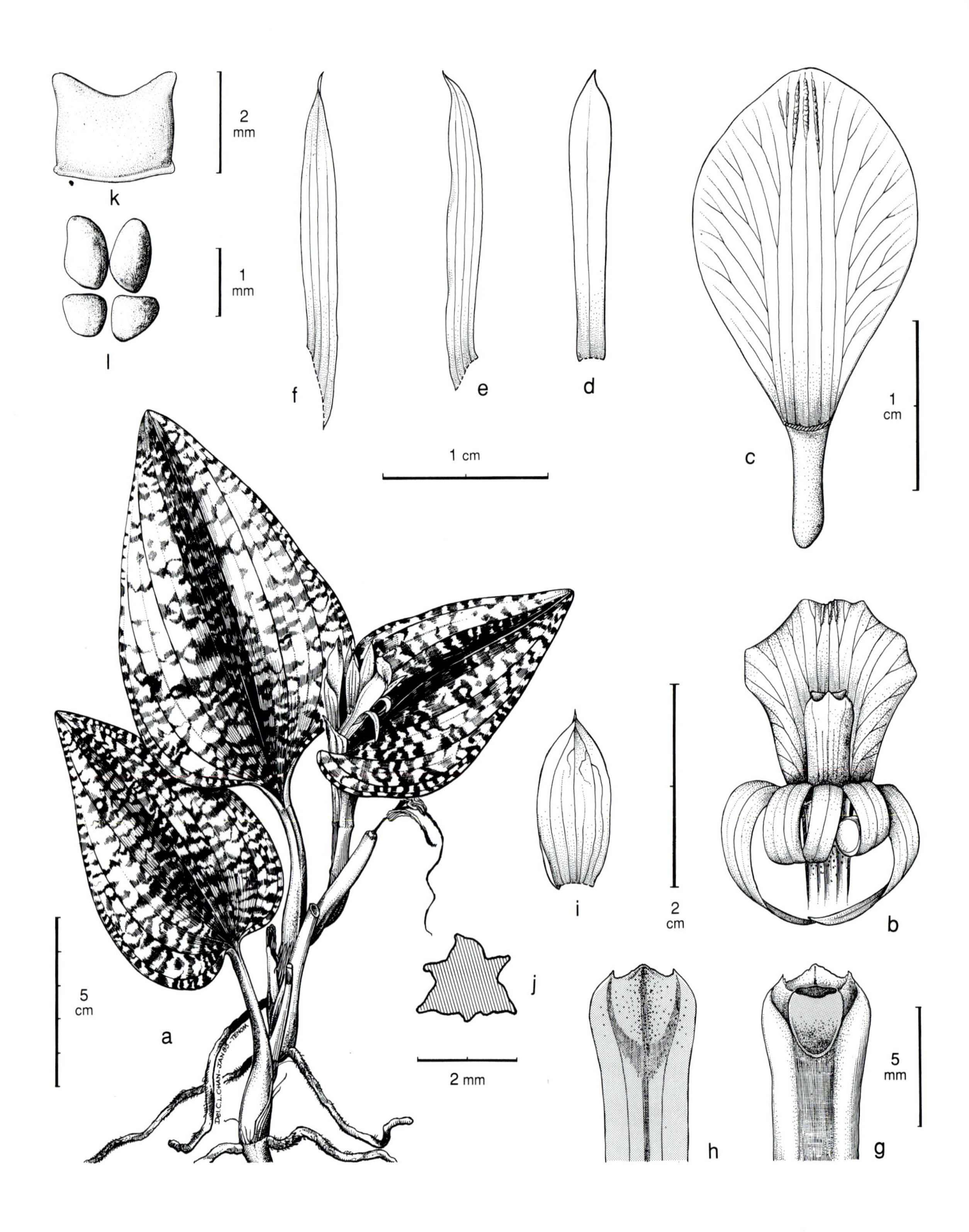

Figure 51. Nephelaphyllum aureum J.J. Wood - A: plant. - B: flower. - C: lip. -D: dorsal sepal. - E: petal. - F: lateral sepal. - G: column, ventral view. - H: column, dorsal view. - I: floral bract. - J: ovary, transverse section. - K: anther. - L: pollinia. All drawn from *Lamb* AL 60/83 by Chan Chew Lun.

51. NEPHELAPHYLLUM AUREUM J.J.Wood

Nephelaphyllum aureum *J.J.Wood* **sp. nov.**, *N. pulchro* Blume bene disperso quam maxime affinis, sed foliis manifestius cordatis, floribus majoribus, labio intense aureo-luteo disco nervis centralibus prominentibus 5 ornato, calcari et columna multo longiore discedit. Typus: Borneo, Sabah, Mt. Alab, Sinsuron Road, 1200 m, May 1982, *Lamb* 230 in SAN 93500 (holotypus K).

Terrestrial ***herb.*** ***Rhizome*** creeping, decumbent, rooting at nodes, 7-10 cm long. ***Pseudobulbs*** 2.5-3.5 x 0.6-0.8 cm, fusiform, fleshy, purplish olive-green, enclosed in pale brown papery sheaths when young. ***Leaves*** 6.5-10.5 x 4.8-6.5 cm, triangular-ovate to cordate, acute, very shortly petiolate, fleshy, pale buff-grey to flesh-coloured with darker purplish brown tessellation and green nerves above, amethyst-purple below; petiole 2-3 mm long. ***Inflorescence*** terminal, 3- to 5-flowered; rachis 1.5-2 cm long; floral bracts 1.2-1.3 x 0.9 cm, ovate, carinate, acute, papery, pale brown speckled and flecked dark purple. ***Flowers*** non-resupinate, faintly sweet-scented; sepals and petals pale translucent greenish cream, flecked and stained pale purple, darker when in bud; lip golden yellow, greenish towards base, nerves darker orange-yellow; spur greenish yellow, faintly flushed pale purple at apex; column pale greenish white flecked purple below and very finely speckled purple above, anther-cap cream with very fine purple speckles and dark purple speckles and a dark purple spot on each projection. ***Pedicel*** with ***ovary*** 1.2 cm long, 6-keeled, green with purple keels. ***Sepals*** and ***petals*** subequal, reflexed. ***Sepals*** 1.7-2.1 x 0.3 cm, linear-ligulate, acute, carinate at base. ***Petals*** 1.6-2 x 0.4 cm, oblong-ligulate, acute. ***Lip*** 1.8-2.5 x 1.6-2.1 cm, 1.3 cm broad at apex, oblong-flabellate, entire, obtuse, the sides embracing the column, apex decurved, with 5 prominent and somewhat raised, smooth to minutely papillose central nerves, the outermost 2 becoming obscure towards the lip apex; spur cylindrical, obtuse, straight or curving inwards. ***Column*** 1.3 x 0.6 cm, oblong, slightly winged; anther-cap 3 x 3 mm, with 2 conical dorsal projections; pollinia 8. Plate 11E.

HABITAT AND ECOLOGY: Mixed oak/chestnut lower montane forest on sandstone ridge soils; in leaf litter in deep shade. Alt. 900 to 1400 m.

DISTRIBUTION IN BORNEO: SABAH: Mt. Alab. SARAWAK: locality uncertain.

GENERAL DISTRIBUTION: Endemic to Borneo.

NOTES: *Nephelaphyllum* is a small genus of a dozen or so species distributed in China and from Indo-China through Thailand and Peninsular Malaysia eastwards through Indonesia to the Philippines. Nine species, including *N. aureum*, have so far been recorded from Borneo. All are terrestrial and have slender, fusiform pseudobulbs, which are rarely distinct from the shorter petioles, and more or less cordate and somewhat variegated, fleshy leaves. The small to medium-sized flowers have an entire or subentire, spurred lip that is borne uppermost, and a footless column. *N. aureum* is closely related to the widespread *N. pulchrum* Blume, but differs in having more distinctly cordate leaves and larger flowers with a golden yellow lip bearing five prominent central nerves on the disc and a much longer spur and column.

DERIVATION OF NAME: The generic name is derived from the Greek *nephela*, cloud, and *phyllos*, leaf, referring to the hazy opaqueness of the upper surface of the leaf. The specific epithet is derived from the Latin *aureus*, golden, referring to the deep golden yellow colour of the lip.

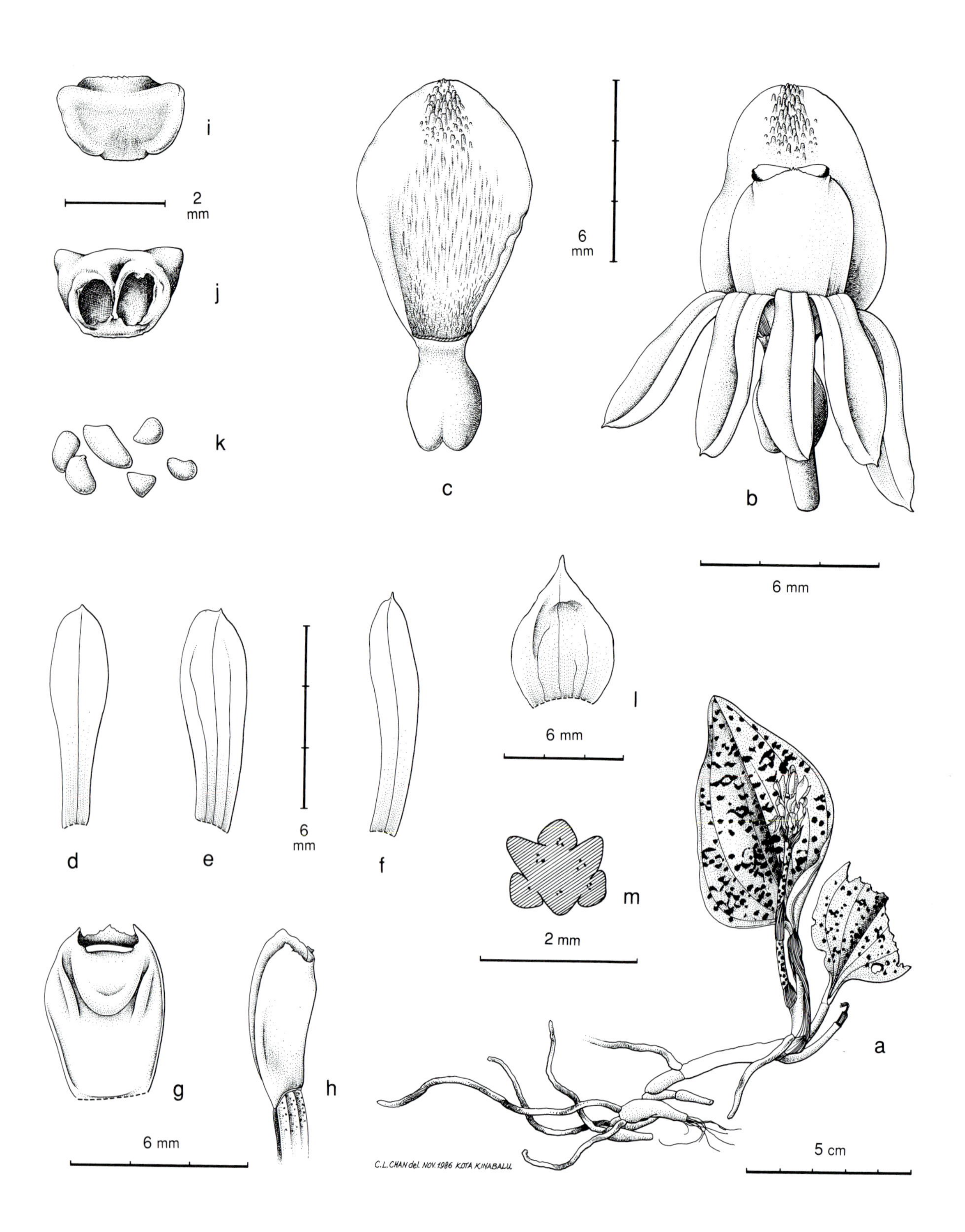

Figure 52. Nephelaphyllum flabellatum Ames & C. Schweinf. - A: plant. - B: flower. - C: lip. - D: dorsal sepal. - E: petal. - F: lateral sepal. - G: column, ventral view. - H: column and ovary, lateral view. - I: anther, abaxially. - J: anther, adaxially. - K: pollinia. - L: floral bract. - M: ovary, transverse section. All drawn from living plant collected from Lohan River, Mt. Kinabalu by Chan Chew Lun.

52. NEPHELAPHYLLUM FLABELLATUM Ames & C.Schweinf.

Nephelaphyllum flabellatum *Ames & C. Schweinf.* in Ames, Orch. 6: 19 (1920). Type: Borneo, Sabah, Mt. Kinabalu, Marai Parai spur, *Clemens* s.n. (holotype AMES).

Terrestrial ***herb***. ***Rhizome*** creeping, rooting at the nodes. ***Pseudobulbs*** up to 2 x 0.4 cm, fusiform, decumbent, partially concealed by loose, papery sheaths, purplish. ***Leaves*** 5.5-6 x 4 cm, ovate-cordate, obtuse and mucronate or subacute, thin-textured, main nerves prominent beneath; petiole 3-5 mm long, conduplicate; blade pale olive to grey-green with darker purplish brown blotches and fine dark green nerves. ***Inflorescence*** 4- to 5-flowered; peduncle 2.5 cm, olive-green with purple streaks; sterile bracts 2-3, 0.9-1.1 cm long, broadly ovate, acute, brown, papery; rachis 2-2.5 cm long; floral bracts 5-7 mm long, ovate, acute, papery, brown. ***Flowers*** erect to spreading; sepals and petals apple-green, tipped pink; lip pale pink with pink nerves and a pale green streaked and spotted purple central area fading to cream, apical callus orange; spur greenish; column pink, spotted darker pink and with a green ventral stripe. ***Pedicel*** with ***ovary*** 0.6-0.8 cm long, narrowly clavate, 6-angled, ramentaceous. ***Sepals*** and ***petals*** connivent, recurved, circinate. ***Sepals*** linear-ligulate, obtuse and mucronate, 1-nerved. ***Dorsal sepal*** 0.8-1 x 0.15-0.2 cm. ***Lateral sepals*** 0.9-1.1 x 0.13-0.17 cm wide, slightly falcate. ***Petals*** 0.8-0.95 x 0.2-0.28 cm, oblong-ligulate, obtuse and mucronate, 3-nerved. ***Lip*** 1-1.3 x 0.6-1.1 cm, elliptic, obtuse, fleshy, papillose, margin often inflexed near base, with a bunch of apical papillae, becoming progressively longer near the apex, hirsute at spur entrance; spur 4-4.5 x 1.5-3 mm, subglobose-cylindric, apex retuse. ***Column*** 0.6 x 0.35 cm, oblong, broadly winged; anther-cap quadrate, somewhat flattened; pollinia 8. Plate 11F.

HABITAT AND ECOLOGY: Hill-forest and lower montane forest; amongst leaf litter in deep shade; Alt. 600 to 1400 m.

DISTRIBUTION IN BORNEO: SABAH: Mt. Kinabalu; Sipitang District.

GENERAL DISTRIBUTION: Endemic to Borneo.

DERIVATION OF NAME: The specific epithet is derived from the Latin *flabellatus*, fan-shaped, referring to the lip shape.

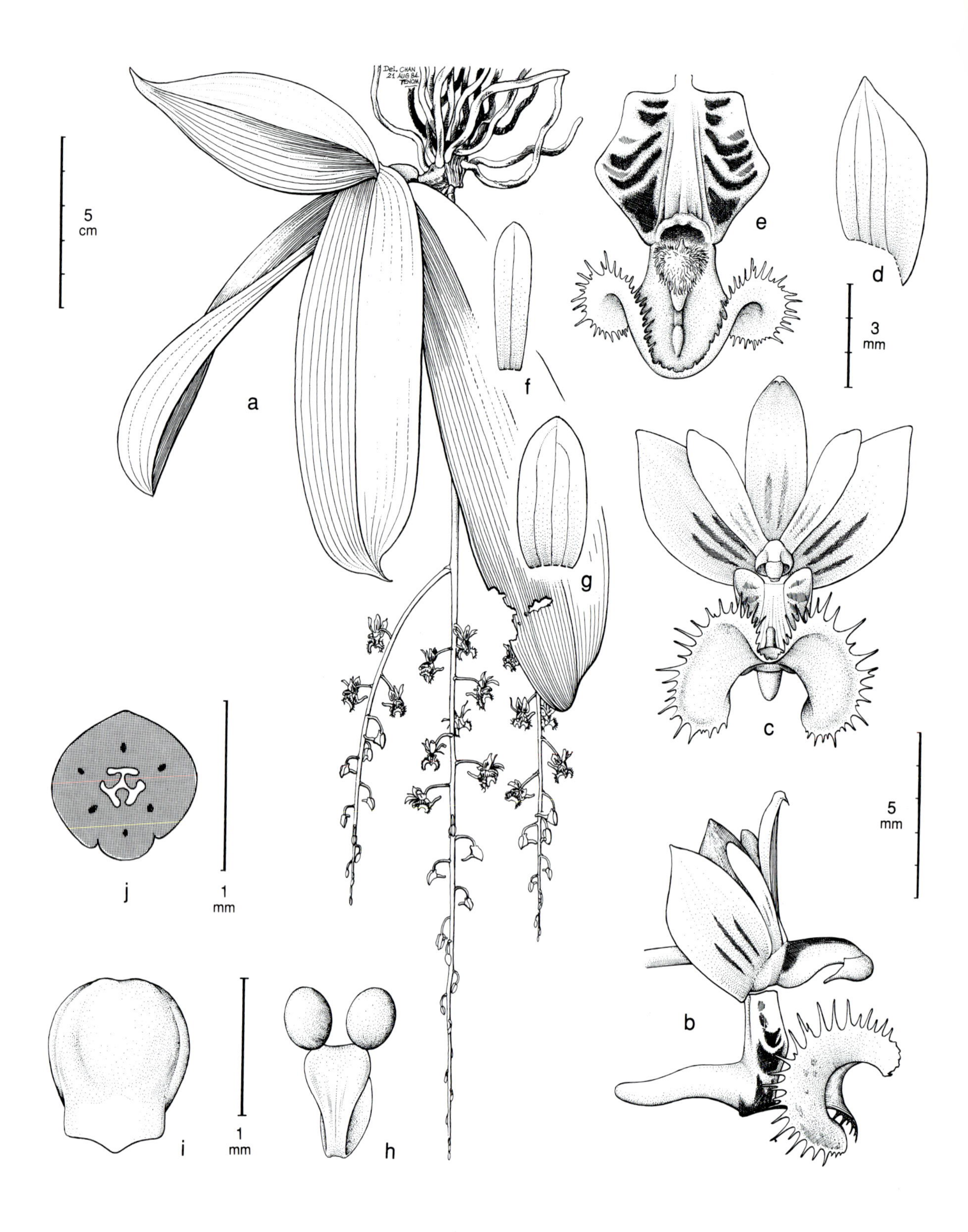

Figure 53. Ornithochilus difformis (Wall. ex Lindl.) Schltr. var. **difformis** - A: plant. - B: flower, lateral view. - C: flower, front view. - D: lateral sepal. - E: lip. - F: petal. - G: dorsal sepal. - H: pollinarium. - I: anther, front view. - J: ovary, transverse section. All drawn from *Lamb* AL 265/84 by Chan Chew Lun.

53. ORNITHOCHILUS DIFFORMIS (Wall. ex Lindl.) Schltr. var. DIFFORMIS

Ornithochilus difformis *(Wall. ex Lindl.) Schltr.* in Feddes Repert., Beih. 4: 277 (1919). Type: Nepal, *Wallich* drawing (holotype K).

Aerides difforme Wall. ex Lindl., Gen. Sp. Orch. Pl.: 242 (1833).

Ornithochilus fuscus Wall. ex Lindl., Gen. Sp. Orch. Pl.: 242 (1833) in synon. of *Aerides difforme* and based on same type.

O. eublepharon Hance in J. Bot. 22: 364 (1884). Type: China, cult. Hong Kong B.G. (holotype K).

Sarcochilus difformis (Wall. ex Lindl.) Tang & Wang in Acta Phytotax. Sin. 1: 92 (1951).

Trichoglottis difformis (Wall. ex Lindl.) Bân & Huyen, Fl. Tayngyuen: 206 (1983).

var. **difformis**

Epiphytic ***herb***. ***Stem*** short, 1-4 cm long, concealed by persistent sheathing leaf bases. ***Leaves*** 2-6, 5-20 x (2-)3-4.5 cm, obliquely elliptic-oblong, acute, attenuate towards base, many-nerved; sheathing bases 0.5-1 cm long. ***Inflorescence*** 15-42 cm long, lateral, racemose or paniculate, loosely many-flowered; peduncle and rachis slender; floral bracts 2-3 mm long, subulate, adpressed to pedicel with ovary. ***Flowers*** c. 1 cm across; sepals and petals greenish-yellow to yellow with purple streaks; lip white, side lobes flushed yellow and striped purple or pale mauve; spur greenish yellow, yellow at the entrance; callus purple; column green, flushed purple; anther-cap yellow. ***Pedicel*** with ***ovary*** 1-1.3 cm long, slender, straight. ***Sepals*** and ***petals*** obtuse, spreading. ***Dorsal sepal*** 5 x 1.5 mm, oblong. ***Lateral sepals*** 5.5 x 3 mm, obliquely-obovate, often cucullate at apex. ***Petals*** 4 x 1 mm, linear. ***Lip*** 3-lobed; side lobes 4 x 2 mm, subquadrate, obtuse; mid-lobe 5-6 x 5-6 mm, clawed, flabellate, fimbriate-pectinate, with a central tooth, inflexed, provided with a velvety hairy flap over the spur entrance; callus keel-like; spur 4-7 mm long, cylindrical to narrowly conical, obtuse, incurved. ***Column*** 3 mm long, column-foot absent; rostellum forcipate; pollinia 4, in two masses. Plate 12A.

HABITAT AND ECOLOGY: Hill-forest and lower montane primary forest on sandstone ridges. Alt. 900 to 1500 m. Flowering observed from December to February.

DISTRIBUTION IN BORNEO: KALIMANTAN TIMUR: Apokayan. SABAH: Penampang District; Sinsuron Road.

GENERAL DISTRIBUTION: N. India, SW. China, Burma, Thailand, Indo-China, Sumatra, Peninsular Malaysia and Borneo.

NOTES: Sabah plants can have either racemose or paniculate inflorescences. Thai plants have a purple lip and spur, whereas Sabah plants have a greenish yellow spur and a white and purple mid-lobe. The spur is distinctly bent down as in Thai and

Peninsular Malaysian plants. The specimen depicted here was cultivated at Tenom Orchid Centre at a much lower altitude than the original collection site.

DERIVATION OF NAME: The generic name is derived from the Greek *ornis*, *-ithos*, bird, and *cheilos*, lip, referring to the bilobed lip resembling a bird in flight. The Latin specific epithet *difformis*, meaning irregular or uneven, probably refers to the irregularly toothed lip margin.

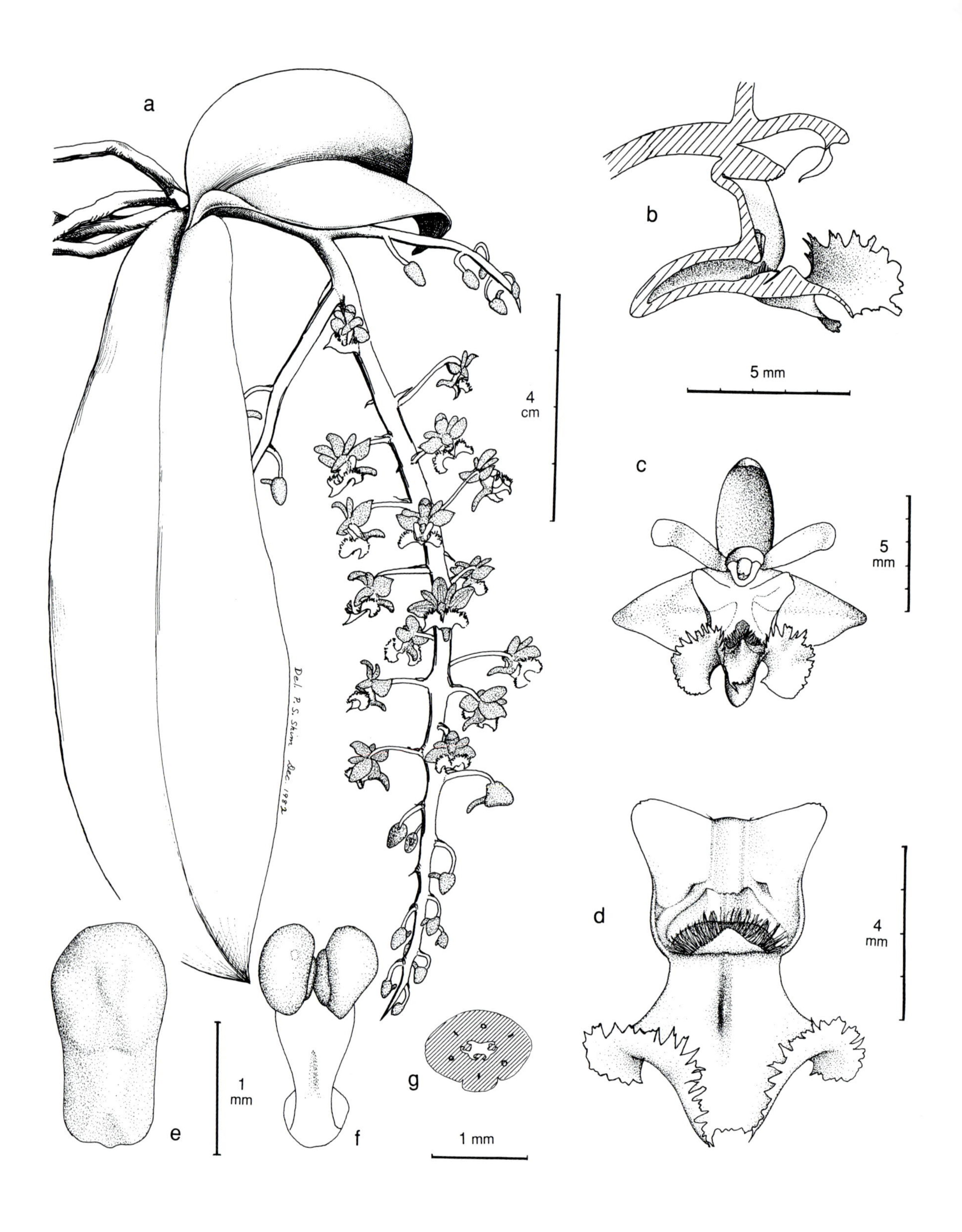

Figure 53a. Ornithochilus difformis var. **kinabaluensis** J.J. Wood, A. Lamb & Shim - A: plant. - B: column and lip, longitudinal section. - C: flower, front view. - D: lip. - E: anther. - F: pollinarium. - G: ovary, transverse section. All drawn from *Jukian & Lamb* in *Lamb* AL 4/82 by Shim Phyau Soon.

53a. ORNITHOCHILUS DIFFORMIS (Wall. ex Lindl.) Schltr. var. KINABALUENSIS J.J. Wood, A. Lamb & Shim

Ornithochilus difformis *(Wall. ex Lindl.) Schltr.* var. **kinabaluensis** *J.J.Wood, A.Lamb et Shim* **var. nov.** a var. typico sepalis petalisque viridi-luteis, labello albo lobo medio breviore fimbriato lobis lateralibus anguste alatis callo minore carinato calcare rectiore differt. Typus: Borneo, Sabah, Mt. Kinabalu, Pinosuk Plateau, 1400-1500 m, 5 December 1982, *Jukian & Lamb* in *Lamb* AL 4/82 (holotypus K).

Leaves up to 15 x 5 cm. ***Inflorescence*** up to 17 cm long, paniculate, 20- to 30-flowered. ***Flowers*** faintly sweet-scented, sepals and petals greenish yellow, turning to yellow; lip white, flap over spur entrance yellow; column white, flushed yellow at apex. ***Dorsal sepal*** 5 x 2-2.8 mm. ***Lateral sepals*** 4.5 x 3-4 mm. ***Petals*** 4 x 1-2 mm. ***Lip*** side lobes 2.5-3 mm long, 2-1.5 mm wide at apex; mid-lobe erose to shortly fimbriate-pectinate; callus a low keel; spur 3-4 mm long, straight or gently incurved, on the same plane as the mid-lobe. ***Column*** thickly conical, 3-4 mm long; anther-cap oblong, 1.7 mm long; pollinia 4, in 2 unequal masses. Plate 12B.

HABITAT AND ECOLOGY: Epiphytic on trees in lower montane forest. Alt. 1300 to 1500 m. Flowering over a long period during the rainy season.

DISTRIBUTION IN BORNEO: SABAH: Mt. Kinabalu.

GENERAL DISTRIBUTION: Endemic to Borneo (Sabah only).

NOTES: Differs from the typical variety in having pure greenish yellow sepals and petals, a pure white lip with shorter fimbriate processes on the mid-lobe, more distinct, narrowly wing-like side lobes, a smaller keel-like callus and a shorter, usually less curved spur.

DERIVATION OF NAME: The specific epithet refers to Mt. Kinabalu, the type locality.

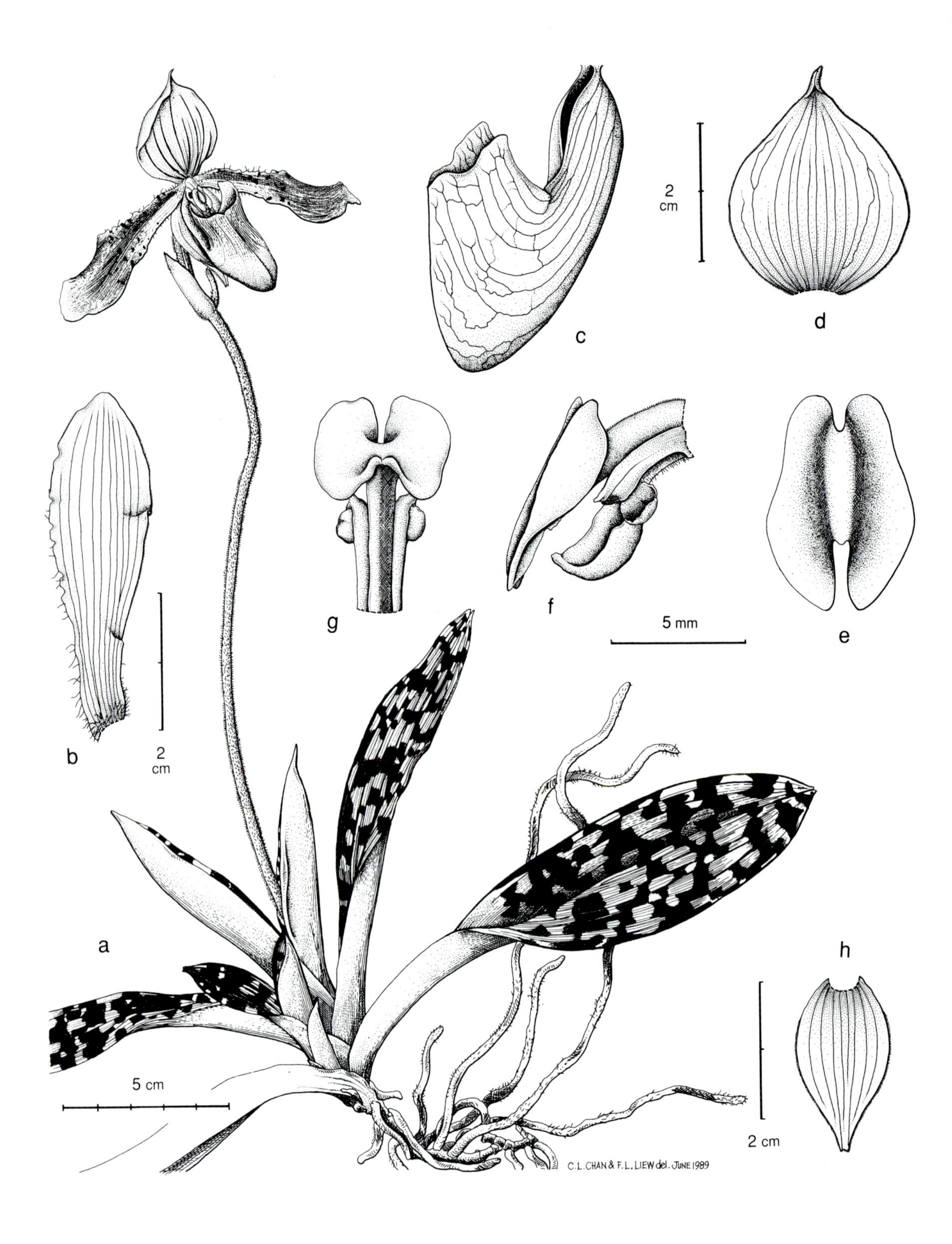

Figure 54. Paphiopedilum bullenianum (Rchb. f.) Pfitzer - A: plant. - B: petal. - C: lip, lateral view. - D: dorsal sepal. - E: staminode, front view. - F: staminode, lateral view. - G: staminode, dorsal view. - H: synsepal. All drawn from cult. TOC by Chan Chew Lun and Liew Fui Ling.

54. PAPHIOPEDILUM BULLENIANUM (Rchb.f.) Pfitzer

Paphiopedilum bullenianum *(Rchb.f.) Pfitzer* in Bot. Jahrb. Syst. 19: 40 (1894). Type: Borneo, hort. *Low* (holotype W).

Cypripedium bullenianum Rchb.f. in Bot. Zeitung (Berlin) 23: 99 (1865).

C. bullenianum Rchb.f. var. *oculatum* Rchb.f. in Gard. Chron. n.s. 15: 563 (1881). Type: Borneo, cult. *Bull* (holotype W).

Paphiopedilum amabile Hallier f. in Natuurk. Tijdschr. Ned.-Indië 54: 450 (1895). Type: Kalimantan, *Hallier* s.n. (holotype BO).

Cordula bulleniana (Rchb.f.) Rolfe in Orchid Rev. 20: 2 (1912).

C. amabilis (Hallier f.) Ames in Merr., Bibl. Enum. Born. Pl.: 135 (1921).

C. bulleniana (Rchb.f.) Rolfe var. *oculata* (Rchb.f.) Ames in Merr., Bibl. Enum. Born. Pl.: 135 (1921).

Paphiopedilum linii Schoser in Orchidee Hamburg 16: 181 (1966). Type: Sarawak, cult. Tuebingen B.G., *Sheridan-Lea* s.n. (holotype TUB).

For full synonymy see Cribb, The Genus Paphiopedilum: 167 (1987).

Terrestrial ***herb***. ***Leaves*** 7-8, elliptic, oblanceolate or oblong-elliptic, tridenticulate at obtuse apex, 7-14 x 2.4 cm, boldly tessellated dark and pale green above, sometimes flushed with purple below. ***Inflorescence*** erect, 1-flowered; peduncle 20-55 cm long, green and purple, pubescent; floral bract 1.5-2.1 cm long, ovate-elliptic, acute, ciliate. ***Flower*** up to 9.5 cm across; sepals white with green veins, often dark purple-marked at base of dorsal sepal; petals green at base, purple above with dark maroon-black spotting on margins and sometimes on the lamina of basal half; lip ochre to greenish. ***Pedicel*** with ***ovary*** 4-6 cm long, pubescent. ***Dorsal sepal*** 2.4-3 x 1.4-2.2 cm, usually concave, acute, shortly pubescent on outer surface. ***Synsepal*** 1.9-2.5 x 1-1.5 cm, lanceolate, acute. ***Petals*** 3.8-5.2 x 0.9-1.4 cm, spathulate to oblanceolate, obtuse, ciliate. ***Lip*** 3-4 cm long, emarginate at apex. ***Staminode*** 1.9 x 0.6-0.8 cm, subcircular to subrhombic, deeply incised at apex, lateral teeth subparallel to spreading, ± with a short tooth in the apical sinus. Plate 12C.

HABITAT AND ECOLOGY: In moss and leaf litter amongst stilt roots of mangroves; coastal swamp forest. At higher altitudes, from 700 to 950 m, it grows in mixed hill/heath forest with small crowns, in leaf litter on poor podsolic soils, and on wet moss-covered rocks. Alt. sea level to 1000 m.

DISTRIBUTION IN BORNEO: KALIMANTAN BARAT: Kapuas River. SARAWAK: Kuching area.

GENERAL DISTRIBUTION: Peninsular Malaysia, Sumatra and Borneo.

NOTES: This is a variable species in leaf shape and markings, petal spotting and staminode shape. It is doubtful if it is specifically distinct from *P. appletonianum* (Gower) Rolfe from mainland SE. Asia. In Borneo its habitat is under threat from coastal development. A distinct variety (var. *celebesense*) is also recognised from Sulawesi and Seram.

DERIVATION OF NAME: The generic name is derived from the Greek *Paphia*, of Paphos; epithet of Venus or Aphrodite (Paphos is a town on Cyprus near where the Goddess of Love was supposed to have been born) and *pedilon*, sandal, from the slipper-shaped lip. The specific epithet honours Mr Bullen, a nurseryman with Messrs Low and Co. of Clapton, London, who first sent material from Borneo to Reichenbach in Vienna.

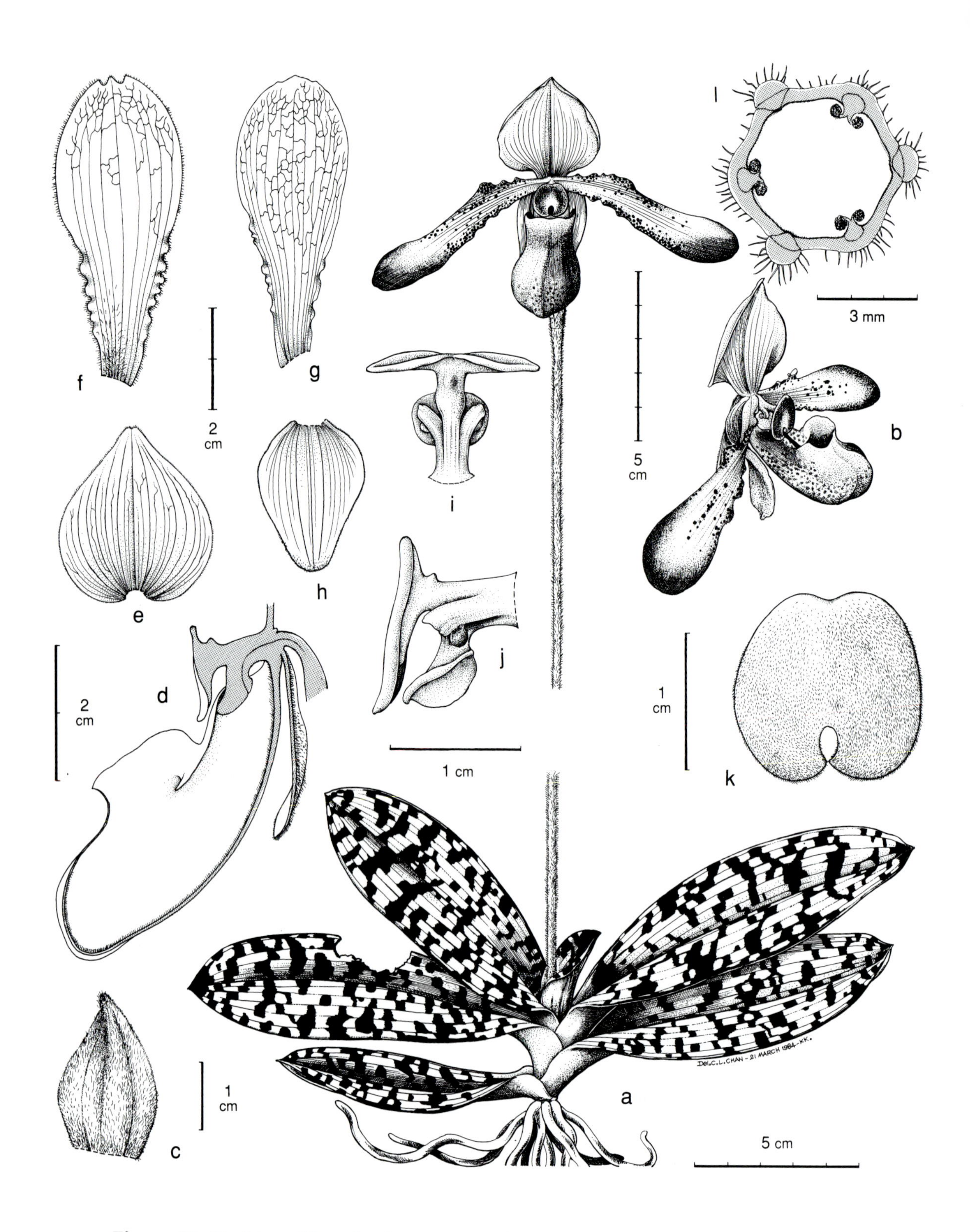

Figure 55. Paphiopedilum hookerae var. volonteanum (Sander ex Rolfe) Kerch. - A: plant. - B: flower, lateral view. - C: floral bract. - D: flower, longitudinal section. - E: dorsal sepal. - F & G: petal. - H: synsepal. - I: staminode, dorsal view. - J: staminode, lateral view. - K: staminode, front view. - L: ovary, transverse section. All drawn from *Beaman* 8990 by Chan Chew Lun.

55. PAPHIOPEDILUM HOOKERAE (Rchb.f.) Stein var. VOLONTEANUM (Sander ex Rolfe) Kerch.

Paphiopedilum hookerae *(Rchb.f.) Stein* var. **volonteanum** (*Sander ex Rolfe*) *Kerch.*, Orch.: 456 (1894). Type: Borneo, *Low* s.n. (holotype K).

Cypripedium hookerae Rchb.f. var. *volonteanum* Sander ex Rolfe in Gard. Chron. ser. 3, 8: 66 (1890).

Paphiopedilum volonteanum (Sander ex Rolfe) Pfitzer in Engler, Pflanzenr. Orch. Pleon.: 80 (1903).

Cordula hookerae (Rchb.f.) Rolfe var. *volonteana* (Sander ex Rolfe) Ames in Merr., Bibl. Enum. Born. Pl.: 136 (1921).

Terrestrial ***herb***. ***Leaves*** 5-6, 7-23 x 2.7-3 cm, oblong-elliptic, obtuse and minutely tridentate at apex, boldly tessellated dark and light green on upper surface, purple-spotted below. ***Inflorescence*** 1-flowered; peduncle to 50 cm long, purple, white-pubescent; floral bract lanceolate, acute, 2-3 x 1.4 cm, pale brownish pubescent. ***Flower*** c. 8 cm across; dorsal sepal cream, flushed bright green in centre; synsepal pale yellow; petals pale green, heavily spotted brown in basal two-thirds, margins and apical third purple; lip brown, brown-warted on side lobes. ***Pedicel*** with ***ovary*** 5 cm long, bright green, pubescent. ***Dorsal sepal*** 3-4 x 2.3-2.9 cm, ovate, acute, with reflexed basal margins. ***Synsepal*** 2-3 x 1.4-1.6 cm, elliptic, bidentate. ***Petals*** 5.5 x 1.5-2.2 cm, deflexed, half twisted in middle, spathulate, subobtuse, ciliate. ***Lip*** 3.8-4.2 x 1.7 cm, slightly constricted below the horizontal mouth, ciliate on slightly reflexed apical margin. ***Staminode*** 1 x 1 cm, circular, apically excised; side lobes at apex incurved-falcate. Plate 12D.

HABITAT AND ECOLOGY: Hill and lower montane forest on ultramafic substrate; sometimes lithophytic on mossy rocks; plants are usually found on banks and cliff ledges, or amongst leaf litter between rocks and boulders or at the base of trees such as *Gymnostoma sumatrana*. Quite often the tree canopy is low and small-crowned, typical of exposed ridges and cliffs, and producing a light shade. The high-altitude forms of *P. hookerae* var. *volonteanum* grow in stunted mossy forest on banks, ledges and amongst serpentine boulders. Alt. 200 to 2600 m. Flowering observed from February to April.

DISTRIBUTION IN BORNEO: SABAH: Mt. Kinabalu.

GENERAL DISTRIBUTION: Endemic to Borneo (Sabah only).

NOTES: *P. hookerae* var. *volonteanum* differs from var. *hookerae*, from Kalimantan and Sarawak, in having proportionately narrower leaves, purple-spotted below, broader and more obtuse petals and the lip a little constricted below the horizontal mouth. The degree of purple mottling on the underside of the leaves varies considerably and seems to be related to the very wide altitudinal range at which this variety is found. Some specimens have an unusually tall scape. A broad-leaved form of this variety is found on the western slopes of Mt. Kinabalu and in Lahad Datu District, whereas the narrow-leaved form is found on eastern Kinabalu, and the Labuk Valley to the east of the mountain.

DERIVATION OF NAME: The specific epithet honours Lady Hooker, wife of Sir William Jackson Hooker, the first Director of the Royal Botanic Gardens, Kew.

The variety is named for M. Volonte, a client of the nursery of Jean Linden of Ghent, Belgium.

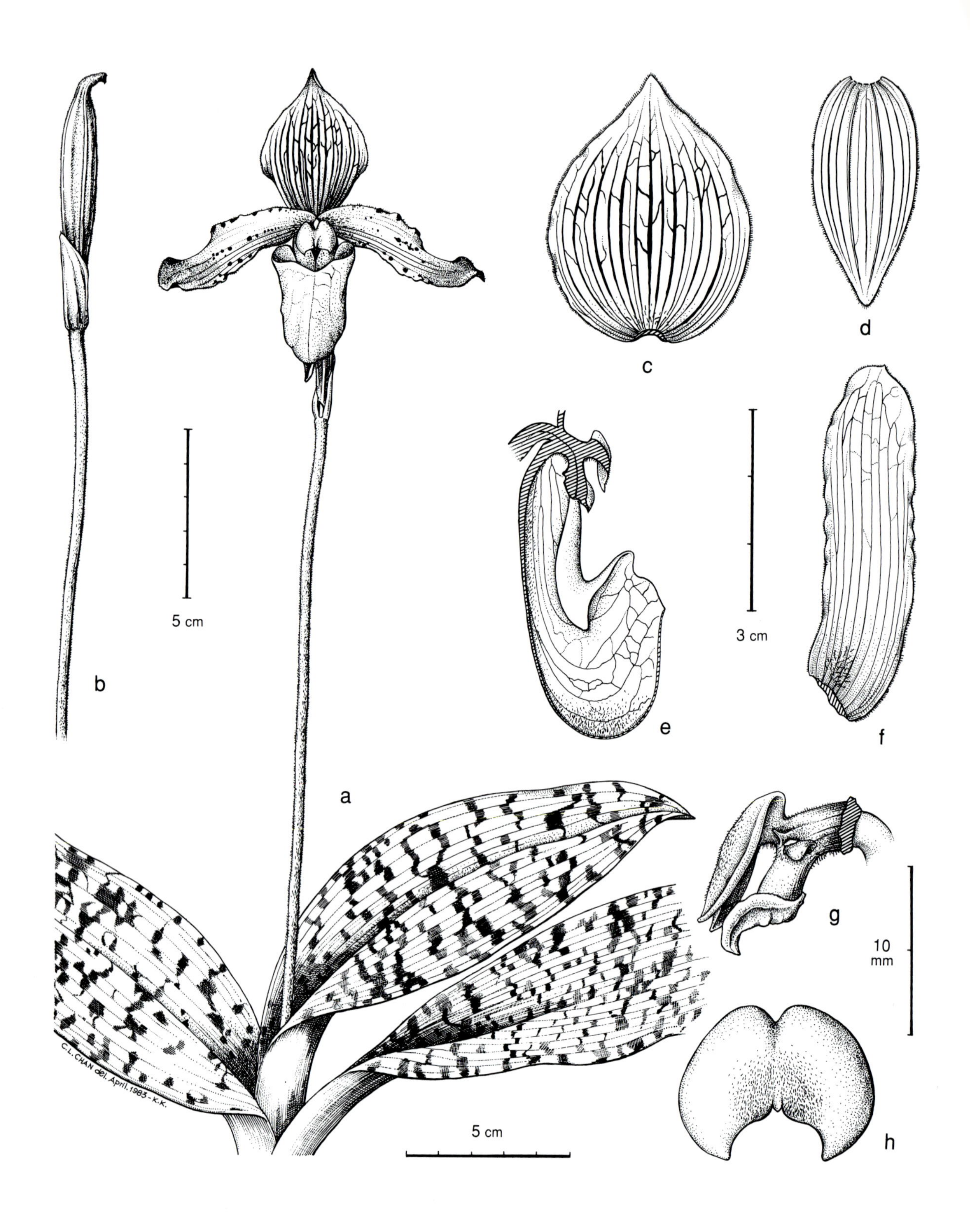

Figure 56. Paphiopedilum javanicum var. virens (Rchb. f.) Stein - A: plant. - B: seed pod. - C: dorsal sepal. - D: synsepal. - E: column and lip, longitudinal section. - F: petal. - G: staminode, lateral view. - H: staminode, front view. All from living plant collected from Mesilau River, Mt. Kinabalu. Drawn by Chan Chew Lun.

56. PAPHIOPEDILUM JAVANICUM (Reinw. ex Lindl.) Pfitzer var. VIRENS (Rchb.f.) Stein

Paphiopedilum javanicum *(Reinw. ex Lindl.) Pfitzer* var. **virens** *(Rchb.f.) Stein*, Orchideenbuch: 471 (1892). Type: Borneo, cult. *Day* (holotype W).

Cypripedium virens Rchb.f. in Bot. Zeitung (Berlin) 21: 128 (1863).

C. javanicum Reinw. ex Lindl. var. *virens* (Rchb.f.) J.J.Veitch, Man. 4: 35 (1881).

Paphiopedilum virens (Rchb.f.) Pfitzer in Bot. Jahrb. Syst. 19: 41 (1896).

Cordula virens (Rchb.f.) Rolfe in Orchid Rev. 20: 2 (1912).

P. purpurascens Fowlie in Orchid Digest 38: 155 (1974). Type: cult. Los Angeles Arb., *Hilberg* H66B1 (holotype UCLA).

Terrestrial ***herb***. ***Leaves*** spreading, 4-5, 12-23 x 3.4 cm, narrowly elliptic, obtuse at minutely tridenticulate apex, pale green, veined and lightly mottled darker green. ***Inflorescence*** 1-flowered; peduncle 16-36 cm long, purple, shortly white-pubescent; floral bract elliptic, obtuse, 1.5-2.5 x 1-1.4 cm, ciliate on margins and mid-vein, pale green, very lightly spotted with purple. ***Flower*** 8-9.5 cm across; dorsal sepal green with darker green veins and a whitish pink margin; synsepal pale green; petals pale green with a pink purple apical quarter, finely spotted with dark maroon in basal half to three-quarters; lip bright green with darker veins, often brown flushed. ***Pedicel*** with ***ovary*** 4.3-5 cm long, pubescent. ***Dorsal sepal*** 3-3.8 x 2.5-2.9 cm, ovate to elliptic, almost acuminate, shortly ciliate, pubescent on outer surface. ***Synsepal*** 2.5-2.6 x 1.1-1.3 cm, lanceolate, acute, pubescent on outer surface. ***Petals*** 4.2-4.8 x 1.3-1.4 cm, usually deflexed at c. 45° to horizontal, narrowly oblong, obtuse, wide, shortly ciliate. ***Lip*** 3.6-4 x 1.8-2 cm, very shortly pubescent on outer surface, verrucose on side lobes. ***Staminode*** 0.8 x 1 cm, reniform, convex, shortly pubescent all over surface. Plate 12E.

HABITAT AND ECOLOGY: Hill-forest to mixed lower montane forest, with oaks, chestnuts and conifers. It prefers a shaded, humid habitat, often growing in leaf litter in the cracks between boulders, and is usually found on steep banks above rivers or near to small streams, particularly on granitic or sandstone soils. Alt. 900 to 1700 m. Flowering is usually sporadic throughout the year, but mainly from January to April.

DISTRIBUTION IN BORNEO: SABAH: Mt. Kinabalu; Crocker Range. SARAWAK: unconfirmed reports from Kelabit Highlands.

GENERAL DISTRIBUTION: Endemic to Borneo.

NOTES: This variety is distinguished from var. *javanicum*, from Java, Sumatra, Bali and Flores, by its brighter green flower, more spathulate, almost horizontal and less heavily spotted petals, almost acuminate dorsal sepal, more pubescent ovary and convex staminode. This slipper orchid has become rare in the wild due to increasing habitat destruction on Mt. Kinabalu. In the Crocker Range National Park, plants can still be found being sold at the roadside stalls along the Sinsuron road.

P. purpurascens Fowlie was described from a single cultivated plant that was collected on Mt. Kinabalu together with a batch of typical *P. javanicum* var. *virens*. However, it falls within the variation of the variable *P. javanicum* and was reduced to synonymy by Cribb (1987).

DERIVATION OF NAME: The specific epithet refers to Java from where it was first described and the varietal epithet is derived from the Latin *virens*, green, referring to the greener flowers of the variety.

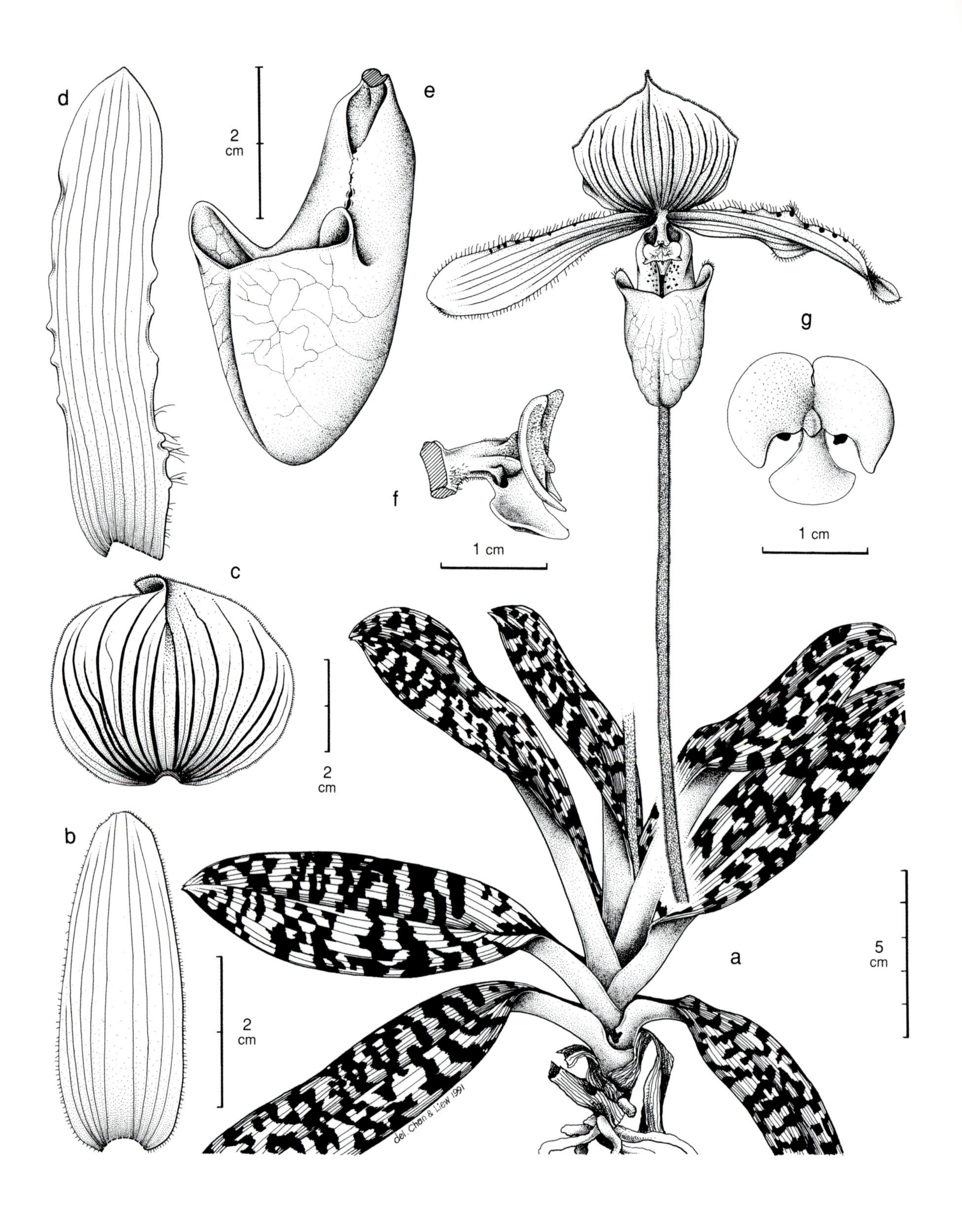

Figure 57. Paphiopedilum lawrenceanum (Rchb. f.) Pfitzer - A: plant. - B: synsepal. - C: dorsal sepal. - D: petal. - E: lip, front view. - F: staminode, lateral view. - G: staminode, front view. A from Hort. Bot. Kew. B-G from *L. de Rothschild* (Kew spirit 22242). Drawn by Chan Chew Lun and Liew Fui Ling.

57. PAPHIOPEDILUM LAWRENCEANUM (Rchb.f.) Pfitzer

Paphiopedilum lawrenceanum *(Rchb.f.) Pfitzer* in Jahrb. Wiss. Bot. 19: 163 (1888). Type: Borneo, cult. Sander, *Burbidge* s.n. (holotype W).

Cypripedium lawrenceanum Rchb.f. in Gard. Chron. n.s. 10: 748 (1878).

Cordula lawrenceana (Rchb.f.) Rolfe in Orchid Rev. 20: 2 (1912).

Paphiopedilum barbatum (Lindl.) Pfitzer subsp. *lawrenceanum* (Rchb.f.) M.W.Wood in Orchid Rev. 84: 352 (1976).

Terrestrial ***herb*** with a short stem and clustered growths on a very short rhizome. ***Roots*** 2-3 mm in diameter, terete, brown-pubescent. ***Leaves*** 5-6 in a fan, 12-19 x 4-6.5 cm, elliptic to narrowly elliptic, obtuse to subacute and minutely tridenticulate at apex, pale green or yellow green, boldly tessellated with dark green. ***Inflorescence*** 25-35 cm long, erect, one-flowered; peduncle terete, purple, densely shortly pubescent; floral bract 1.6-2 cm long, ovate, acute, green veined with maroon. ***Flower*** 10-11.5 cm across; sepals white veined with green and purple; petals greenish with a purple apex, with large purple warts and cilia on both margins; lip green flushed with maroon or brown; staminode green with darker venation and a purple margin. ***Pedicel*** with ***ovary*** 4-6.5 cm long, pubescent, green with maroon ridges. ***Dorsal sepal*** 5.8-6.5 x 5.8-6.5 cm, erect, flat, broadly ovate-subcircular, obtuse. ***Synsepal*** 4 x 1.4-1.6 cm, lanceolate, acuminate to bifid at apex. ***Petals*** 5.5-6 x 0.9-1.1 cm, horizontal, linear-ligulate, subacute, warted and ciliate on both margins. ***Lip*** 5.5-6.5 x 2.5-3.2 cm, deeply saccate, 3-lobed; side lobes incurved, acute, warted. ***Column*** 0.8-0.9 cm long; staminode 0.9-1.1 x 1.3-1.45 cm, lunate, 3-toothed in front. Plate 12F.

HABITAT AND ECOLOGY: Lowland forest; usually along river banks in yellow clay and alluvial soils and sometimes on mossy limestone. Alt. 300 to 500 m. Flowering time unknown.

DISTRIBUTION IN BORNEO: SABAH: unconfirmed reports. SARAWAK: locality unknown.

GENERAL DISTRIBUTION: Endemic to Borneo.

NOTES: This species is closely related to *P. barbatum* (Lindl.) Pfitzer, from Peninsular Malaysia, but is distinguished by its much larger flower, particularly the very large dorsal sepal, petals with warts on both margins, differences in the shape of the staminode, and the boldly tessellated leaves. During 1989 and 1990 plants were smuggled out of Sarawak and sold in Germany. This is probably the rarest and most endangered Bornean slipper orchid.

DERIVATION OF NAME: Named in honour of Sir Trevor Lawrence of Burford, Dorking, Surrey, an eminent Victorian orchid grower and President of the Royal Horticultural Society.

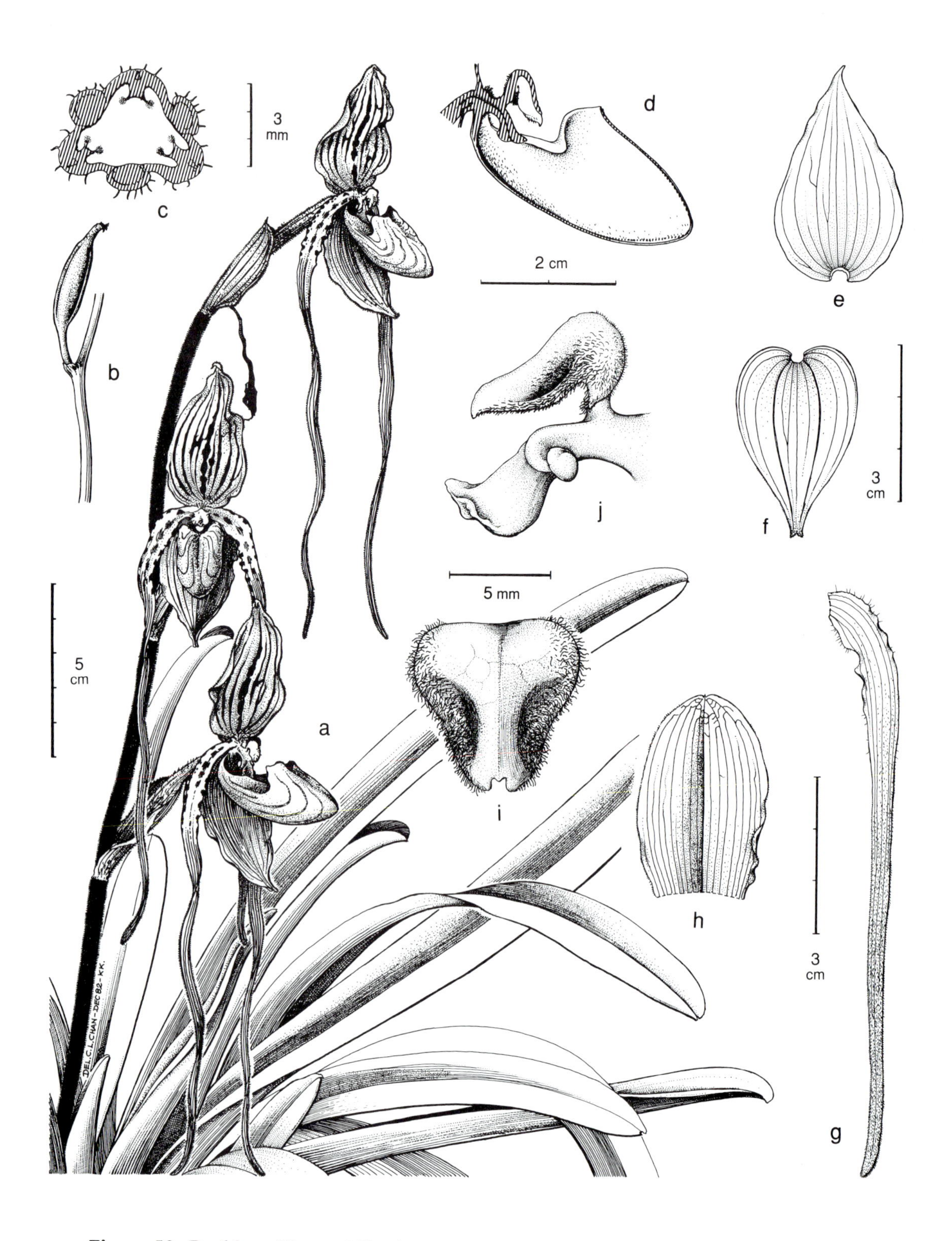

Figure 58. Paphiopedilum philippinense (Rchb. f.) Stein - A: plant. - B: seed pod. - C: ovary, transverse section. - D: column and lip, longitudinal section. - E: dorsal sepal. - F: synsepal. - G: petal. - H: floral bract. - I: staminode, front view. - J: staminode, lateral view. A from plant cultivated by Chan Kwong Choi, Sandakan, B-J from cult. Shim Phyau Soon. Drawn by Chan Chew Lun.

58. PAPHIOPEDILUM PHILIPPINENSE (Rchb.f.) Stein

Paphiopedilum philippinense *(Rchb.f.) Stein*, Orchideenbuch: 482 (1892). Type: without provenance (holotype W).

Cypripedium philippinense Rchb.f. in Bonplandia 10: 335 (1862).

C. laevigatum Bateman in Bot. Mag. 91: t. 5508 (1861). Type: hort. *Veitch* (illustration in Bot. Mag. t.5508).

Selenipedium laevigatum (Bateman) May in Rev. Hort. 1885: 301 (1885).

Cypripedium cannartianum Linden in Lindenia 3: 93, t. 141 (1888). Types: Philippines, hort. *Cannart d'Hamale* & hort. *Wallaert* (not found).

C. roebelenii Rchb.f. var. *cannartianum* Linden in Lindenia 3: 93, t. 141 (1888) in synon.

Paphiopedilum laevigatum (Bateman) Pfitzer in Engler & Prantl, Nat. Pflanzenf. 2, 6: 84 (1889).

P. philippinense (Rchb.f.) Stein var. *cannartianum* (Linden) Pfitzer in Engler, Pflanzenr. Orch. Pleon.: 62 (1903).

Terrestrial or lithophytic ***herb***. ***Leaves*** 20-50 x 2-5.5 cm, coriaceous, up to 9, ligulate, rounded at asymmetric apex, V-shaped in cross-section, very thick in texture. ***Inflorescence*** erect, 2-4-flowered, up to 50 cm long; peduncle purple-pubescent; floral bracts up to 5 x 2 cm, elliptic, acute, pubescent. ***Flowers*** rather variable in size; sepals white, dorsal striped with maroon; petals white or yellow at base, maroon above with marginal dark maroon warts in basal half; lip and staminode yellow. ***Pedicel*** with ***ovary*** 4.5-6.5 cm long, purple, pubescent. ***Dorsal sepal*** 4-5 x 2-2.5 cm wide, ovate, acute. ***Synsepal*** 4.5-5.3 x 2 cm, similar to dorsal sepal. ***Petals*** 6-30 x 0.5-0.6 cm, linear, tapering to apex, ciliate. ***Lip*** 3.8 x 1.4 cm, small, rather ovoid in shape. ***Staminode*** convex, rather cordate-subquadrate, emarginate, yellow, veined with green, purple-pubescent on sides. Plate 13A.

HABITAT AND ECOLOGY: In peaty leaf litter in pockets or ledges on limestone cliffs, and on andesitic rocks, usually close to the sea. Usually found in sunny to lightly shaded sites shaded by shrubs or low trees. Alt. sea level to 500 m.

DISTRIBUTION IN BORNEO: SABAH: offshore islands.

GENERAL DISTRIBUTION: Philippines (Luzon, Mindoro, Mindanao, Palawan) and Borneo.

NOTES: Plants in cultivation at Tenom Orchid Centre, Sabah, normally flower from February until March. Seedlings growing on andesitic rocks collected by Phillipps in 1981 flowered in cultivation, and produced a 52-cm-long peduncle and large flowers with petals to 17.5 cm long, compared to 10 cm long for those from limestone areas. Visits in recent years to both localities confirmed that all remaining plants had been removed. *P. philippinense* var. *roebelenii* (Veitch) P.J.Cribb from Luzon, with larger flowers and petals up to 30 cm long closely resemble the Sabahan plants growing on andesitic rocks.

DERIVATION OF NAME: The specific epithet refers to the Philippines, from where it was originally described.

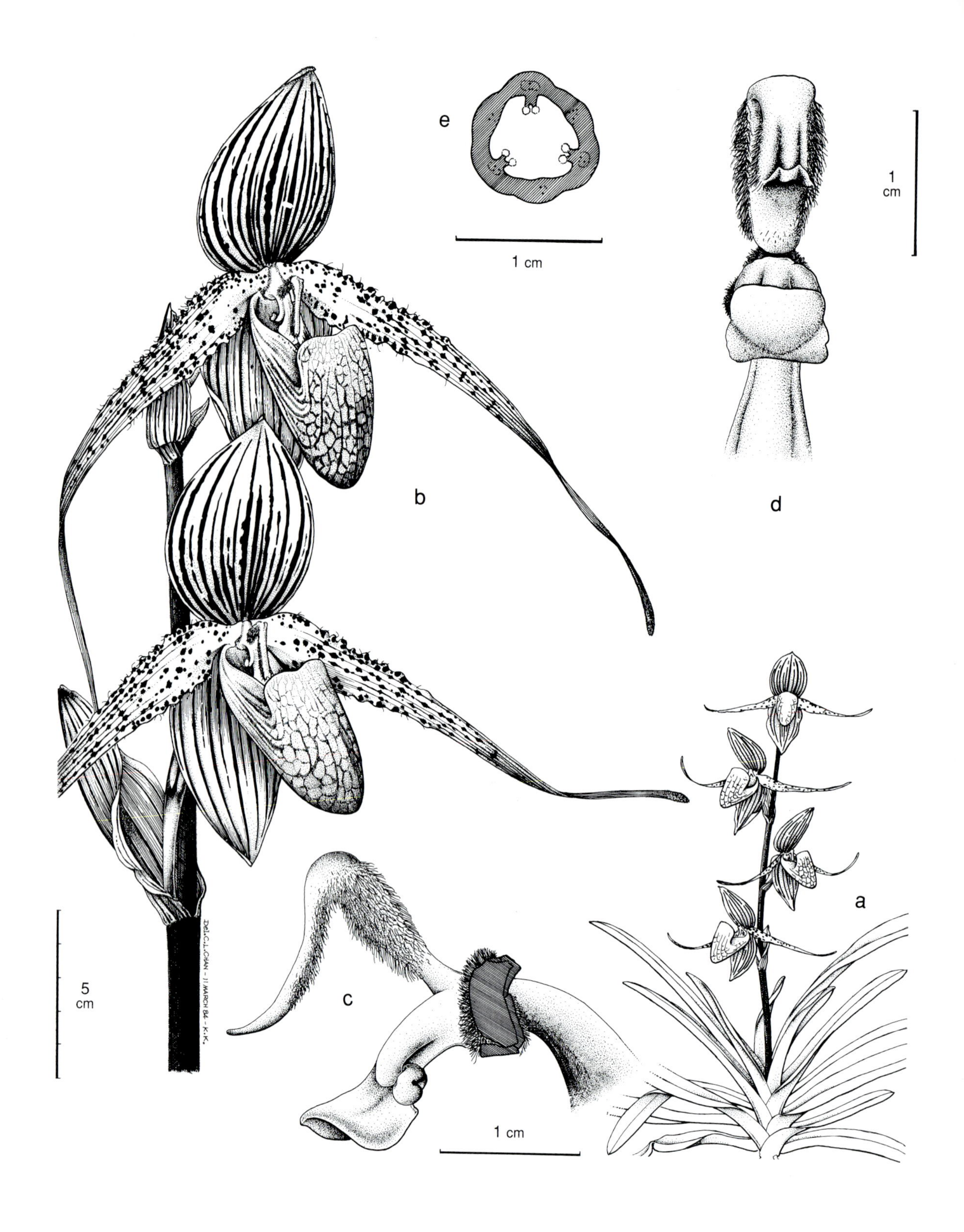

Figure 59. Paphiopedilum rothschildianum (Rchb. f.) Stein - A: plant. - B: inflorescense. - C: staminode, lateral view. - D: staminode, front view. - E: ovary, transverse section. A from photograph by A. Lamb. B-E from *Beaman* 8855. Drawn by Chan Chew Lun and Liew Fui Ling.

59. PAPHIOPEDILUM ROTHSCHILDIANUM (Rchb.f.) Stein

Paphiopedilum rothschildianum *(Rchb.f.) Stein*, Orchideenbuch: 482 (1892). Type: hort. *Sander* (holotype W).

Cypripedium rothschildianum Rchb.f. in Gard. Chron. ser. 3, 3: 457 (1888).

C. elliottianum O'Brien in Gard. Chron. ser. 3, 4: 501 (1888). Type: hort. *Sander* (holotype K).

Paphiopedilum elliottianum (O'Brien) Stein, Orchideenbuch: 466 (1892).

P. rothschildianum (Rchb.f.) Stein var. *elliottianum* (O'Brien) Pfitzer in Engler, Pflanzenr. Orch. Pleon.: 59 (1903).

Cordula rothschildiana (Rchb.f.) Ames in Merr., Bibl. Enum. Born. Pl.: 137 (1921).

Terrestrial or lithophytic ***herb*** often growing in large clumps. ***Leaves*** several, up to 60 x 4-5 cm, linear to narrowly oblanceolate-acute, sparsely ciliate at base, green. ***Inflorescence*** 2- to 4-flowered, erect; peduncle up to 45 cm long, purple, shortly pubescent; floral bracts ovate-elliptic, obtuse, up to 5.5 cm long, ciliate and hairy on mid-vein, pale green or yellow, purple-striped. ***Flowers*** very large, up to 30 cm across; dorsal sepal, synsepal and petals ivory-white or yellow with maroon veins and markings; lip golden, heavily purple-suffused; staminode pale yellow-green. ***Pedicel*** with ***ovary*** 7.5 cm long, pale green sparsely spotted purple, glabrous. ***Dorsal sepal*** 6.6 x 3.7-4.1 cm, ovate, acute to acuminate. ***Synsepal*** c. 5.7 x 3.3 cm, similar but smaller. ***Petals*** up to 12.4 x 1.4 cm, narrowly tapering to rounded apex, ciliate, papillose towards apex. ***Lip*** c. 5.7 x 2.2 cm, subporrect, grooved on back; side lobes not auricular. ***Staminode*** 1.4-1.6 x 0.4-0.5 cm, linear, bifid at apex, geniculate, densely glandular pubescent on margins and at base. Plate 13B.

HABITAT AND ECOLOGY: On peaty detritus and moss accumulated on rock ledges and steep slopes or cliffs in hill-forest with *Gymnostoma sumatrana*, usually lightly shaded by low trees or shrubs. It prefers to grow near seepages in these rocky habitats and is so far only recorded from ultramafic rocks. Alt. 500 to 1800 m. Flowering March and April.

DISTRIBUTION IN BORNEO: SABAH: Mt. Kinabalu.

GENERAL DISTRIBUTION: Endemic to Borneo (Sabah only).

NOTES: The population in the known locality within the Kinabalu National Park has been seriously depleted by collectors. The other site, degazetted from the Park in recent years, has been destroyed by fire. Attempts to re-introduce plants to the Park by Grell et al. (1988) have not been successful, due to the lack of follow-up attention. *P. rothschildianum* is considered by many to be the aristocrat of slipper orchids with its graceful petals held out like the arms of the local Sumazau dancers. The curious glandular hairs on the bent staminode have been shown by Atwood (1985) to mimic an aphid colony, which is the normal brood site of syrphid flies (*Dideopsis aegrota*). When these flies are fooled into laying eggs on the staminode, they often fall into the pouch-like lip. Escape is only possible through a gap between the base of the lip and the column, passing beneath the stigma and pollinia and thereby effecting pollination.

DERIVATION OF NAME: The specific epithet honours Baron Ferdinand de Rothschild, an eminent Victorian orchid grower.

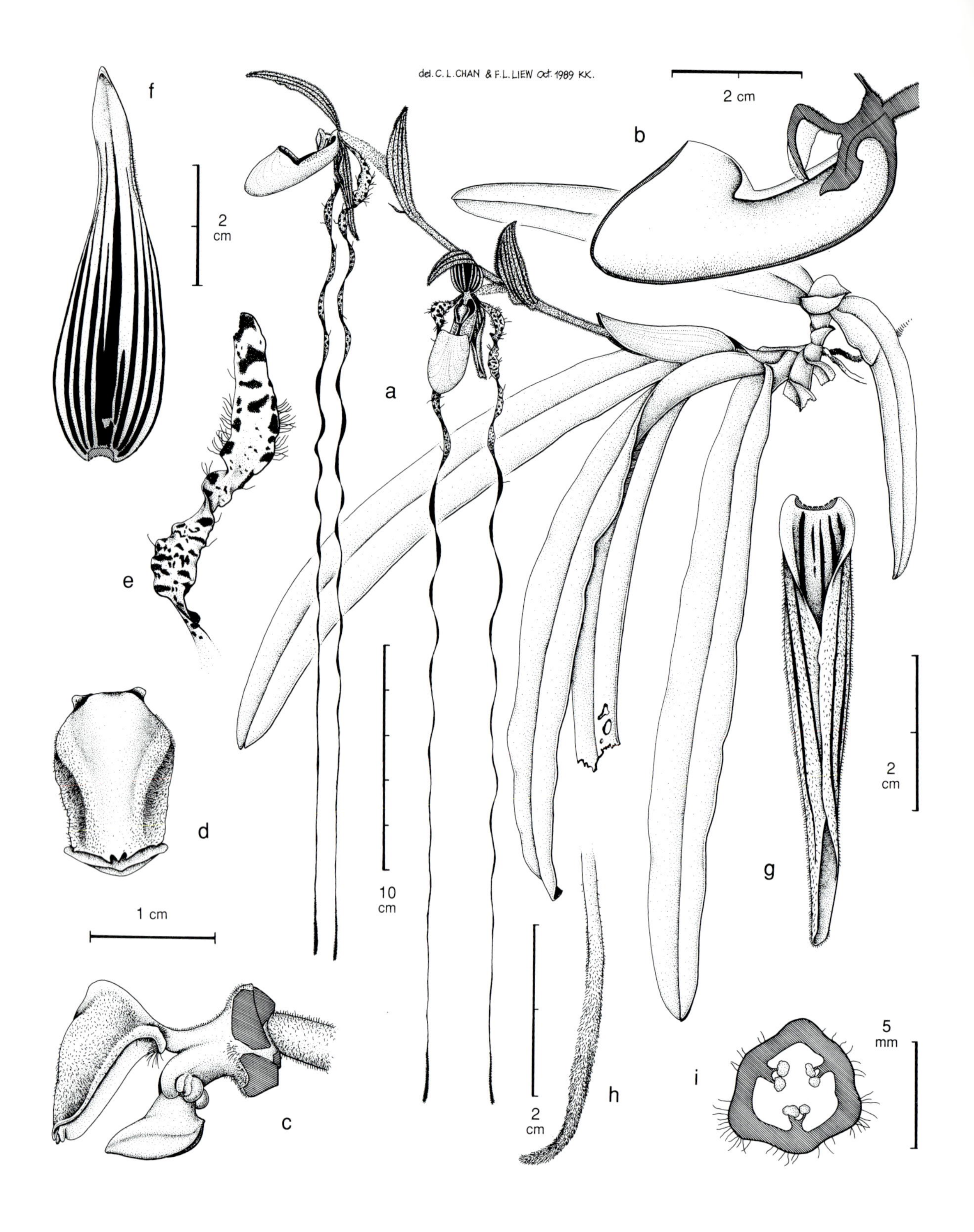

Figure 60. Paphiopedilum sanderianum (Rchb. f.) Stein - A: plant. - B: column and lip, longitudinal section. - C: staminode, lateral view. - D: staminode, front view. - E: petal, base. - F: dorsal sepal. - G: synsepal. - H: petal, apex. - I: ovary, transverse section. A after a painting by Chan Chew Lun. B - I from cult. A. Lamb. Drawn by Chan Chew Lun and Liew Fui Ling.

60. PAPHIOPEDILUM SANDERIANUM (Rchb.f.) Stein

Paphiopedilum sanderianum *(Rchb.f.) Stein*, Orchideenbuch: 482 (1892). Type: Borneo, Sarawak, *Foerstermann* s.n., cult. *Sander* (holotype W).

Cypripedium sanderianum Rchb.f. in Gard. Chron. n.s. 25: 554 (1886).

Medium-sized to large lithophytic ***herb*** with a short stem and russet-hairy roots. ***Leaves*** 4-5 in a fan, 35-50 x 4-5.5 cm, coriaceous, arcuate to suberect, linear, obtuse, green or yellowish green. ***Inflorescence*** up to 60 cm long, spreading to arcuate, laxly 2-5-flowered; peduncle and rachis terete, purple, pubescent; floral bracts 4-8 cm long, elliptic to obovate, acute, purple. ***Flowers*** narrow but very long; sepals yellow boldly striped with maroon; petals pale yellow or white spotted with maroon in the basal part and maroon above; lip pale yellow flushed with brown on pouch. ***Pedicel*** with ***ovary*** 6-7 cm long, white, densely pubescent. ***Dorsal sepal*** 4.8-6.5 x 1.3-2.5 cm, lanceolate, acuminate. ***Synsepal*** 3.5-6 x 1-2 cm, lanceolate, acuminate, 2-keeled on outer side. ***Petals*** 30-90 x 0.5 cm, ribbon-like, linear- tapering, rounded and fleshy at apex, spreading for basal 3 cm, then pendent, strongly twisted, shortly pubescent, ciliate. ***Lip*** 5.5-6.5 x 2-2.5 cm, strongly dilated-subfusiform, with acute auricles within. ***Column*** 9 mm long; staminode 1.2-1.3 x 1 cm, oblong-tapering, obtuse, densely pilose on margins. Plate 13C.

HABITAT AND ECOLOGY: Sheer south-east facing limestone cliffs especially where there are seepages and the rock face is moist. It likes to be shaded, but with some morning sunshine, and is often found near streams where conditions are cool and moist. The locality receives over 3000 mm of rain fairly evenly distributed throughout the year. Alt. 100 to 600 m. Flowering in April.

DISTRIBUTION IN BORNEO: SARAWAK.

GENERAL DISTRIBUTION: Endemic to Borneo.

NOTES: *P. sanderianum* was for long considered extinct but was rediscovered in Sarawak by a Danish botanist in 1979. Unfortunately, this colony was discovered by collectors in 1986 and the populations have subsequently been heavily depleted. Consequently, it is considered to be on the verge of extinction and urgent action to secure its future in the wild is necessary. Lamb observed that after flowering a new fan of leaves developed and took two years to flower. The opening of the flowers and development of the petals was observed. About 8 cm of petal in 2 cm long folds fell out on opening and then expanded mainly overnight to 22 cm long. A day later this had expanded to 36 cm, at which stage the pendulous petals were straight. On the fourth full day the petals expanded another 12 cm to 48 cm long, and started to twist. Subsequently growth slowed to 2 cm a day and finally reached 60 cm after 10 days. Other more robust plants in the wild have petals measuring 90 cm long. The tips of these extraordinary dark purple petals are hairy and scented, as if to attract insects.

DERIVATION OF NAME: Named in honour of Frederick Sander, the eminent Victorian orchid nurseryman whose collector J. Foerstermann first discovered this spectacular orchid.

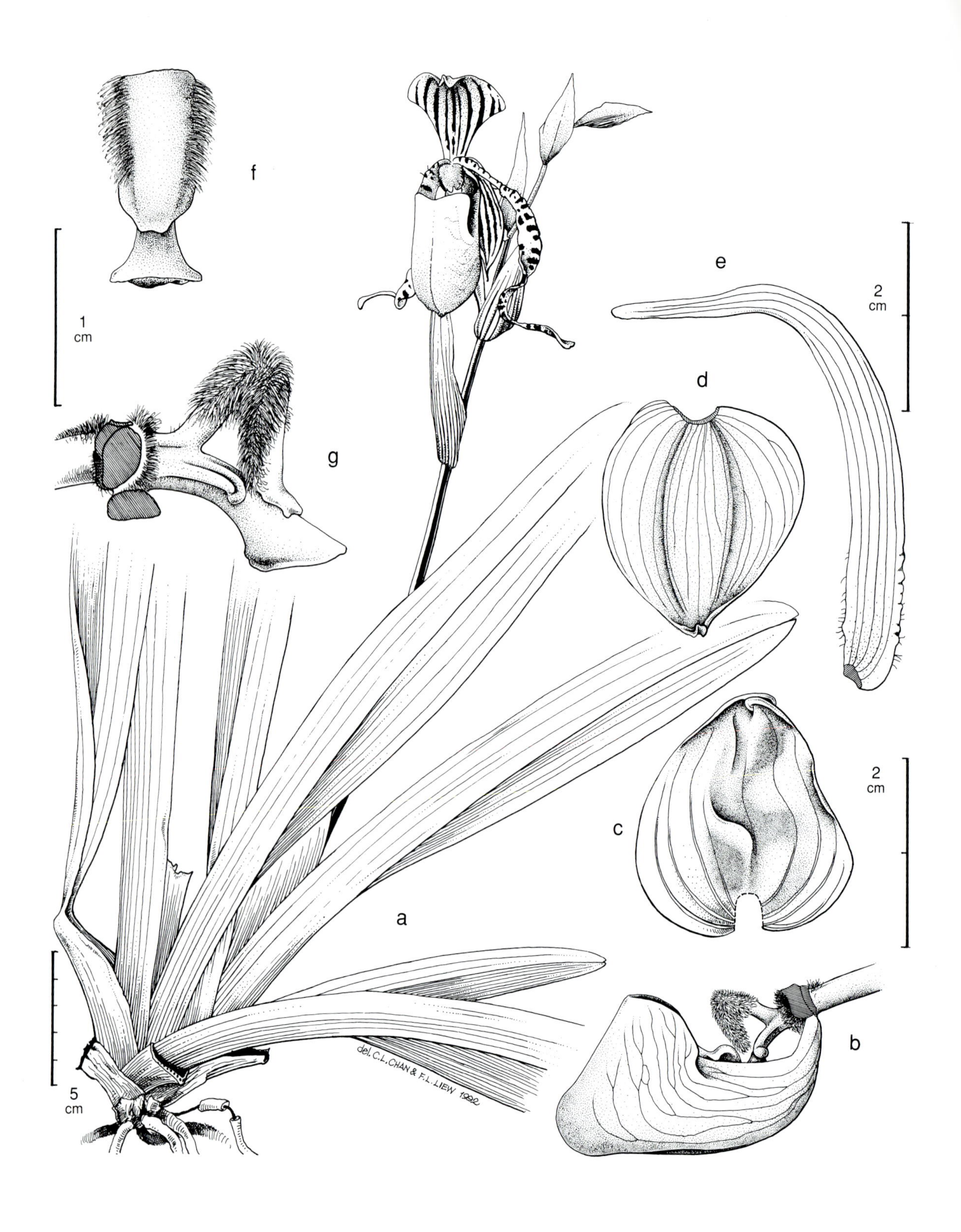

Figure 61. Paphiopedilum supardii Braem & Loeb - A: plant. - B: column and lip, longitudinal section. - C: dorsal sepal (deformed). - D: synsepal. - E: petal. - F: staminode, front view. - G: staminode, lateral view. A from cult. Hort. Bot. Kew and photographs. B - G from *Grell* s.n. Drawn by Chan Chew Lun and Liew Fui Ling.

61. PAPHIOPEDILUM SUPARDII Braem & Loeb

Paphiopedilum supardii *Braem & Loeb* in Orchidee (Hamburg) 36(4): 142 (1985). Type: Borneo, Kalimantan, *Supard* in *Braem* GB 585 (holotype Herb. Braem).

Paphiopedilum 'victoria' de Vogel, Panda Nieuws 12: 117 (1975).

Medium-sized to large lithophytic ***herb*** with a very short stem. ***Roots*** 3-5 mm in diameter, red-pubescent. ***Leaves*** 7-9 in a fan, 30-55 x 3.5-5.5 cm, coriaceous, linear-ligulate, obtuse or rounded at unequally bilobed apex, green. ***Inflorescence*** 30-50 cm long, spreading arcuate to suberect, rather densely 4- to 5-flowered; peduncle and rachis terete, purple, pubescent; floral bracts 4.5-5.5 cm long, elliptic-ovate, acute, yellow striped with purple. ***Flowers*** 5-6 cm across, opening simultaneously; sepals yellow boldly longitudinally striped with maroon; petals pale yellow spotted with maroon; lip rich glossy purple; staminode yellow with a fringe of purple hairs. ***Pedicel*** with ***ovary*** 5.5-6 cm long, glabrous. ***Dorsal sepal*** 5-5.5 x 2.4-2.6 cm, elliptic, acute. ***Synsepal*** 5-5.5 x 2-2.2 cm, ovate, acute or acuminate. ***Petals*** 8-9 x 0.7-0.9 cm, linear, acute, incurved, undulate and twisted in apical half. ***Lip*** 4-4.6 x 1.5-1.8 cm, deeply pouched, dilated at apex, with acute auricles within. ***Column*** 0.6-0.7 cm long; staminode 0.8-0.9 x 0.8 cm, circular-subquadrate, tapering to acutely bifid apex. Plate 13D.

HABITAT AND ECOLOGY: In leaf-mould-filled hollows, half shaded in forest on limestone. Alt. 600 to 1000 m.

DISTRIBUTION IN BORNEO: KALIMANTAN SELATAN.

GENERAL DISTRIBUTION: Endemic to Borneo (Kalimantan only).

NOTES: This strange orchid with a rather grotesquely contorted flower is closely allied to *P. rothschildianum* (Rchb.f.) Stein. It was discovered in 1972 in Kalimantan by Dr E. de Vogel of the Rijskherbarium, Leiden, in the Netherlands, who published a photograph of it, under the name *P. 'victoria'* in 1975. Schoser & Deelder intended describing it shortly afterwards as *'P. devogelii'*, but failed to do so, although this name became widely used in the horticultural trade. Braem & Loeb described it as *P. supardii* in 1985.

DERIVATION OF NAME: Named after Mr Supard, a collector for A. Kolopaking of the Simanis Nursery in Lawang, E. Java, who collected the type material.

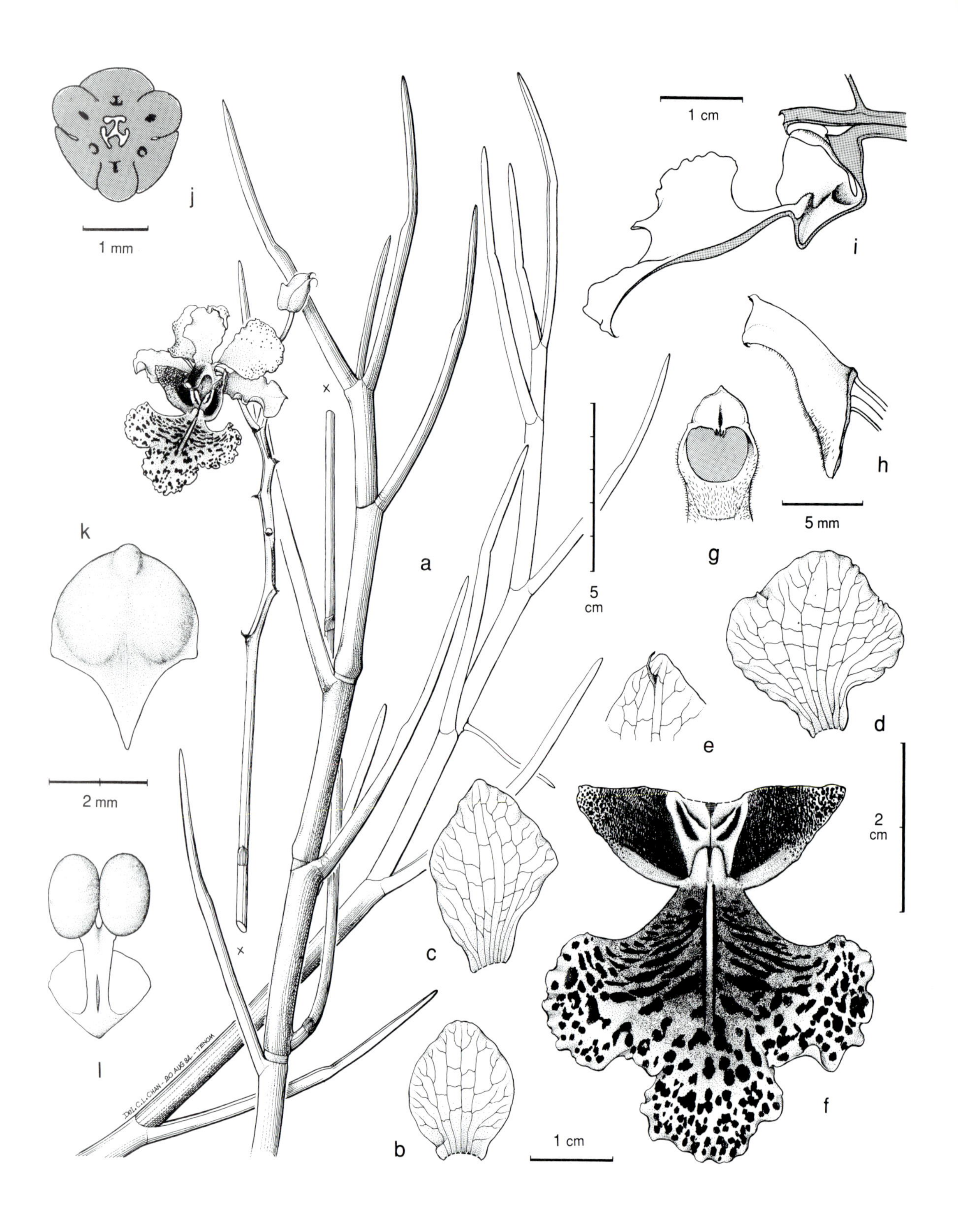

Figure 62. Papilionanthe hookeriana (Rchb. f.) Schltr. - A: plant. - B: dorsal sepal. - C: lateral sepal. - D: petal. - E: lateral sepal, apex. - F: lip. - G: column, ventral view. - H: column, lateral view. - I: column and lip, longitudinal section. - J: ovary, transverse section. - K: anther. - L: pollinarium. All drawn from cult. TOC by Chan Chew Lun.

62. PAPILIONANTHE HOOKERIANA (Rchb.f.) Schltr.

Papilionanthe hookeriana *(Rchb.f.) Schltr.* in Orchis 9: 80 (1915). Type: Borneo, Labuan, *Motley* 347 (holotype K).

Vanda hookeriana Rchb.f. in Bonplandia 4: 324 (1856).

Large scrambling terrestrial ***herb***. ***Stems*** slender, terete, up to 3 m long; internodes 4-5 cm long. ***Leaves*** 7-10 x 0.3-0.5 cm, terete, nearly straight, sulcate, constricted some 2 cm below the apex, mucronate. ***Inflorescence*** up to 30 cm long, 2- to 15-flowered. ***Flowers*** large, showy; dorsal sepal and petals white to pale mauve, tessellated with darker mauve and somewhat spotted; lateral sepals almost white; lip rich purple, purple-coloured near the base of the side lobes, and pale mauve marked with rich purple spots on the mid-lobe; column purple. ***Pedicel*** with ***ovary*** 1.8-3.5 cm long. ***Dorsal sepal*** 1.2-2 x 1.3-1.5 cm, erect, obovate-oblong, obtuse, with crisped undulate margins. ***Lateral sepals*** to 2.5 x 1.6-1.8 cm, spreading, less undulate. ***Petals*** 2.2 x 1.5 cm, twisted at the base, broadly elliptic, obtuse, with strongly undulate crisped margins. ***Lip*** deflexed, 3-lobed, 2-3.3 cm long, 3.5-5.3 cm wide when flattened; side lobes erect to spreading, oblong to triangular-falcate, 1.3-1.5 x 0.7-1.1 cm; mid-lobe reniform-flabellate, somewhat trilobulate, 2.8-3.2 x 4-4.5 cm; callus fleshy, 2-ridged, small, basal; spur conical, about 2 mm long. ***Column*** about 1 cm long, terete, incurved, pubescent around the stigma and at the base. ***Fruit*** with pedicel to 2.5 cm and ovary to 7 cm long. Plate 13E.

HABITAT AND ECOLOGY: Swamps, usually in quite deep peaty soils, often growing with other shrubs and tall grasses for support. These swampy areas are subject to seasonal flooding or inundation and the roots become submerged, but it has adapted very successfully to this habitat; usually exposed to full sunlight. Alt. near sea level. Flowering April and May.

DISTRIBUTION IN BORNEO: KALIMANTAN BARAT: Pontianak area. KALIMANTAN SELATAN: Banjarmasin area. SABAH: Papar area; Weston area. SARAWAK: Baram District; 3rd Division.

GENERAL DISTRIBUTION: Thailand, Vietnam, Peninsular Malaysia, Sumatra and Bangka.

NOTES: A white-flowered form has been collected in Peninsular Malaysia, but is not recorded for Borneo. At one time this showy orchid was common in swamps along the railway line in Sabah from Kota Kinabalu to Papar but, as in other areas, eg. Labuan, it has almost disappeared through collecting. However, it must still be abundant in the vast swampy areas of West and Central Kalimantan and possibly Sarawak. It was one of the parents of the earliest hybrid raised in Singapore, *Vanda* Miss Joaquim (*P. teres x P. hookeriana*) which is now Singapore's national flower. In Peninsular Malaysia, this orchid is known as the 'Kinta Weed' from the Kinta Valley in Perak where it was once abundant.

DERIVATION OF NAME: The generic name derives from the Latin *papilio*, a butterfly, and the Greek *anthe*, a flower. This species is named in honour of Sir William Jackson Hooker, a former Director of the Royal Botanic Gardens, Kew.

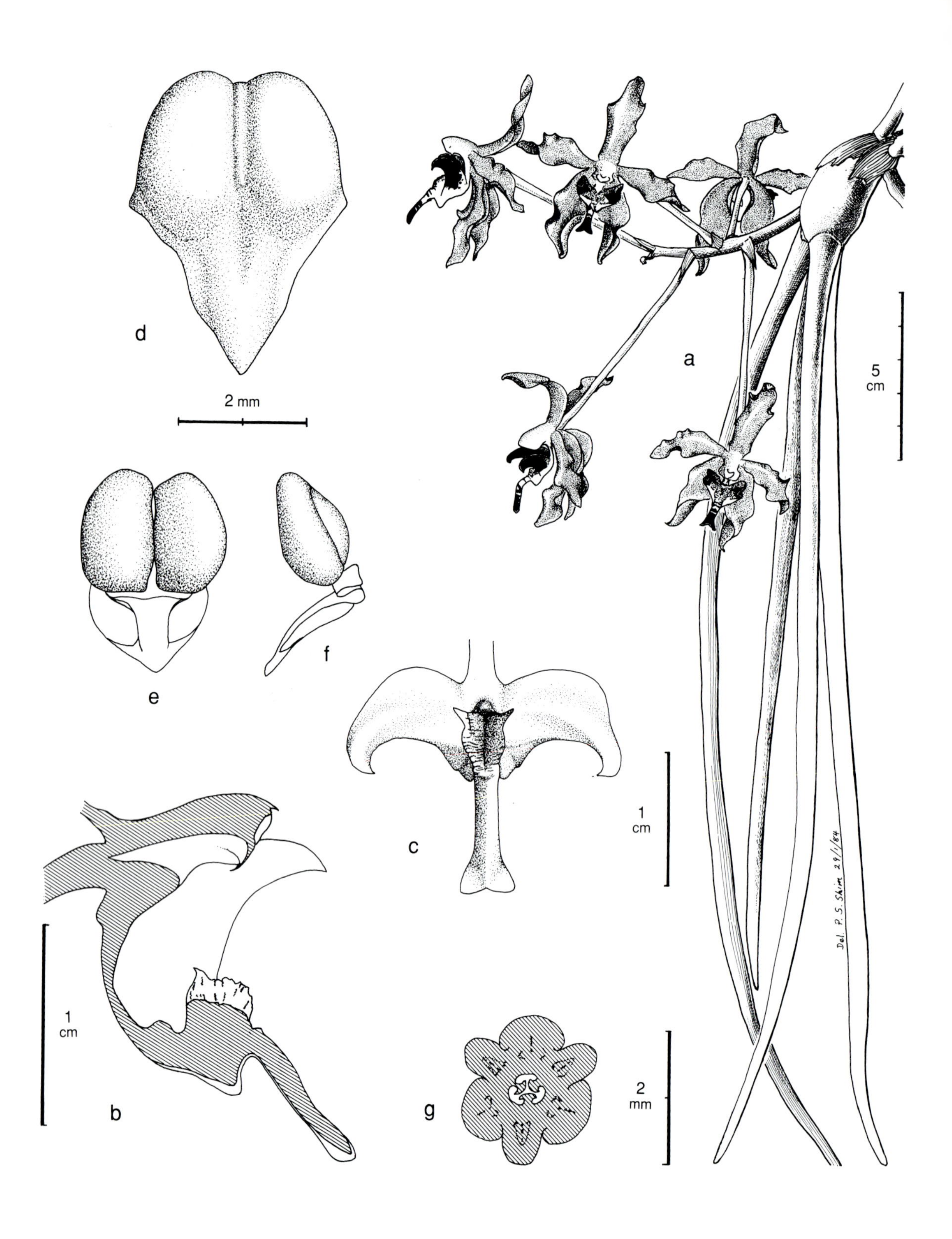

Figure 63. Paraphalaenopsis denevei (J.J. Sm.) A.D. Hawkes - A: plant. - B: column and lip, longitudinal, section. - C: lip, spread out. - D: anther, front view. - E: pollinarium, front view. - F: pollinarium, lateral view. - G: ovary, transverse section. All drawn from cult. TOC by Shim Phyau Soon.

63. PARAPHALAENOPSIS DENEVEI (J.J.Sm.) A.D.Hawkes

Paraphalaenopsis denevei *(J.J.Sm.) A.D.Hawkes* in Orquidea (Rio de Janeiro) 25: 212 (1963). Type: Borneo, West Kalimantan, near Pontianak, Nanga Djetah Plantation, *de Neve* s.n., cult. hort. Bogor, under no. 484 (holotype L).

Phalaenopsis denevei J.J.Sm. in Recueil Trav. Bot. Néerl. 22: 264 (1925).

Vanda denevei (J.J. Sm.) Zurowetz in Orchid Rev. 41: 76 (1933), nomen.

Epiphytic ***herb*** with branched aerial roots. ***Stem*** short, 3- to 6-leaved, up to 3 cm long. ***Leaves*** up to 70 x 1 cm, fleshy, elongate, terete with a narrow groove on the upper surface. ***Inflorescences*** produced in succession; peduncle short, up to 2.5 cm long; rachis rather congested, up to 5-flowered (11-13 in wild plants); floral bracts triangular, adpressed, up to 5 mm long. ***Pedicel*** with ***ovary*** up to 6 cm long, terete. ***Flowers*** with spreading sepals and petals, 4 cm in diameter; sepals and petals greenish yellow to yellow or golden yellow; lip white with large crimson spots and bars on the mid-lobe, apex deep magenta to beetroot purple, apex of side lobes deep magenta, base white, spotted magenta; callus yellow with brownish red bars. ***Dorsal sepal*** up to 2.6 x 1 cm, ovate-elliptic to narrowly elliptic, acute. ***Lateral sepals*** up to 2.8 x 1.3 cm, obliquely ovate-elliptic, acute. ***Petals*** up to 2.4 x 0.7-1 cm, from a fleshy, cuneate base, ovate-lanceolate with a falcate acuminate apex. ***Lip*** 1.7 x 2 cm, 3-lobed; side lobes obliquely oblong falcate to falcate triangular, acuminate, anterior margin thickened, 1-1.2 x 0.6 cm; mid-lobe recurved, linear spathulate, papillose, apex retuse, 0.3 cm; disc 3-ribbed, ribs confluent with plate-like subquadrate, rugose denticulate callus. ***Column*** white, up to 1 cm long, cylindric, papillose; anther-cap 4.5-5 x 3.5 x 4 mm; pollinia 2, cleft. Plate 13F.

HABITAT AND ECOLOGY: Riverside trees in lowland primary forest. Alt. below 300 m.

DISTRIBUTION IN BORNEO: KALIMANTAN BARAT: Pontianak area.

GENERAL DISTRIBUTION: Endemic to Borneo (Kalimantan only).

NOTES: The flowers are very sweetly scented and according to Sweet (1980) they appear in the rainy season. In cultivation they flower three to four times per year. *P. denevei* has been crossed with species in several other genera, including *Aerides*, *Arachnis*, *Papilionanthe*, *Renanthera* and *Vanda*. The four species of *Papilionanthe* are very distinct because of their long terete leaves, but hybrids often produce erect, semi-terete leaves, eg. those involving *Vanda*.

DERIVATION OF NAME: The generic name is derived from the Greek *para*, and *Phalaenopsis*, the orchid genus of that name, referring to the close affinity of the two genera. The specific epithet honours T.A. de Neve who collected the type.

Figure 64. Paraphalaenopsis labukensis Shim, A. Lamb & C.L. Chan - A & B: plant. - C: stipes. - D: pollinia. - E: anther. - F: lip. - G: column and lip, longitudinal section. - H: lip sidelobe, apex. - I: ovary, transverse section. All drawn from cult. TOC by Chan Chew Lun and Shim Phyau Soon.

64. PARAPHALAENOPSIS LABUKENSIS Shim, A.Lamb & C.L.Chan

Paraphalaenopsis labukensis *Shim, A. Lamb & C.L. Chan* in Orchid Digest 45: 139 (1981). Type: Borneo, Sabah, Kuala Labuk near Pamol, *Lamb* SAN 91503 (isotypes K, UCLA).

Phalaenopsis labukensis (Shim, A.Lamb & C.L.Chan) Shim in Malayan Nat. J. 36(1): 21 (1982).

Epiphytic ***herb***. ***Roots*** few, thick. ***Stems*** short, enclosed by leaf sheaths 3 cm long. ***Leaves*** 3-5, terete, 0.6-0.9 x 165(-210) cm, slightly constricted 1.3-2.5 cm from acute apex. ***Inflorescence*** slightly erect to pendulous, more than one per internode, bearing 5-15 cinnamon-scented flowers; peduncle 3 cm; rachis 4 cm; floral bracts 0.7 x 0.9 cm, triangular-ovate, apex obtuse. ***Flowers*** 6.2 cm wide; sepals and petals purplish cinnamon, speckled yellow, edges greenish yellow, inner halves of the lateral sepals with purple dots; lip externally orange grading to greenish white basally, cleft at base purple, inner surface white apically and yellow basally with purple dots and streaks, hook white, mid-lobe yellowish white with small purple dots on inner surface and orange bands down outer surface; column pale green apically, turning reddish purple basally, column-foot purple, anther-cap purple and yellow; whole flower turning more golden yellow with age. ***Pedicel*** with ***ovary*** 6-12.5 cm long, yellowish-green, spotted red at base. ***Dorsal sepal*** 1.3 x 2.8 cm, edges recurved, oblong-elliptic, apex obtuse, unevenly bilobed, twisted. ***Lateral sepals*** 1.5-3 cm, partly attached to column-foot, edge strongly convoluted, obliquely ovate-elliptic, apex acute, twisted, spreading. ***Petals*** 0.9-3.1 cm, lanceolate-elliptic, apex acute, edge wavy, at 90° to dorsal sepal. ***Lip*** base almost at right angle to column-foot; side lobes spreading, narrow, almost rectangular, 0.4 x 1 cm, convex, longitudinally carinate on external surface with a hook at apex; mid-lobe with a low rounded median keel, bent at 90° to base of lip, 0.25 x 1.2 cm, dilating to blunt rounded apex 0.7 cm wide; callus at junction of mid-lobe and side lobes conduplicate, 0.5-0.4 x 0.6 cm long at base, anterior edges crenate, decurrent with edges of mid-lobe. ***Column*** 0.8 cm long, with a fairly long column-foot; pollinia 2, ovoid grooved; viscidium narrow, hastate; stipes sigmoid. Plate 14A.

HABITAT AND ECOLOGY: Hill-forest; recorded on *Gymnostoma sumatrana* on ultramafic rocks above valleys. Alt. 500 to 1000 m. Flowering observed in March, April and August.

DISTRIBUTION IN BORNEO: SABAH: Kuala Labuk; Mt. Kinabalu.

GENERAL DISTRIBUTION: Endemic to Borneo (Sabah only).

NOTES: Plants collected by estate managers in the Kuala Labuk area, clearing land for agriculture, including some low hills on ultramafic substrate, at first confused the species with *P. denevei*. Many plants have been collected from along the Labuk River, as most of the area has been logged or cleared for agricultural development. The species is now endangered unless some reserves are created to preserve it. In the wild, the terete leaves do not normally exceed 120 cm but in cultivation they have exceeded 2 metres.

DERIVATION OF NAME: Named after the Labuk Valley in eastern Sabah, from where the type was collected.

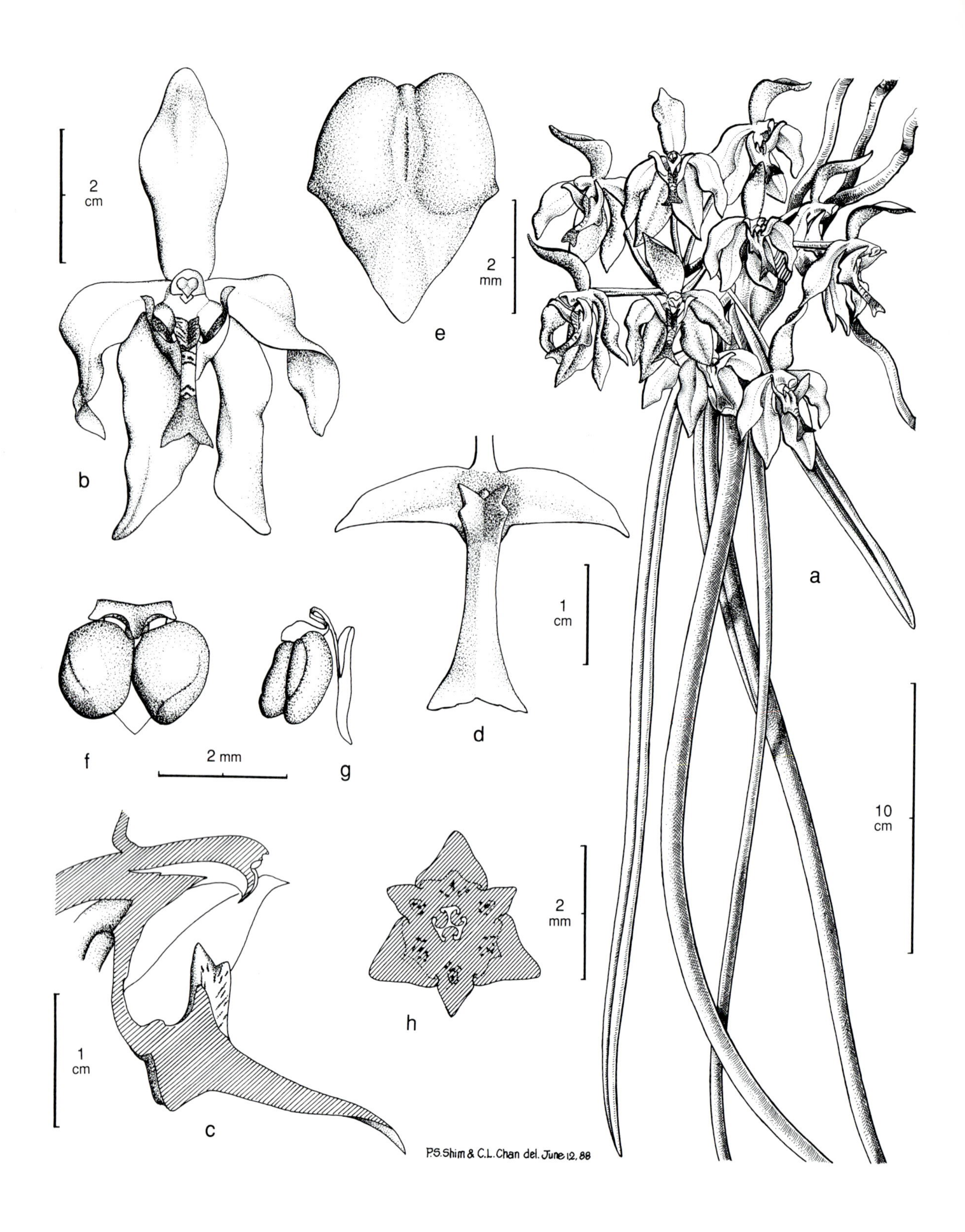

Figure 65. Paraphalaenopsis laycockii (M.R. Hend.) A.D. Hawkes - A: plant. - B: flower, front view. - C: column and lip, longitudinal section. - D: lip, spread out. - E: anther. - F: pollinarium, front view. - G: pollinarium, lateral view. - H: ovary, transverse section. All drawn from cult. TOC by Shim Phyau Soon and Chan Chew Lun.

65. PARAPHALAENOPSIS LAYCOCKII (M.R.Hend.) A.D.Hawkes

Paraphalaenopsis laycockii *(M.R.Hend.) A.D.Hawkes* in Orquidea (Rio de Janeiro) 25: 212 (1964). Type: Borneo, Kalimantan, cult. *Laycock* s.n. (holotype SING).

Phalaenopsis laycockii M.R.Hend. in Orchid Rev. 43: 108 (1935).

Pendent, epiphytic ***herb***. ***Roots*** branching. ***Stem*** short, few leaved, up to 5 cm long. ***Leaves*** up to 1 m long, 1 cm in diameter; terete with a distinct groove. ***Inflorescence*** very short with a congested rachis, several-flowered, rarely up to 15; floral bracts ovate, acute, up to 1 cm long. ***Flowers*** pinkish mauve to lilac with yellow and brown markings on lip; sepals and petals white, centrally pale pink on dorsal sepal, pinkish lilac to pansy-violet on petals and lateral sepals; side lobes of lip pale brown at apex, white with reddish brown spots on basal portion; callus yellow, barred and spotted with reddish to orange spots and bars at base to pale orange brown over whole apex; column white. ***Pedicel*** with ***ovary*** up to 6 cm long, cylindric. ***Dorsal sepal*** up to 4 x 1.5 cm, narrowly lanceolate-elliptic, acute to subacuminate. ***Lateral sepals*** up to 4.5 x 1.7 cm, obliquely ovate-lanceolate, acute. ***Petals*** up to 4 x 1.4 cm, from a fleshy cuneate base, subfalcate-lanceolate, acute. ***Lip*** 3-lobed; side lobes erect, linear-oblong, obliquely truncate to rounded at apex, up to 1.8 x 0.7 cm; mid-lobe porrect, linear-spathulate with a divaricately bilobed apex, lobes triangular, acute, up to 1.5 x 1.2 cm across tip; disc with a conduplicate, quadrate callus with entire margins, pale yellow with transverse brown stripes. ***Column*** up to 1 cm long; pollinia 2, cleft. Plate 14C.

HABITAT AND ECOLOGY: Lowland and hill-forest. Alt. not known.

DISTRIBUTION IN BORNEO: KALIMANTAN TIMUR: locality unknown.

GENERAL DISTRIBUTION: Endemic to Borneo (Kalimantan only).

NOTES: The flowers of this species are strongly scented but with a sweet, musty, slightly unpleasant perfume, reported by Sweet (1980) as similar to crushed lemon grass. In the late 1970s logging company employees apparently rediscovered this species in East Kalimantan and many plants appeared for sale along the East Coast, in Sabah and in Singapore.

DERIVATION OF NAME: Named after John Laycock who first imported to and cultivated the plant in Singapore.

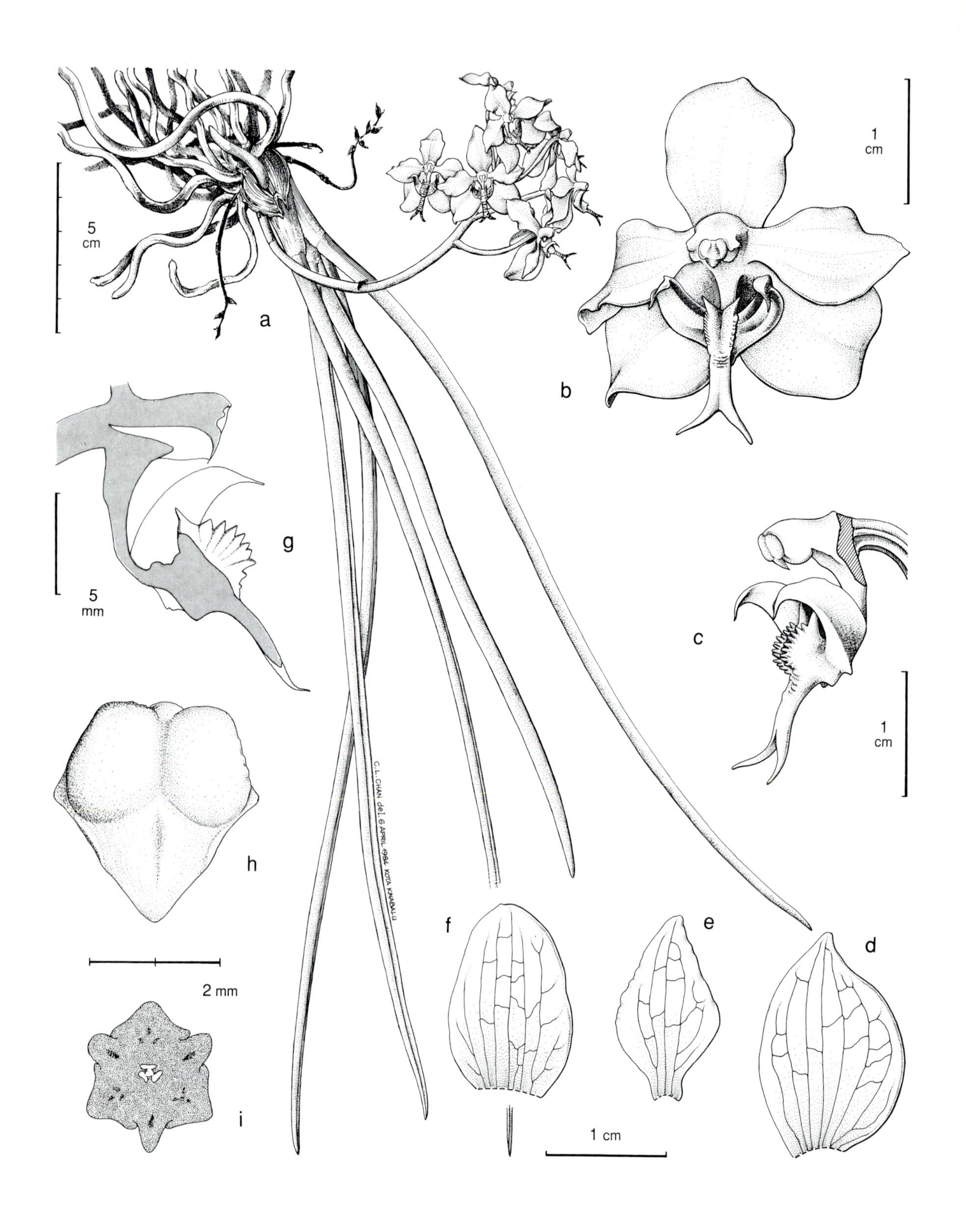

Figure 66. Paraphalaenopsis serpentilingua (J.J. Sm.) A.D. Hawkes - A: plant. - B: flower, front view. - C: column and lip, lateral view. - D: lateral sepal. - E: petal. - F: dorsal sepal. - G: column and lip, longitudinal section. - H: anther. - I: ovary. All drawn from cult. Lee Shong Mai by Chan Chew Lun.

66. PARAPHALAENOPSIS SERPENTILINGUA (J.J.Sm.) A.D.Hawkes

Paraphalaenopsis serpentilingua *(J.J.Sm.) A.D.Hawkes* in Orquidea (Rio de Janeiro) 25: 212 (1964). Types: Borneo, West Kalimantan, Singkawang and Sintang, cult. Bogor B.G. (holotype L, isotype BO).

Phalaenopsis serpentilingua J.J.Smith in Orchid Rev. 41: 147 (1933).

P. denevei J.J.Smith var. *alba* Price, Orchid Culture in Ceylon and the East, ed. 2: 90 (1933). Type: cult. Sri Lanka (specimen not preserved).

P. simonsei Simonse in Amer. Orchid Soc. Bull. 29: 531 (1960), nomen nudum.

Pendent, epiphytic ***herb***. ***Roots*** aerial, thick, unbranched. ***Stem*** short, few-leaved, up to 3 cm long. ***Leaves*** up to 30 x 0.7 cm, terete, with a distinct groove. ***Inflorescence*** up to 35 cm long, up to 7-flowered, erect to ascendant, elongate; rachis compact. ***Flowers*** small to medium-sized, scented; sepals and petals white inside; lip lemon-yellow, transversely striated with purple. ***Pedicel*** with ***ovary*** up to 6 cm long. ***Dorsal sepal*** 1.6 x 1 cm, elliptic to obovate-elliptic, acute to subapiculate. ***Lateral sepals*** 1.8 x 1.1 cm, obliquely elliptic to ovate-elliptic, acute. ***Petals*** 1.6 x 0.7 cm, from a fleshy, cuneate base, rhombic-lanceolate, acute to subobtuse. ***Lip*** three-lobed; side lobes falcate-linear, subulate, up to 1 cm long; mid-lobe recurved, linear with a distinct bifid apex, the segments of which are subulate and somewhat diverging, up to 0.9 cm long; disc with a conduplicate, dentate callus. ***Column*** up to 0.6 cm long; pollinia 2, cleft. Plate 14B.

HABITAT AND ECOLOGY: Reports indicate that it occurs in lowland seasonally swampy forest or rarely on moss-covered rocks. Alt. below 1000 m.

DISTRIBUTION IN BORNEO: KALIMANTAN BARAT: Singkawang area; Kapuas River, Sanggau area; Sintang area.

GENERAL DISTRIBUTION: Endemic to Borneo (Kalimantan only).

NOTES: There is little information on flowering time, but it is probably during the rainy season. This is the smallest species in the genus, with short leaves up to 30 cm long. The flowers, however, are very striking. It has become extremely rare in cultivation in the Malesian region. Plants were originally collected along with *P. denevei* near Sintang in West Kalimantan. Sweet (1980) reports it growing on moss-covered rocks.

DERIVATION OF NAME: The specific epithet is derived from the Latin *serpentinus*, snake-like and *lingua*, tongue, referring to the forked mid-lobe of the lip which resembles the tongue of a snake.

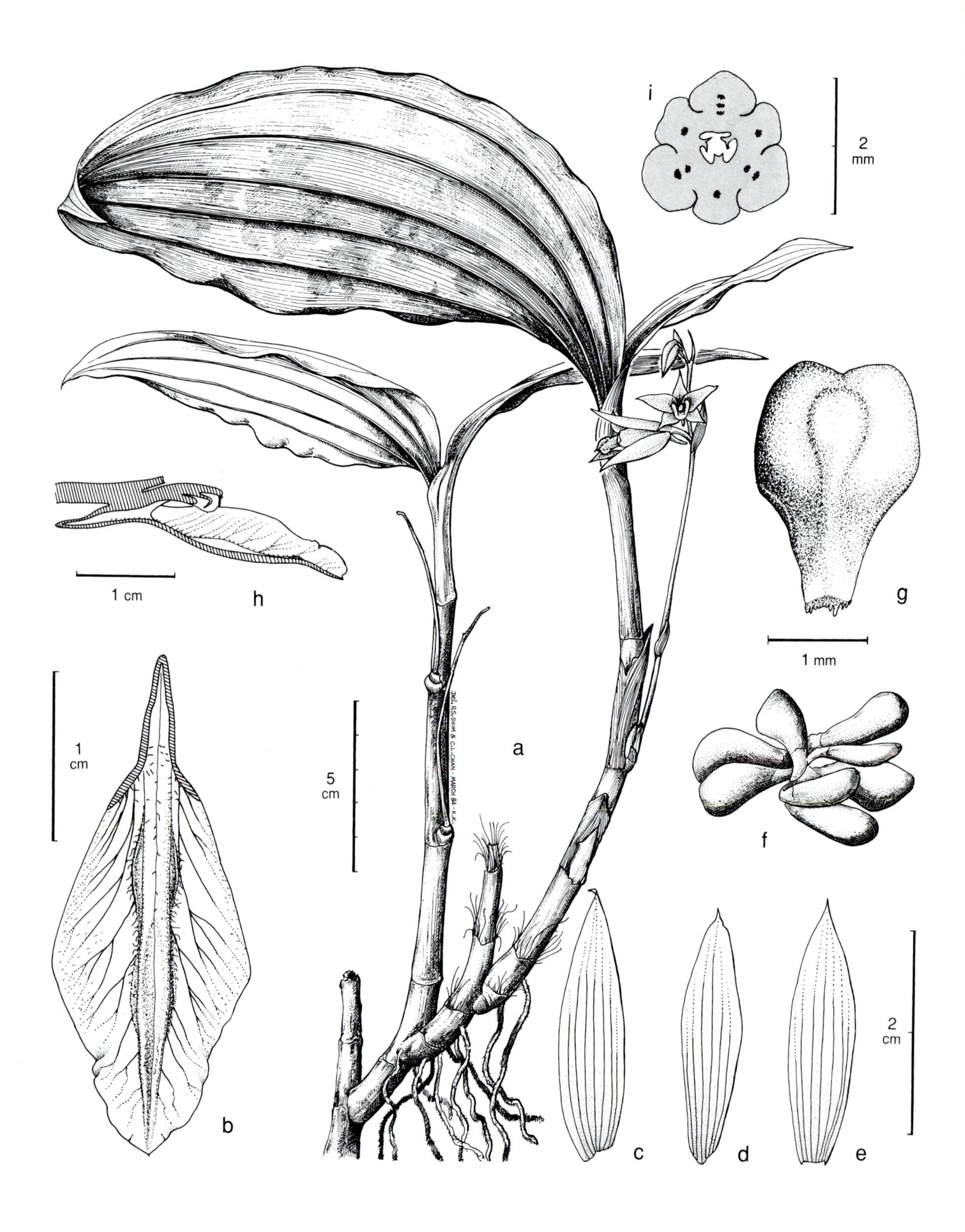

Figure 67. Phaius baconii J.J. Wood & Shim - A: plant. - B: lip. - C: lateral sepal. - D: petal. - E: dorsal sepal. - F: pollinarium. - G: anther. - H: column and lip, longitudinal section. - I: ovary, transverse section. All from *Lamb* AL 39/83 by Shim Phyau Soon and Chan Chew Lun.

67. PHAIUS BACONII J.J.Wood & Shim

Phaius baconii *J.J.Wood et Shim* **sp. nov.** *P. paucifloro* (Blume) Blume arcte affinis, sed differt caulibus basin versus, angustis, nec tumidis pseudobulbos parvos basi efficientibus, foliis tantum duobus pro caulem evolutis, florum labio angustiore integro calcari breviore apice emucronato disco verruculoso-papilloso quasi duobus superciliis angustis prope basin praedito; a *P. corymboidi* Schltr. statura multo breviore, foliis paucioribus, petalis longioribus, labio oblongo-elliptico calcari breviore distincta; a *P. stenocentro* Schltr. foliis longioribus, labio angustiore oblongo-elliptico lamellis brevibus parallelis mediis deficientibus, calcari multo breviore praeterea differt. Typus: Borneo, Sabah, Mt. Kinabalu, Penibukan, 1500 m, January 1983, *Lamb* AL 39/83 (holotypus K).

Erect terrestrial ***herb***. ***Stems*** to 29 x 0.7-1 cm, fleshy, branched and rooting at the nodes towards the base, dark green, flushed purple; internodes 2.5-5.5 cm long, with 3 to 4 tubular, acute, scarious sheaths below, the remains of which are persistent. ***Leaves*** 2 at apex of stem, rarely with a third much smaller lower leaf, 16-28 x 4.5-7 cm, narrowly elliptic to elliptic, acuminate, plicate, narrowed into a sheathing petiole; petiole 5-9 cm long. ***Inflorescence*** a 3- to 6-flowered, lax, porrect raceme borne from the middle to lower nodes of the stem and piercing the enclosing tubular sheath; peduncle 8-18 cm long, with 3 ovate, acute, scarious 6-8 mm long sheaths below, naked above; rachis 6-8.5 cm long, somewhat fractiflex, white; floral bracts 1.5-1.8(-3) x 0.5-0.8 cm, ovate-elliptic to narrowly elliptic, acute to acuminate, scarious, concave, white, flushed green. ***Flowers*** c. 5 cm in diameter, white, the lip with purple stripes and a yellow disc sometimes speckled with red. ***Pedicel*** with ***ovary*** 1-1.2 cm long, clavate, straight or slightly curved. ***Dorsal sepal*** 2.3-3.2 x 0.7-0.8 cm, narrowly elliptic, slightly oblique, acute, slightly carinate at apex. ***Lateral sepals*** 2.3-3.2 x 0.7-8 cm, narrowly elliptic, slightly oblique, acute, slightly carinate at apex. ***Petals*** 2.3-3.1 x 0.7-0.8 cm, narrowly elliptic, apex mucronate and reflexed. ***Lip*** 2.4-2.7 x 1.1-1.3 cm, oblong-elliptic, entire, obtuse, undulate-crenulate distally, sides erect and with slightly raised minutely papillose nerves, becoming flat along the apical portion; disc consisting of a raised fleshy, verruculose-papillose area provided with 2 low flanges near the base, hirsute at and towards the base, the thickened area narrowing abruptly and continuing to just below the apex of the lip; spur 0.5-0.6 cm long, cylindrical to slightly conical, straight. ***Column*** 0.75-0.9 cm long, 3-4 mm broad at apex, slender, broadened at apex, straight; anther-cap 2 x 1 mm, ovate, cucullate, pollinia 8, in 2 groups of 4, spathulate-clavate. Plate 14D.

HABITAT AND ECOLOGY: Lower montane forest on ultramafic substrate; in leaf litter over rocks in bamboo thickets. Alt. 1200 to 1500 m. Flowering observed in January and April.

DISTRIBUTION IN BORNEO: SABAH: Mt. Kinabalu.

GENERAL DISTRIBUTION: Endemic to Borneo (Sabah only).

NOTES: *P. baconii* is most closely allied to *P. pauciflorus* (Blume) Blume distributed in Peninsular Malaysia, Sumatra and Java, but differs in having stems that are narrow throughout and not swollen to form small pseudobulbs at the base. There are usually only two leaves per stem and the flowers have a slightly narrower, entire and shorter-spurred lip lacking an apical mucro. The verruculose-papillose disc is provided with two low basal flanges near the base. It is distinguished from the Sumatran *P. corymboides* Schltr. by its much shorter stature, fewer leaves, longer petals and longer, narrower oblong-elliptic lip with a shorter spur. It also differs from *P. stenocentron* Schltr. from Sulawesi by its longer leaves and narrower oblong-elliptic lip lacking any short, parallel lamellae and having a much shorter spur.

DERIVATION OF NAME: The generic name is derived from the Greek *phaios*, grey or swarthy, referring to the flowers which turn dark grey-blue with age or if damaged. The specific epithet honours Dr Andrew Bacon, formerly a veterinary officer with the Sabah Government and a keen orchid collector who discovered this species on Mt. Kinabalu.

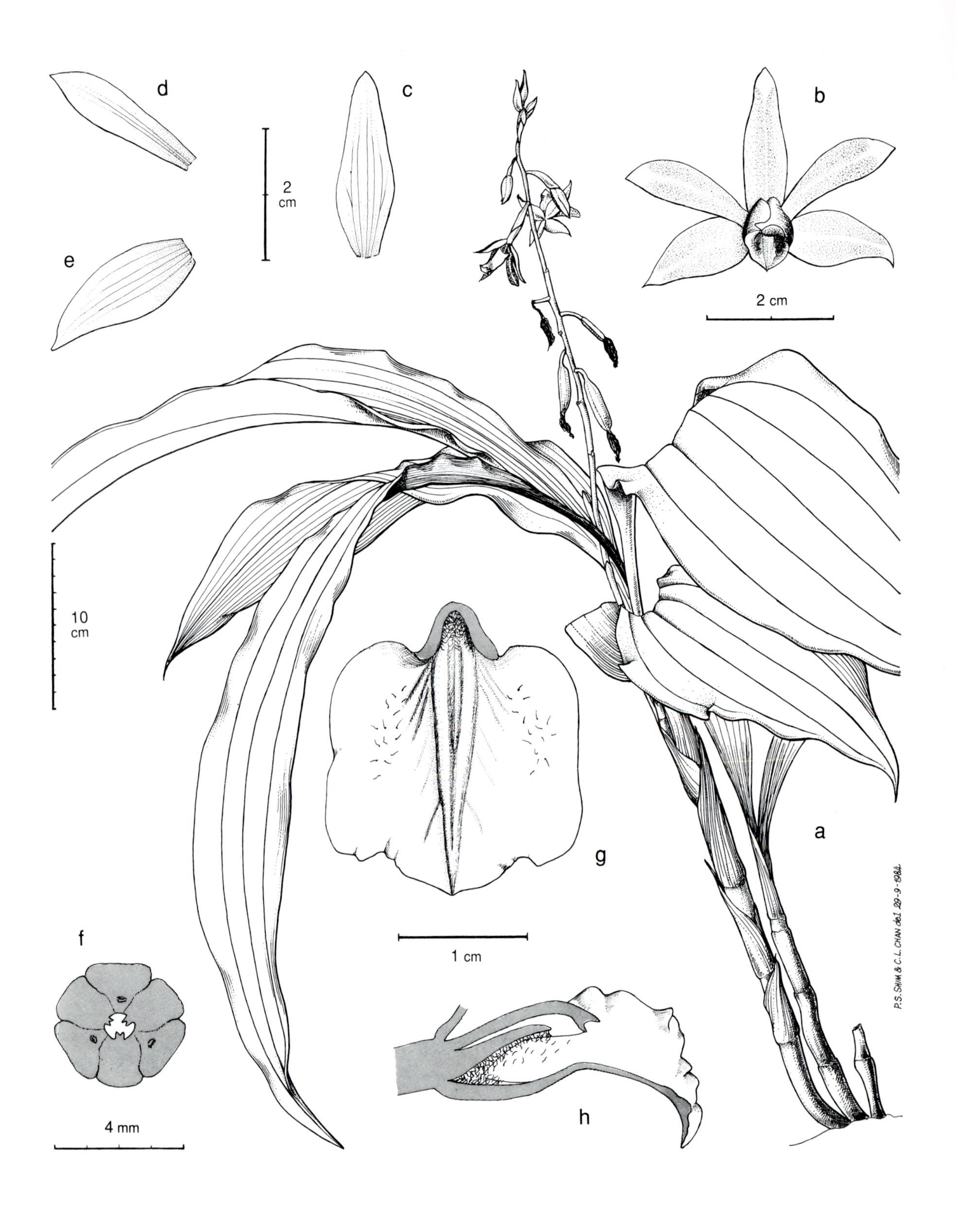

Figure 68. Phaius borneensis J.J. Sm. - A: plant. - B: flower, front view. - C: dorsal sepal. - D: petal. - E: lateral sepal. - F: ovary, transverse section. - G: lip. - H: column and lip, longitudinal section. All drawn from cult. TOC by Shim Phyau Soon and Chan Chew Lun.

68. PHAIUS BORNEENSIS J.J.Sm.

Phaius borneensis *J.J.Sm.*, Icon. Bogor. 2: 6, t. 3, fig. C (1903). Type: Borneo, Kalimantan, Bukit Kasian, *Nieuwenhuis* s.n. (holotype BO).

Terrestrial ***herb***. ***Stem*** to 25-30 cm long, covered with 3-4 acute, tubular sheaths which are shed from older stems, square in cross section towards apex; internodes to 5-7 cm long. ***Leaves*** c. 6, 35-50 x 11-13 cm, plicate, elliptic to lanceolate, acuminate; petiole 10-20 cm long. ***Inflorescence*** 45-53 cm long, with up to 17 resupinate flowers; floral bracts 2.5 x 1.3 cm, falling soon after flowers open. ***Flowers*** spreading; sepals and petals yellowish green with pinkish red to red bands covering most of the surface, tips green; lip pale yellow to yellow with pink to red lines of spots on the side lobes, apex spotted red, central disc yellow, column white. ***Pedicel*** with ***ovary*** 2-2.5 cm long. ***Dorsal sepal*** 2.7 x 0.7 cm, 5-veined, lanceolate, acute. ***Lateral sepals*** 2.5 x 0.9 cm, ovate, acute, 5-veined. ***Petals*** 2.3 x 0.6 cm, 3-veined, oblanceolate, acute. ***Lip*** 2.2 x 2 cm, subquadrate, spurless, hairy at saccate base, 3-lobed; side lobes enveloping column, oblong, 1.5 cm wide, sparsely hairy; mid-lobe broadly triangular, acute; disc oblong-lanceolate, the 2-basal keels uniting on the mid-lobe to form a raised median keel, acute at apex. ***Column*** 1.3 cm long, base with white hairs, attached to lip for 3 mm at base; anther-cap and pollinia deformed in all flowers examined. Plate 14E.

HABITAT AND ECOLOGY: Mixed hill-forest and lower montane forest on ultramafic and other soils; in deep shade, usually near streams, in humid valleys. Alt. near sea level to 1500 m.

DISTRIBUTION IN BORNEO: KALIMANTAN: Bukit Kasian. KALIMANTAN TIMUR: Sungai Susuk. SABAH: Mt. Kinabalu.

GENERAL DISTRIBUTION: Endemic to Borneo.

NOTES: This was the first species of *Phaius* to be discovered in Borneo and is the only spurless species of *Phaius* on the island.

DERIVATION OF NAME: Named after the island of Borneo.

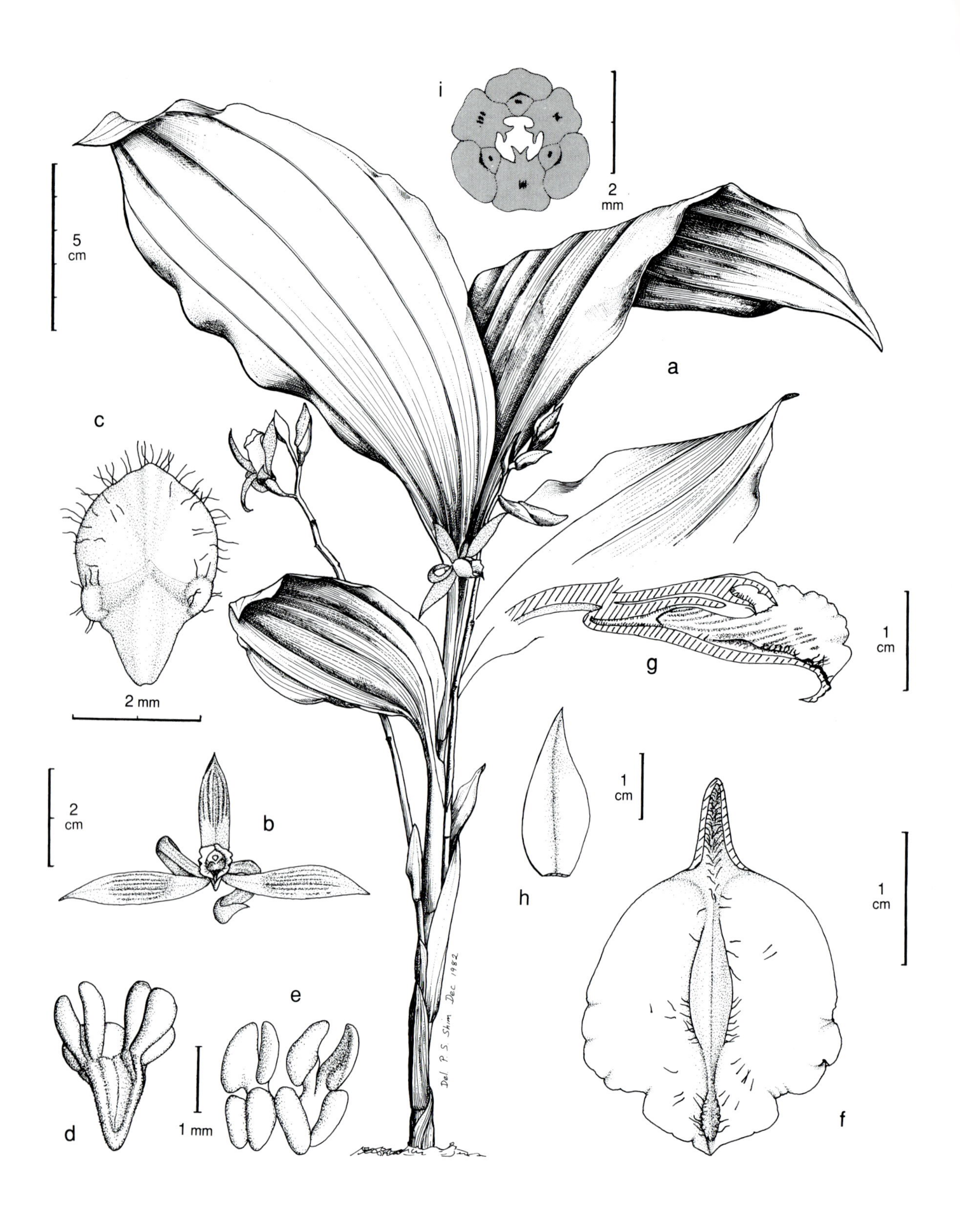

Figure 69. Phaius reflexipetalus J.J. Wood & Shim - A: plant. - B: flower, front view. - C: anther. - D & E: pollinarium. - F: lip. - G: column and lip, longitudinal section. - H: floral bract. - I: ovary. All drawn from *Lamb* AL 25/82 by Shim Phyau Soon.

69. PHAIUS REFLEXIPETALUS J.J.Wood & Shim

Phaius reflexipetalus *J.J.Wood et Shim*, **sp. nov**. *P. borneensi* J.J.Sm. affinis sed foliis paucis parvis, petalis angustioribus et valde reflexis, callo medio labelli latiore, elliptico, complanato, antice attenuato, apice dilatato, ruguloso, parte libera calcaris 1-2 mm longa distinguenda. Typus: Borneo, Sabah, Mt. Kinabalu, Penataran Ridge, 1100 m, 1982, *Lamb* AL 25/82 (holotype K).

Erect, terrestrial ***herb***. ***Stems*** fleshy, simple, dark green, 20-30 x 1 cm; internodes 4-4.5 cm long, enclosed by 4 to 5 tubular, acute to acuminate scarious sheaths 2.5-10 cm long. ***Leaves*** 3-4, blade 14-25 x 4.5-8 cm, elliptic, acuminate, plicate, narrowed into a sheathing petiole 8-10 cm long. ***Inflorescence*** up to 12-flowered, usually two arising from axils of lower stem sheaths; peduncle 11-14 cm long, bearing 3 tubular, acute sheaths 2-2.5 cm long; rachis 10-13 cm long; floral bracts narrowly elliptic, acuminate, concave, deciduous. ***Flowers*** c. 5 cm in diameter; sepals and petals pale greenish yellow, stained pink or brown; lip yellow with a cream margin, disc with reddish brown streaks, callus speckled reddish brown at apex; column cream. ***Pedicel*** with ***ovary*** 1.5-2.2 cm long, clavate, curved. ***Sepals*** spreading, recurved at tips. ***Dorsal sepal*** 3-3.5 x 0.8-0.9 cm, narrowly elliptic, acute to acuminate. ***Lateral sepals*** 3-3.5 x 0.7-0.8 cm, oblong-elliptic to narrowly elliptic, acute to acuminate. ***Petals*** 2.9-3.4 x 0.3-0.4 cm, linear, oblanceolate or ligulate, acute, strongly reflexed. ***Lip*** 1.6-2.2 x 1.9-2.2 cm, entire or subentire, broadly ovate, apiculate, sometimes with an obscure mid-lobe, margins undulate, with a few sparse long hairs, particularly above apical callus; disc with a broad median fleshy, flattened elliptical callus commencing above the base, attenuated distally before becoming raised, dilated and rugulose at apex of lip; spur fused to column for 7-9 mm, free portion obtuse, 1.2 mm long, minutely pubescent, particularly along upper sides and inside. ***Column*** 1.6-1.8 x 0.5-0.6 cm, slightly curved, minutely pubescent below stigmatic cavity; anther-cap 4 x 2.5 mm, ovate, acute to acuminate, hirsute; pollinia 8. Plate 14F.

HABITAT AND ECOLOGY: Mixed hill-forest on ultramafic substrate with serpentine boulders; in deep humus amongst limestone boulders; in deep shade in humid forest near rivers. Alt. 1100 m. Flowering in May, June, November and December.

DISTRIBUTION IN BORNEO: SABAH: Mt. Kinabalu.

GENERAL DISTRIBUTION: Endemic to Borneo (Sabah only).

NOTES: *P. reflexipetalus* differs from the related *P. borneensis* J.J.Sm. by its fewer, smaller leaves, narrower, sharply reflexed petals, elliptic median callus which is broader and flattened, narrowed at the front, expanded and somewhat wrinkled at the apex and the free portion of the spur which is 1-2 mm long.

DERIVATION OF NAME: The specific epithet is derived from the Latin *reflexus*, reflexed, bent abruptly backwards, referring to the sharply reflexed petals.

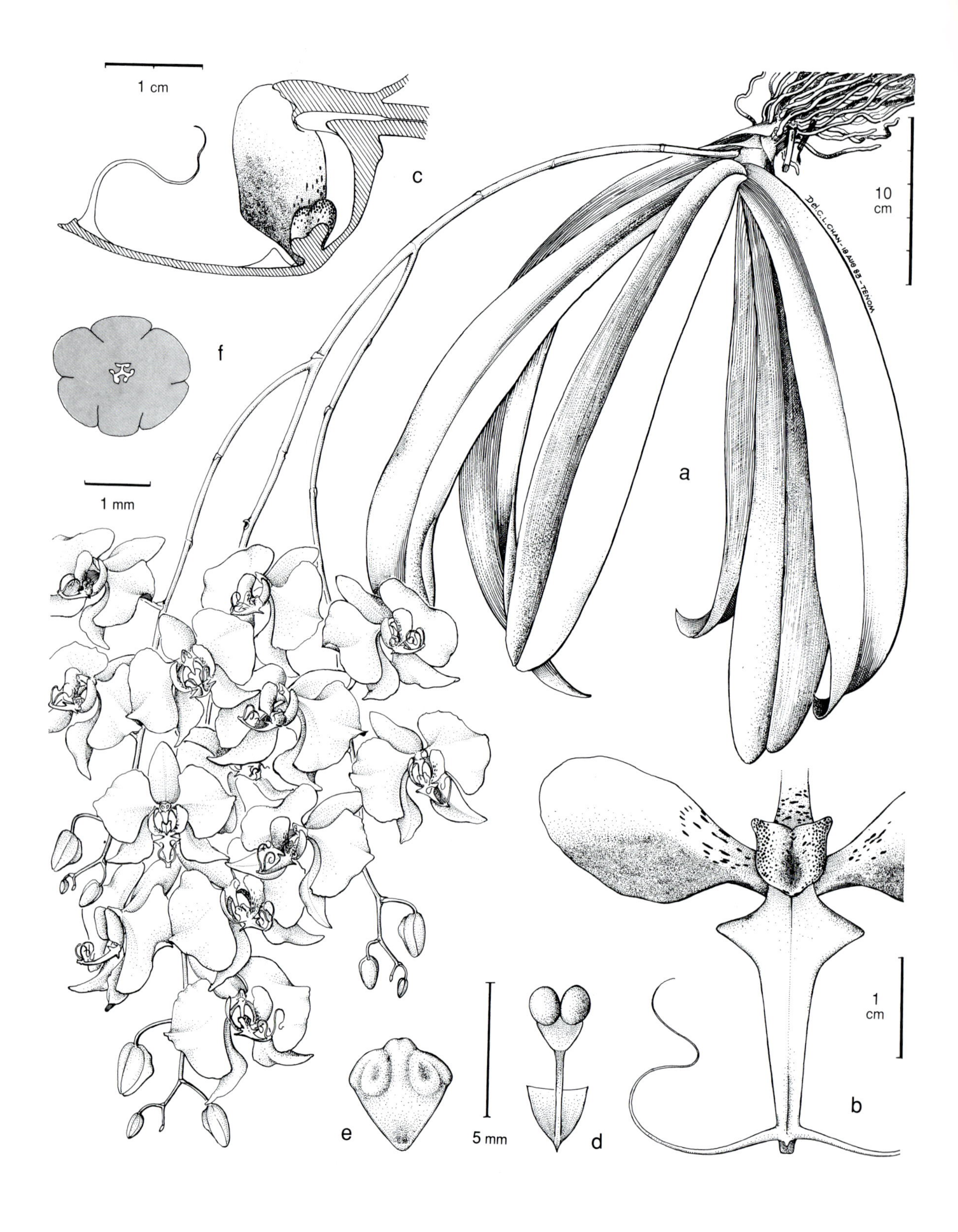

Figure 70. Phalaenopsis amabilis (L.) Blume - A: plant. - B: lip, spread out. - C: column and lip, longitudinal section. - D: pollinarium. - E: anther. - F: ovary, transverse section. All drawn from *Lamb* AL 120/83 by Chan Chew Lun.

70. PHALAENOPSIS AMABILIS (L.) Blume

Phalaenopsis amabilis (*L.*) *Blume*, Bijdr.: 294 (1825). Type: Java, *Osbeck* s.n. (holotype LINN).

Epidendrum amabile L., Sp. Plant.: 953 (1753).

Phalaenopsis amabilis (L.) Blume var. *fuscata* Rchb.f. in Bot. Zeitung (Berlin) 20: 214 (1862). Type: Borneo, cult. *Low* (holotype:W).

P. grandiflora Lindl. var. *aurea* Hort. in Proc. Roy. Hort. Soc. London 4:135 (1864). Type: Borneo, cult. *Warner* (neotype: Select Orch.Pl., ser.2: t.7,1869).

P. grandiflora Lindl. var. *fuscata* (Rchb.f.) Burb. in The Garden (London) 22: 118(1882).

P. amabilis (L.) Blume var. *aurea* (Hort.) Rolfe in Gard. Chron. n.s. 26: 212 (1886).

For full synonymy see Sweet, The genus *Phalaenopsis*: 21 (1980).

Epiphytic ***herb***. ***Roots*** extensive, to 3 m or more, fleshy, often branched, flexuous, glabrous. ***Stems*** short, robust, completely enclosed by imbricate leaf-sheaths. ***Leaves*** up to 50 x 10 cm, few, fleshy or leathery, elliptic, ovate-elliptic to obovate-oblong or oblong-lanceolate, apex obtuse, rarely acute. ***Inflorescence*** racemose, to 1 m or more long; floral bracts 5 mm long, scarious, triangular, cucullate. ***Flowers*** faintly sweet-scented, white, margins of lip and tendrils (cirrhi) yellow, callus dark yellow with crimson markings. ***Pedicel*** with ***ovary*** 4-5 cm long. ***Dorsal sepal*** 3 - 4.5 x 1.5-2.7 cm, erect, elliptic or elliptic-ovate, obtuse. ***Lateral sepals*** 3-5 x 1-2.5 cm, ovate, ovate-elliptic or ovate-lanceolate, somewhat oblique, acute, rarely subacuminate. ***Petals*** 3-4.5 x 2.5-5 cm, broadly ovate to elliptic, contracted into a cuneate base. ***Lip*** long-unguiculate, fleshy, 3-lobed; side lobes 1-1.2 x 1.3-1.5 cm, erect, obovate-oblanceolate, rounded above, laterally emarginate below apex; mid-lobe from a cuneate base, cruciform, apex terminating in two elongate, filiform tendrils (cirrhi); callus at junction of mid-lobe and side lobes peltate with entire margins. ***Column*** 0.8 cm long, cylindric; anther-cap 4 x 4 mm, rounded, front margin drawn out, triangular, acute; pollinia 2, 1.7 mm long. ***Fruit*** to 10 cm long. Plate 15A.

HABITAT AND ECOLOGY: Mixed lowland and hill-forest on all soil types. It is most commonly found on the drier eastern slopes of the Crocker Range in Sabah and some of the valleys of the interior. It is also found on the drier offshore islands, particularly those off the north east coast of Sabah and is widely scattered elsewhere in Borneo. It grows on *Vitex pubescens* trees in the hills of south-west Sabah. It thrives in areas where the rainfall ranges between 1500 and 2000 mm per year, avoiding areas of high rainfall, preferring a climate experiencing distinct but short dry seasons. Alt. sea level to 1500 m. Flowering January to March and July.

DISTRIBUTION IN BORNEO: KALIMANTAN: locality unknown. SABAH: Mt. Kinabalu; Tenom District; Banggi Island. SARAWAK: localities unknown.

GENERAL DISTRIBUTION: Sumatra, Java, Sulawesi, Seram and other Indonesian islands east to New Guinea (including the Bismarck Archipelago), Australia (Queensland), north to the Philippines.

NOTES: This beautiful orchid, which is Indonesia's national flower, was at one time a common sight on the offshore islands all around Sabah. Plants from the Kinabatangan River have a pure white lip with a pale yellow callus. Those from some of the offshore islands along the east coast have a mauve-brown margin to the lip. Similar plants are also found in the Sulu Islands (Philippines). Many plants from the interior regions have the backs of the sepals flushed with pink. Flowering starts after the two main monsoon periods. Flowers are most commonly visited by large blue-black carpenter bees (*Xylocopa spp.*) and pollination is over 50 percent successful. If pollination is prevented, the arching sprays continue flowering for nearly three months, and inflorescences of 1.5 metres in length often occur. *P. amabilis* is placed in section *Phalaenopsis* by Sweet (1980). The species of this section have petals larger than the sepals and distinctive 'tendrils' or cirrhi at the apex of the lip. Several varieties of *P. amabilis* have been recognised including var. *aurea* (Hort.) Rolfe which has a deep yellow lip. The round white flowers have given rise to the popular name "Moon Orchid". *P. amabilis* is by far the most popular native orchid collected and grown by local people who commonly plant it in ironwood baskets or coconut husks, with charcoal and chicken feather manure and hung under the eaves of shady verandahs.

DERIVATION OF NAME: The generic name is derived from the Greek *phalaina*, moth, and *opsis*, appearance, referring to the delicate moth-like flowers, particularly of the white-flowered species. The specific epithet is derived from the Greek *amabilis*, lovely, referring to the beautiful arching sprays of white flowers.

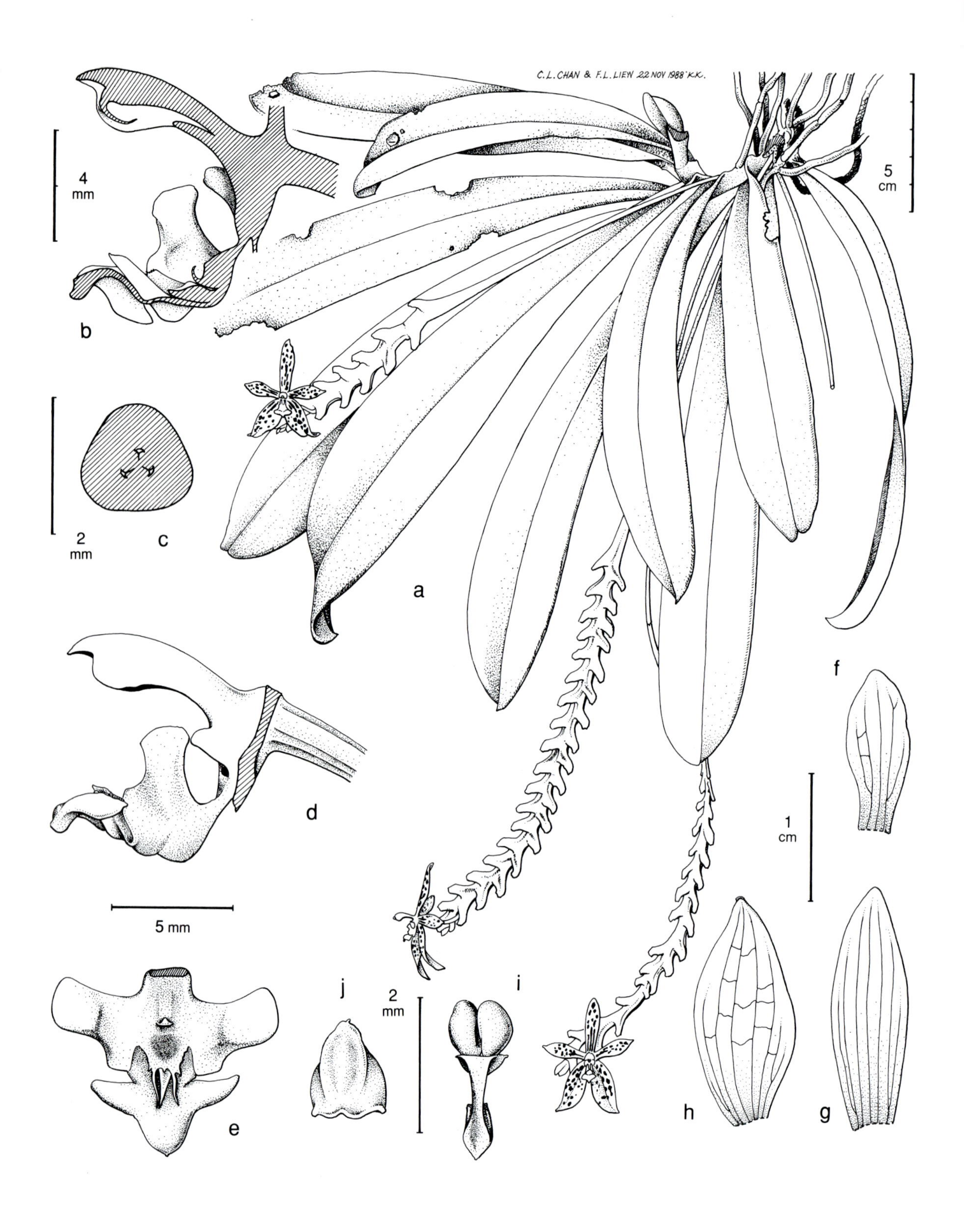

Figure 71. Phalaenopsis cornucervi (Breda) Blume & Rchb. f. - A: plant. - B: column and lip, longitudinal section. - C: ovary, transverse section. - D: column and lip, lateral view. - E: lip, spread out. - F: petal. - G: dorsal sepal. - H: lateral sepal. - I: pollinarium. - J: anther. All drawn from cult. TOC by Chan Chew Lun and Liew Fui Ling.

71. PHALAENOPSIS CORNUCERVI (Breda) Blume & Rchb.f.

Phalaenopsis cornucervi (*Breda*) *Blume & Rchb. f.* in Hamburger Garten-Blumenzeitung 16:116 (1860). Type: Java, Bantam Province, *van Hasselt* s.n. (holotype L).

Polychilos cornucervi Breda in Kuhl & van Hasselt, Gen. Sp. Orch.: t. 1 (1827).

Polystylus cornucervi (Breda) Hassk. in Natuurk. Tijdschr. Ned.-Indië 10:3 (1856). Type: Java, *Lobb* s.n. (holotype K).

Phalaenopsis devriesiana Rchb.f. in Hamburger Garten-Blumenzeitung 16:116 (1860). Type: Java, *de Vries* s.n. (holotype W).

Often rather robust epiphytic ***herb***. ***Roots*** profusely produced from rhizome-like stem, fleshy, flexuous, often branched, glabrous. ***Stem*** robust, completely enclosed by imbricate leaf sheaths. ***Leaves*** 3-6 per plant, 12-22 x 2.5-4 cm, fleshy, oblong-ligulate to oblong-oblanceolate, obtuse, sometimes bilobed at apex. ***Inflorescences*** 9-42 cm long, one to several; peduncle terete, with one or two small cauline sheaths; rachis simple or branched, laterally compressed, flexuous, commonly several-flowered; floral bracts 5 cm long, alternate, distichous, ovate-cucullate, dorsally keeled. ***Flowers*** fleshy, waxy, with spreading floral segments, several opening simultaneously, yellow to yellowish green; sepals marked with red brown bars, spots and blotches; lip yellow, with apex of lobes and blade white; column yellow. ***Pedicel*** with ***ovary*** 1.5-3 cm long, yellowish green. ***Dorsal sepal*** 1.8-2.3 x 0.5-0.8 cm, obovate-elliptic to oblanceolate-elliptic, prominently carinate toward apex, margins slightly recurved. ***Lateral sepals*** 1.7-2.3 x 0.7-0.9 cm, elliptic to elliptic-lanceolate, somewhat oblique, acute or subacuminate, distinctly carinate toward apex. ***Petals*** 0.7-1.8 x 0.5-0.6 cm, lanceolate or elliptic-lanceolate, acute or obtuse, dorsally somewhat thickened at apex. ***Lip*** 0.8 x 1 cm, fleshy, 3-lobed; side lobes porrect, subquadrate with truncate apex below which there is a fleshy callus, confluent with base of mid-lobe forming a flat semi-circular gibbosity; mid-lobe in front of gibbosity suddently constricted then expanding into an anchor-shaped callus, the acute or obtuse apex provided beneath with a fleshy hook-like protuberance, centre of gibbosity provided with an erect oblong linear, laterally compressed appendage, distal end of gibbosity provided with another plate-like convex bicirrhous structure, the margin of which on each side is provided with an acuminate tooth, with an additional retrorse lanceolate acuminate projection at base of plate. ***Column*** 0.8 cm, fleshy, somewhat arcuate, cylindric, somewhat dilated towards apex, with a small basal protuberance on each side; pollinia 2, cleft. Plate 15B.

HABITAT AND ECOLOGY Lowland and hill-dipterocarp forest, riverine and swamp forest usually on soils derived from sandstone, mudstone and shale. The thick, leathery leaves indicate that the plants are only lightly shaded and normally receive plenty of sunlight high up in the canopy. Alt. sea level to 500 m. Flowering throughout the year except in prolonged dry periods.

DISTRIBUTION IN BORNEO: KALIMANTAN BARAT: Pontianak area. KALIMANTAN SELATAN: Banjarmasin area. KALIMANTAN TIMUR:

Mahakam River. SABAH: Ulu Dusan. SARAWAK: Bukit Pait; Lawas River; Batang River.

GENERAL DISTRIBUTION: India, Nicobar Islands, Burma, Thailand, Laos, Peninsular Malaysia, Sumatra, Java and Borneo.

NOTES: *P. cornucervi* is placed by Sweet (1981) in section *Polychilos*, with three representatives in Borneo. Species of the section have fleshy flowers with similar sepals and petals, a lobed lip with a gibbose cavity in the centre, and anchor-shaped apical lobes. *Polychilos* was previously considered a separate genus by J.J. Smith and Shim (1982) supported Smith's view that all the "star-shaped" *Phalaenopsis* should be reinstated in that genus. However, in the horticultural world, this would cause considerable confusion to breeders and Shim's view is not upheld here.

The variety *picta* (Hassk.) Sweet, with well developed transverse coloured bars instead of spots on the sepals and petals, has not been seen by the authors in Borneo. The species has been hybridised many times with other *Phalaenopsis* species and hybrids to pass on the yellow colouration. Crosses have also been made with species of *Doritis* and *Kingidium*.

Bornean plants, especially in the north, have vividly coloured sepals and petals, generally of a deep yellow strongly spotted with red to reddish brown. Plants can tolerate a wide range of conditions from deep shade to full sun. In Borneo, the more exposure to sunlight, the thicker and more leathery the leaves become. The flattened nature of the rachis and the shape of the floral bracts, have led to the common name of "Crocodile's Tail" orchid. In the Miri area of Sarawak it is sometimes referred to as the "Star of Sarawak", and has often been confused with *P. pantherina*.

DERIVATION OF NAME: The specific epithet is derived from the Latin *cornu*, horn or horn-like, and *cervus*, deer, in reference to the forked appendages on the lip which resemble a deer's forked antlers.

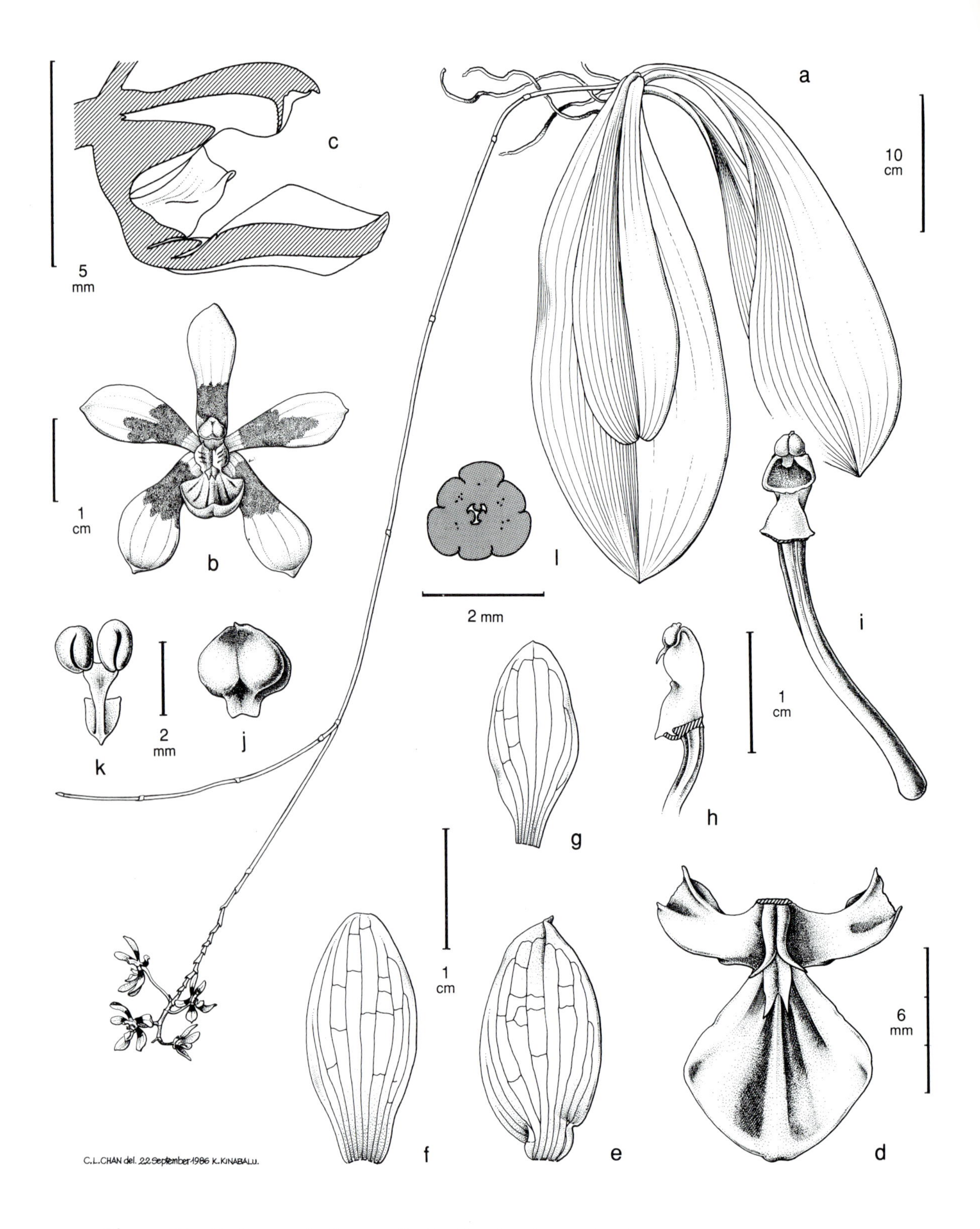

Figure 72. Phalaenopsis fuscata Rchb. f. - A: plant. - B: flower. - C: column and lip, longitudinal section. - D: lip, spread out. - E: lateral sepal. - F: dorsal sepal. - G: petal. - H: column and ovary, lateral view. - I: column and pedicel, ventral view. - J: anther. - K: pollinarium. - L: ovary, transverse section. All drawn from cult. TOC by Shim Phyau Soon and Chan Chew Lun.

72. PHALAENOPSIS FUSCATA Rchb. f.

Phalaenopsis fuscata *Rchb. f.* in Gard. Chron. n.s. 2:6 (1874). Type: Peninsular Malaysia, without precise locality, cult. *Bull* s.n. (holotype W).

Phalaenopsis denisiana Cogn. in Gard. Chron. ser. 3, 26:82 (1899). Type: *Philippines*, without precise locality, introduced in 1897, cult. *M. Fernald Denis* s.n. (holotype BR).

Polychilos fuscata (Rchb.f.) Shim in Malayan Nat. J. 36:23 (1982).

Epiphytic ***herb***. ***Roots*** slightly flexuous, glabrous. ***Stem*** completely enclosed by imbricate leaf sheaths. ***Leaves*** 30 x 10 cm, spreading, deflexed, from a cuneate base, obovate-oblong, thick, acute. ***Inflorescence*** suberect, arcuate, simple or branched, rather stout, few-flowered, occasionally with up to 12 flowers, as long as or longer than leaves. ***Flowers*** fleshy, with spreading segments; sepals and petals cream to yellow to greenish yellow with a brown to orange-brown blotch or two in the basal half; lip pale cream to pale yellow with several pale brown to brownish orange lines on the mid-lobe, apex of side lobes white, base yellowish orange and mid portion with pale brown bars and brownish purple spots, callus appendage cream with pale orange spots; column pale yellow, anther-cap cream. ***Dorsal sepal*** 1.4-1.7 x 0.7-1 cm, ovate or ovate-elliptic, obtuse, rarely acute with revolute margins. ***Lateral sepals*** 1.4-2.5 x 0.7-1.1 cm, from an oblique base, elliptic, obtuse, margins revolute. ***Petals*** 1.2-1.6 x 0.6-0.8 cm, obovate-oblong to elliptic-oblong, obtuse, margins revolute. ***Lip*** 1.1-1.4 x 1-1.4 cm, fleshy, 3-lobed; side lobes subquadrate, somewhat falcate with a truncate apex to 6 x 2.5 mm; mid-lobe ovate or elliptic, fleshy, flat, with a fleshy, heavy, median keel; disc at junction of mid-lobe and side lobes provided with a fleshy bifurcate appendage, superimposed upon which is another bipartate, fleshy callus blending into the main tissue of the disc. ***Column*** 0.7-0.8 cm long, erect, fleshy, cylindric; pollinia 2, cleft. Plate 15C.

HABITAT AND ECOLOGY: Lowland dipterocarp and mixed hill-dipterocarp forest, on all types of soils, including ultramafic. It appears to favour a shady habitat near streams. Alt. sea level to 1000 m. Flowering May-August.

DISTRIBUTION IN BORNEO: KALIMANTAN TIMUR: Kutai, Long Ibok. SABAH: Tenom District; Rundum area.

GENERAL DISTRIBUTION: Peninsular Malaysia, Borneo and the Philippines.

NOTES: Sweet (1980) has made *P. fuscata* the type for the section *Fuscatae* characterised by the sepals and petals being similar, with recurved margins, and a lip that is continuous with the column-foot. The mid-lobe of the lip is flat to concave with one to several keels, the central one fleshy. Like many *Phalaenopsis*, herbarium specimens are often not preserved, the plants more often being kept alive for cultivation, hence the paucity of material in herbaria particularly of the rarer Bornean species. In common with many species in the genus it is becoming increasingly scarce, mainly because of logging activities and agricultural development. Its habitats are under the greatest pressure for timber and land development.

DERIVATION OF NAME: The specific epithet is derived from the Latin *fuscatus*, darkened, referring to the flower colour.

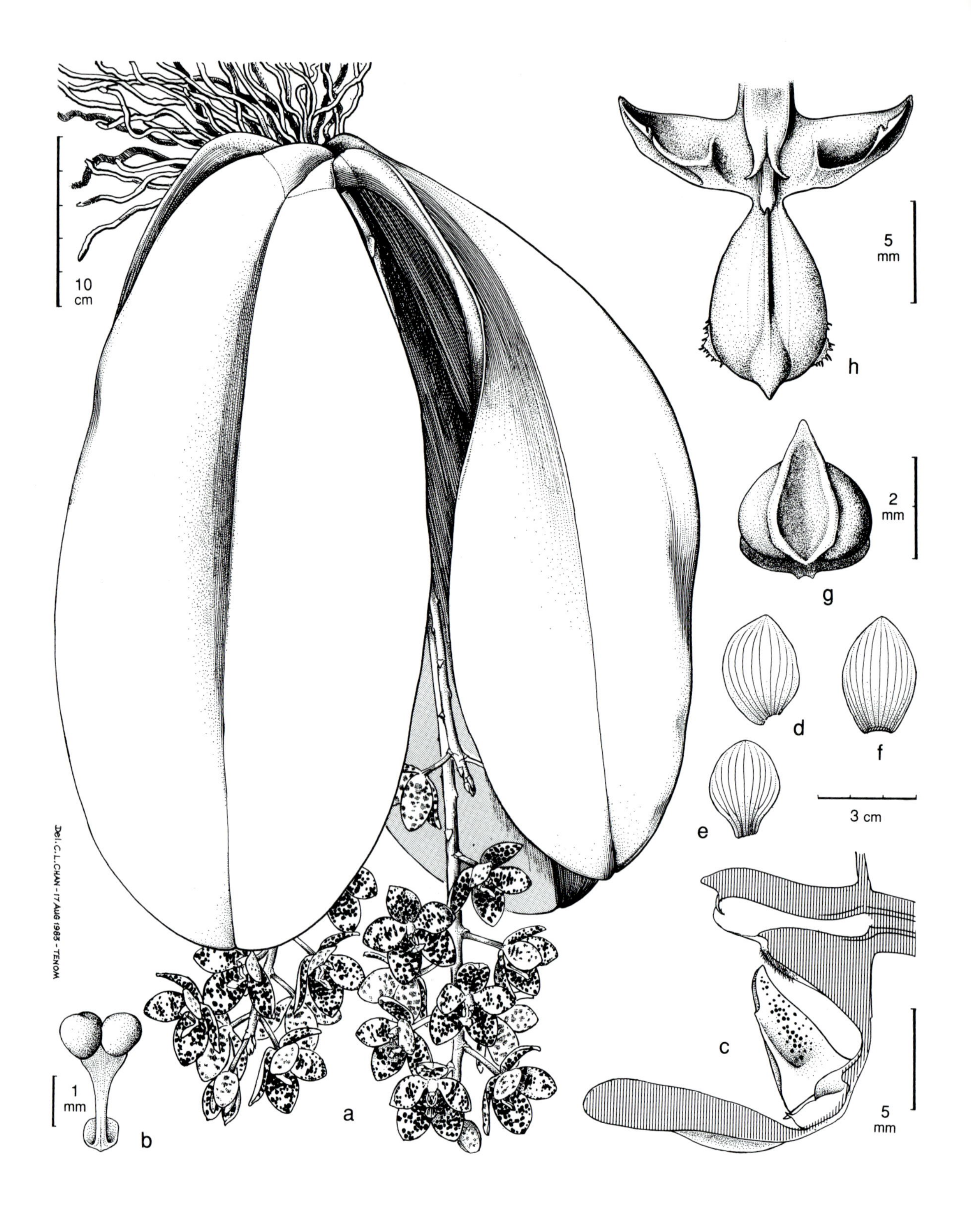

Figure 73. Phalaenopsis gigantea J.J. Sm. - A: plant. - B: pollinarium. - C: column and lip, longitudinal section. - D: lateral sepal. - E: petal. - F: dorsal sepal. G: anther. - H: lip, spread out. A from cult. TOC. B - H from *Lamb* AL 119/83. Drawn by Chan Chew Lun.

73. PHALAENOPSIS GIGANTEA J.J. Sm.

Phalaenopsis gigantea *J.J. Smith* in Bull. Dép. Agric. Indes. Néerl. 22:45 (1909). Type: Borneo, Kalimantan, without precise locality, collected during Nieuwenhuis Expedition by *Jaheri* s.n. (holotype BO).

Epiphytic ***herb***. ***Roots*** abundant, rather fleshy, glabrous. ***Stem*** very short, rather heavy in appearance, completely enclosed by approximate conduplicate leaf bases. ***Leaves*** 5-6 per plant, up to 50-68.5 x 20-25.5 cm, pendulous, rather leathery, shiny on both sides, oblong-ovate to elliptic, obtuse. ***Inflorescence*** up to 40 cm long, from axil of leaves, lateral pendulous, a densely many-flowered raceme, sometimes paniculate; floral bracts 5-6 mm long, triangular, acute, concave. ***Flowers*** c. 5 cm across, fleshy, all open simultaneously, sweetly scented; sepals and petals off-white, cream, or greenish yellow, waxy, covered with reddish brown, maroon, purple or violet brown blotches and bars, often concentrically arranged; lip white, often with three bright magenta or carmine lines or stripes on each side of median keel on mid-lobe, side lobes have orange calli and apical half of each is orange-yellow dotted with magenta, disc between side lobes orange; column white. ***Dorsal sepal*** 2.9-3.5 x 1.8-2.5 cm, spreading, elliptic or ovate-elliptic, obtuse. ***Lateral sepals*** 3.1-3.9 x 2.1-2.6 cm, fleshy, spreading, obliquely elliptic or ovate-elliptic, obtuse, base adnate to column-foot. ***Petals*** 2.7-3 x 1.8-2.2 cm, elliptic or rhombic-elliptic, acute or obtuse. ***Lip*** 1.3-1.6 x 1.3-1.5 cm, 3-lobed, rather small, fleshy; side lobes 0.6-0.8 x 0.4 cm, triangular to subfalcate, with a cushion-like callus toward centre; mid-lobe 0.9-1.1 x 0.5-0.7 cm, ovate or ovate-subrhombic in outline with a few teeth on lateral margins and with an ovoid callus at apex, at junction with side lobes adorned with a fleshy, bidentate, plate-like callus in front of which there is a thin keel in centre; disc between side lobes provided with a fleshy subcylindric callus, the apex of which bifurcates in a falcate manner. ***Column*** 0.9-1.1 cm long, fleshy, cylindric, dilated toward column-foot; pollinia 2, cleft. Plate 15D.

HABITAT AND ECOLOGY: An understorey epiphyte in lowland dipterocarp forest to hill-dipterocarp forest on sandstone or basalt-derived soils. Alt. sea level to 400 m. Flowering mainly in July and August, also in February.

DISTRIBUTION IN BORNEO: KALIMANTAN TIMUR. SABAH.

GENERAL DISTRIBUTION: Endemic to Borneo.

NOTES: Sweet (1980) places *P. gigantea* in section Amboinenses which has rotuliform flowers, broadly elliptic petals and a fleshy lip mid-lobe which is always convex and has a single median keel. The huge leaves of this plant have given it the local name "elephant ears".

DERIVATION OF NAME: The specific epithet is derived from the Latin *giganteus*, gigantic or very large, referring to the enormous leaves of this species.

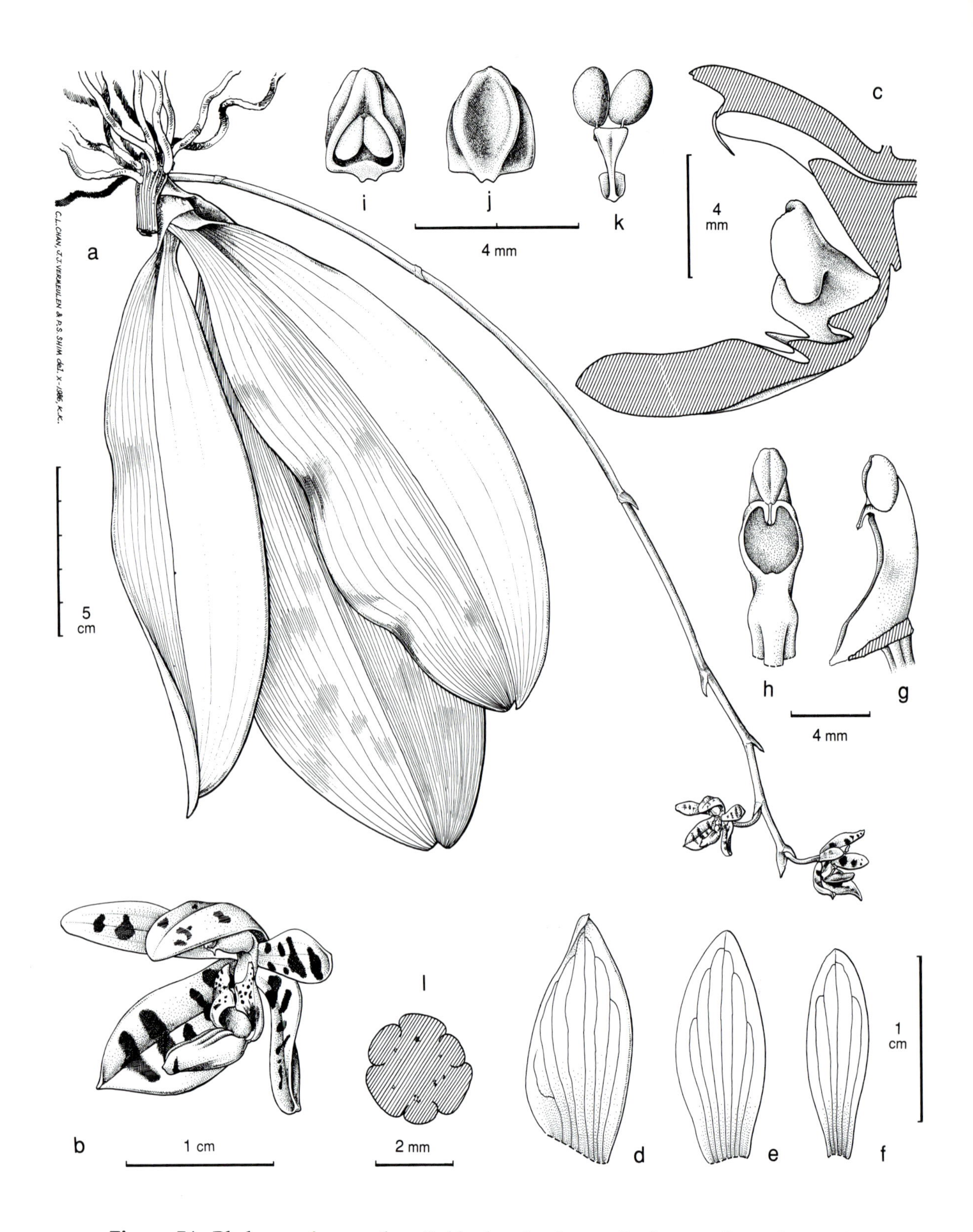

Figure 74. Phalaenopsis maculata Rchb. f. - A: plant. - B: flower , front view. - C: column and lip, longitudinal section. - D: lateral sepal. - E: dorsal sepal. - F: petal. - G: column, lateral view. - H: column, ventral view. - I: anther with pollinia, adaxially. - J: anther, abaxially. - K: pollinarium. - L: ovary, transverse section. All drawn from cult. TOC by Chan Chew Lun and Shim Phyau Soon.

74. PHALAENOPSIS MACULATA Rchb. f.

Phalaenopsis maculata *Rchb. f.* in Gard. Chron. n.s. 16:134 (1881). Type: Borneo, without precise locality, *Curtis* s.n. (holotype W).

Phalaenopsis muscicola Ridl. in Trans. Linn. Soc. London, Bot. 3: 373 (1983). Type: Peninsular Malaysia, Pahang State, Sungai Tahan, *Ridley* s.n. (holotype BM).

P. cruciata Schltr. in Feddes Repert. 8: 457 (1910). Type: Borneo, Kalimantan Timur, Koetai, near Long Sele, *Schlechter* 13480 (holotype B, destroyed).

Epiphytic ***herb***. ***Roots*** fleshy, flexuous, glabrous. ***Stem*** very short, completely enclosed by imbricate leaf sheaths. ***Leaves*** 2 or 3 per plant, 15-21 x 2.5-4 cm, distichous, fleshy, upper surface waxy, oblong-ligulate to oblong-elliptic or oblanceolate, acute at canaliculate apex, gradually narrowing toward base. ***Inflorescence*** lateral, suberect to arcuate, as long as the subtending leaves, rarely exceeding, racemose, rarely branched, few-flowered. ***Flowers*** small, delicate in texture, distichous; sepals and petals waxy cream to white often with a green suffusion blotched with brownish red to purplish red; lip mid-lobe red, side lobes white with red to reddish purple near the apex and red spots at the base, callus yellow; column white. ***Dorsal sepal*** 1.3-1.8 x 0.4-0.6 cm, oblong-lanceolate to oblong-elliptic, acute or obtuse, mucronate at tip. ***Lateral sepals*** 1.3-1.5 x 0.5-0.6 cm, somewhat oblique, ovate-oblong to elliptic, acute or obtuse, dorsally carinate, mucronate at tip. ***Petals*** 1.2 x 0.4 cm, narrowly oblong-lanceolate to oblanceolate, acute or obtuse. ***Lip*** 1 x 0.8 cm, 3-lobed, fleshy, cross-shaped; side lobes subquadrate, apical half fleshy-incrassate into a horseshoe-shaped grooved callus; disc between side lobes with a pair of approximate calli, the anterior part diverging, hence bifid; mid-lobe elliptic or oblong-elliptic, strongly convex, with indistinct longitudinal grooves, without either a prominent keel or an apical swelling or callus; disc at base in front of junction with side lobes with a short abortive bidentate callus. ***Column*** 0.7 cm long, cylindric, slightly arcuate, hardly or not at all dilated toward base; pollinia 2, cleft. Plate 15E.

HABITAT AND ECOLOGY: Normally epiphytic in lowland to mixed hill-forest on all soil types including limestone and ultramafic, preferring shaded cool, moist habitats near to streams. Alt. sea level to 1000 m. Flowering at the end of the rainy seasons.

DISTRIBUTION IN BORNEO KALIMANTAN TIMUR: Kutai, Long Sele. SABAH: Tawau District; Ranau District; Mt. Trus Madi; Danum Valley. SARAWAK: Bidi Cave; Bukit Rawan; Bau; Bukit Batu Tiban.

GENERAL DISTRIBUTION: Peninsular Malaysia and Borneo.

NOTES: *P. maculata* is placed in section Zebrinae by Sweet (1980), characterised by star-shaped flowers, with narrowly obovate petals which are twice as long as wide. This section is further divided into subsections on account of the differences in the type of callosities between the side lobes of the lip. *P. maculata* belongs in subsection Glabrae on account of its glabrous lip with a single bifid callus between the side lobes. This is one of the smallest species of *Phalaenopsis*, and is easily distinguished by its size, the few, rather small flowers on the scape and glabrous red lip. It is widespread along rivers in the foothills of Sabah and Sarawak.

DERIVATION OF NAME: The specific epithet is derived from the Latin *maculatus*, spotted or blotched, referring to the spots and blotches on the sepals and petals.

Figure 75. Phalaenopsis modesta J.J. Sm. - A: plant. - B: flower, front view. - C: column and lip, longitudinal section. - D: pollinarium, front view. - E: pollinarium, lateral view. - F: anther. - G: lip, spread out. - H: ovary, transverse section. All drawn from cult. TOC by Shim Phyau Soon and Chan Chew Lun.

75. PHALAENOPSIS MODESTA J.J. Sm.

Phalaenopsis modesta *J.J. Sm.* in Icon. Bogor. 3:47, t. 218 (1906). Type: Borneo, without precise locality, *Nieuwenhuis* s.n., cult. Bogor (holotype BO).

Polychilos modesta (J.J. Sm.) Shim in Malayan Nat. J. 36: 25 (1982).

Epiphytic ***herb***. ***Roots*** fleshy, compressed, flexuous, glabrous. ***Stem*** short, completely enclosed by imbricate leaf sheaths. ***Leaves*** 1-4 per plant, 11-20(-23) x 4-6 cm, distichous, fleshy, obovate to obovate-lanceolate or obovate-elliptic, acute, tapering to a subpetiolate base, articulate with the sheaths. ***Inflorescence*** slender, arcuate, commonly racemose, rarely subpaniculate, shorter than subtending leaves, few- to many-flowered; floral bracts distichous on a semi-fractiflex rachis, ovate-lanceolate, cucullate, to 3 mm long. ***Flowers*** rather delicate, with spreading segments, strongly scented in the early morning; sepals and petals white, pale red or purple with fine purple or green transverse bars or stripes on basal halves, turning greenish with age; lip mid-lobe white, mauve- and purple-blotched or speckled near the apex, apex white, side lobes white with dark yellow blotches in midsection and at base, callosities pale yellow; column white. ***Pedicel*** with ***ovary*** 1.5-2 cm, pale greenish white. ***Dorsal sepal*** 1.2-1.6 x 0.5-0.7 cm, oblong-elliptic to obovate, obtuse or arcuate. ***Lateral sepals*** 1.2-1.6 x 0.7-0.9 cm, oblique, ovate or ovate-elliptic, acute, dorsally carinate, mucronate at apex. ***Petals*** 1.1-1.5 x 0.4-0.5 cm, narrowly elliptic or ovate-elliptic, acute, rarely subobtuse, usually shorter than sepals. ***Lip*** 1.2-1.4 x 1.2-1.4 cm, 3-lobed; side lobes subquadrate-oblong, narrowing toward subtruncate bifid apex, 0.5-0.6 x 0.15-0.2 cm, with cushion-like callus in middle; mid-lobe oblong-elliptic to ovate-elliptic, rounded at apex, convex, with a lamella-like keel in centre from base to middle, at apex provided with an elongate cushion-like glabrous callus, very rarely decorated with a few short hairs, the disc between the side lobes provided with a pair of superimposed fleshy apically bifid calli, the anterior one grooved in middle and situated at junction of mid-lobe and side lobes. ***Column*** 0.7-0.9 cm, fleshy, cylindric, somewhat arcuate, slightly dilated toward base; pollinia 2, cleft. Plate 15F & 16A.

HABITAT AND ECOLOGY: On trees and lianas in mixed hill-dipterocarp forest. Often found along rivers in hill-forest on sandstone ridge to ultramafic substrate. Usually epiphytic very low down, even at the base of trees or on exposed roots over streams. Alt. 50 to 900 m. Flowering May and June, and September to November.

DISTRIBUTION IN BORNEO: KALIMANTAN SELATAN: locality unknown. KALIMANTAN TIMUR: Kutai. SABAH: Mt. Kinabalu; Tawau District; Tenom District; Sungai Sapulut; Papar District; Beluran District; Mt. Trus Madi.

GENERAL DISTRIBUTION: Endemic to Borneo.

DERIVATION OF NAME: The specific epithet is derived from the Latin *modestus*, unassuming or unpretentious, referring to the rather small size of the flowers, often on scapes hidden below the leaves.

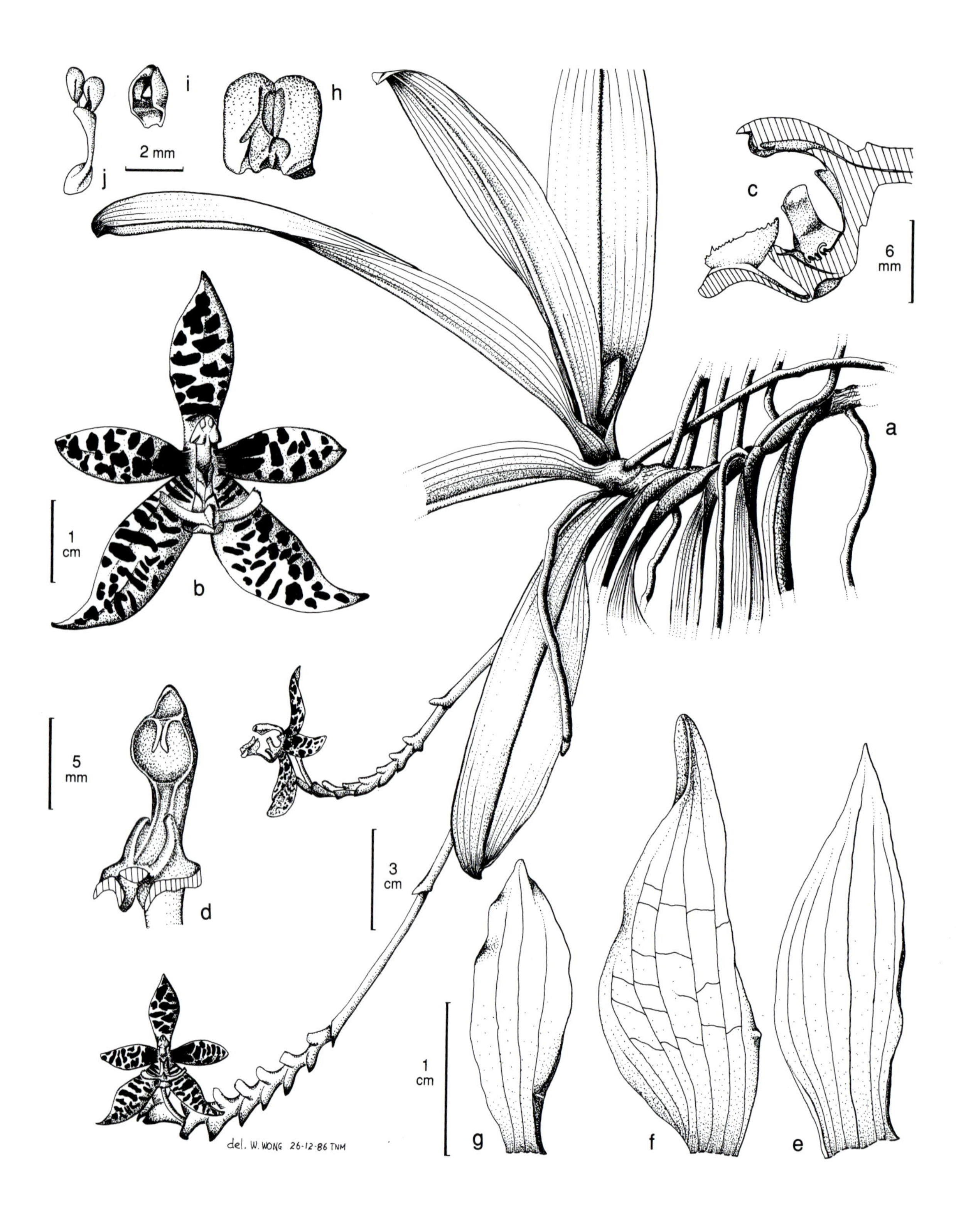

Figure 76. Phalaenopsis pantherina Rchb. f. - A: plant. - B: flower, front view. - C: column and lip, longitudinal section. - D: column, ventral view. - E: dorsal sepal. - F: lateral sepal. - G: petal. - H & I: anther. - J: pollinarium. All drawn from cult. TOC by William W.W. Wong.

76. PHALAENOPSIS PANTHERINA Rchb. f.

Phalaenopsis pantherina *Rchb.* f. in Bot. Zeitung (Berlin) 22:298 (1864). Type: Borneo, cult. *Low* (holotype W).

Phalaenopsis luteola Burb., The Garden of the Sun: 258 (1880), nomen.

Polychilos pantherina (Rchb.f.) Shim in Malayan Nat. J. 36:25 (1982).

Epiphytic ***herb***. ***Stem*** short, enclosed by imbricate leaf sheaths. ***Leaves*** 15-20 x 3-4 cm, elliptic to obovate-elliptic or oblong-lanceolate, apex obtuse. ***Inflorescence*** to 25 cm long, arcuate,terete, with a short laterally compressed, somewhat flexuous rachis, racemose to paniculate; raceme distichous, few-flowered; floral bracts ovate-cucullate, hamate, up to 5 mm long. ***Flowers*** firm, with spreading floral segments, waxy, slightly sweetly scented; sepals and petals yellow covered with red spots; lip with white to yellow side lobes, speckled and barred brownish red, with a yellow callus, and white appendages, some with reddish lilac tips; column yellow with red at the base. ***Pedicel*** with ***ovary*** to 1.5 cm long. ***Dorsal sepal*** 2.5-3.3 x 0.8-0.9 cm, narrowly lanceolate or lanceolate-elliptic, acute, dorsally distinctly carinate toward apex, margins revolute. ***Lateral sepals*** 2.5-3 x 0.8-1.1 cm, obliquely ovate-lanceolate, acute, dorsally distinctly carinate toward apex. ***Petals*** 1.5-1.9 x 0.6 x 0.7 cm, at right angles to dorsal sepal, lanceolate-elliptic, acute, dorsally distinctly carinate at apex, basally margins revolute, thus appearing long-stalked. ***Lip*** 1.5 cm long, 1 cm wide across side lobes, 1.2 cm wide across apex of mid-lobe, fleshy, 3-lobed; side lobes 0.5-0.6 x 0.2-0.3 cm, subquadrate with truncate apex below which is a fleshy callus, confluent with base of mid-lobe forming a flat, semicircular gibbosity, from which the mid-lobe extends into a prominent isthmus and abruptly expands into a transversely reniform anchor-shaped apical lobe with erose-dentate margins, 1.1-1.2 x 0.4-0.5 cm, disc at apex provided with an ovoid callus, the base of which is sparsely covered with hairs; centre of gibbosity provided with an erect, oblong-linear, laterally compressed, truncate appendage and projecting over it originating from the distal end of gibbosity is another bidentate plate-like fleshy structure, with a small tooth at each side. ***Column*** 0.8 cm long, cylindric, arcuate, somewhat dilated towards apex, basally and laterally provided with a small protuberance on each side; pollinia 2, cleft. Plate 16B.

HABITAT AND ECOLOGY: Lowland to mixed hill-dipterocarp forest on soils derived from sandstone and shales. Usually exposed to light shade on branches, in the canopy of tall trees. Alt. sea level to 800 m. Flowering January and February, July and August.

DISTRIBUTION IN BORNEO: KALIMANTAN. SABAH: Labuan; Tenom District. SARAWAK: Marudi & Miri area.

GENERAL DISTRIBUTION: Endemic to Borneo.

NOTES: Flowering once commenced extends over two months. This species is placed in section *Polychilos* together with two other Bornean species, *P. cornucervi* (Breda) Blume & Rchb. f. and *P. lamelligera* Sweet.

Like many *Phalaenopsis* it is frequently collected by local people and cultivated in coconut shells under the eaves of houses. In Sabah, plants are often collected in the Crocker Range from Tenom to Kota Belud, and in the Paling-Paling Hills. Elsewhere it appears to be scattered and rare. Similarly in Sarawak it has been collected in the hills between Miri and Marudi. This species is very close to *P. cornucervi*, with its flattened rachis, and is also called the "Crocodile's Tail" orchid.

DERIVATION OF NAME: The specific epithet is derived from the Latin *pantherinus*, panther-like, spotted like a panther, referring to the spotted flowers.

Figure 77. Phalaenopsis sumatrana Korth. & Rchb. f. - A: plant. - B: flower, front view. - C: lip, spread out. - D: pollinarium, lateral and front view. - E: ovary, transverse section. - F: anther. - G: column and lip, longitudinal section. All drawn from *Lamb* AL 174/84 by Shim Phyau Soon and Chan Chew Lun.

77. PHALAENOPSIS SUMATRANA Korth. & Rchb. f.

Phalaenopsis sumatrana Korth. & Rchb. f. in Hamburger Garten-Blumenzeitung 16:115 (1860). Type: Sumatra, without precise locality, *Korthals* drawing no. 443 (holotype K, isotype L).

Phalaenopsis zebrina Witte in Ann. Hort. Bot. Leiden 4:145, t. (1860); Teijsm. & Binn. in Natuurk. Tijdschr. Ned.-Indië 24: 319 (1862). Type: Sumatra, Palembang, *Gersen* s.n. (holotype BO).

Polychilos sumatrana (Korth. & Rchb. f.) Shim in Malayan Nat. J. 36:26 (1982).

Robust epiphytic ***herb***. ***Roots*** many, flexuous, glabrous. ***Stem*** short, completely covered by persistent leaf sheaths. ***Leaves*** 15-32 x 4-11 cm, fleshy, arcuate or pendent, oblong-elliptic to obovate, acute or rounded, slightly tapering toward articulate base. ***Inflorescence*** erect or slightly arcuate, a loosely, several- to many-flowered raceme, occasionally with a few branches, up to 30 cm long including a slightly fractiflex rachis; floral bracts triangular, cucullate, up to 6 mm long, shorter than pedicel with ovary. ***Flowers*** fleshy, 4-5 cm across, unscented; sepals and petals waxy, spreading, white or cream with red bars and white hairs or blotches; lip white with yellow margins on the side lobes and magenta, purple to red bars on the mid-lobe; column white. ***Pedicel*** with ***ovary*** 2.5 cm, white. ***Dorsal sepal*** (1.3-)2-4 x (6-)1-1.5 cm, oblong-lanceolate to elliptic-lanceolate, acute, canaliculate above, carinate dorsally. ***Lateral sepals*** 2.5-2.8 x 1-1.5 cm, similar to dorsal sepal in size and shape, but slightly oblique. ***Petals*** (1.3-)2.2-3.5 x (6-)9-1.2 cm, lanceolate, acute, somewhat smaller than sepals. ***Lip*** (1.5-)2.5 x (1.2-)1.5 cm, 3-lobed; side lobes linear-oblong with obliquely truncate semi-falcate tip; mid-lobe convex, with a median keel, oblong-elliptic, heavily thickened toward apex, cushion-like, densely hirsute; disc between side lobes with a variable multi-digitate callus; junction of mid-lobe and side lobes with a fleshy bifurcate plate-like callus. ***Column*** (0.8-)1.5 cm long, fleshy, arcuate; clinandrium well-developed, cucullate, with pronounced erose-dentate margins; pollinia 2, cleft. Plate 16C, D & E.

N.B. Figures in brackets refer to *Grell* s.n. (K), from Sabah which has smaller flowers than other material at Kew and measurements given by Sweet (1980).

HABITAT AND ECOLOGY: Lowland to mixed hill-forest, epiphytic low down on understorey trees or on tree trunks, usually in deep shade, particularly beside streams or rivers. Alt. sea level to 700 m. Flowering sporadic, recorded in March and October.

DISTRIBUTION IN BORNEO: KALIMANTAN BARAT: Pontianak area. SABAH: Keningau District; Tenom District. SARAWAK: Lawas River.

GENERAL DISTRIBUTION: Burma, Thailand, Peninsular Malaysia, Sumatra and Borneo.

NOTES: *P. sumatrana* is the type of section Zebrinae and subsection Zebrinae, which contains four other Borneo species, viz. *P. corningiana* Rchb.f., *P. maculata* Rchb.f., *P. modesta* J.J. Sm. and *P. violacea* Witte. Logging and clearance for agriculture have seriously depleted its known habitats in Borneo. The closely related *P. corningiana* is distinguished from *P. sumatrana* by the apical halves of the sepals and petals which are barred or striped, longitudinally with cinnamon to red or crimson. The bars and stripes are crosswise in *P. sumatrana*.

DERIVATION OF NAME: The specific epithet refers to the island of Sumatra from where the type orginated.

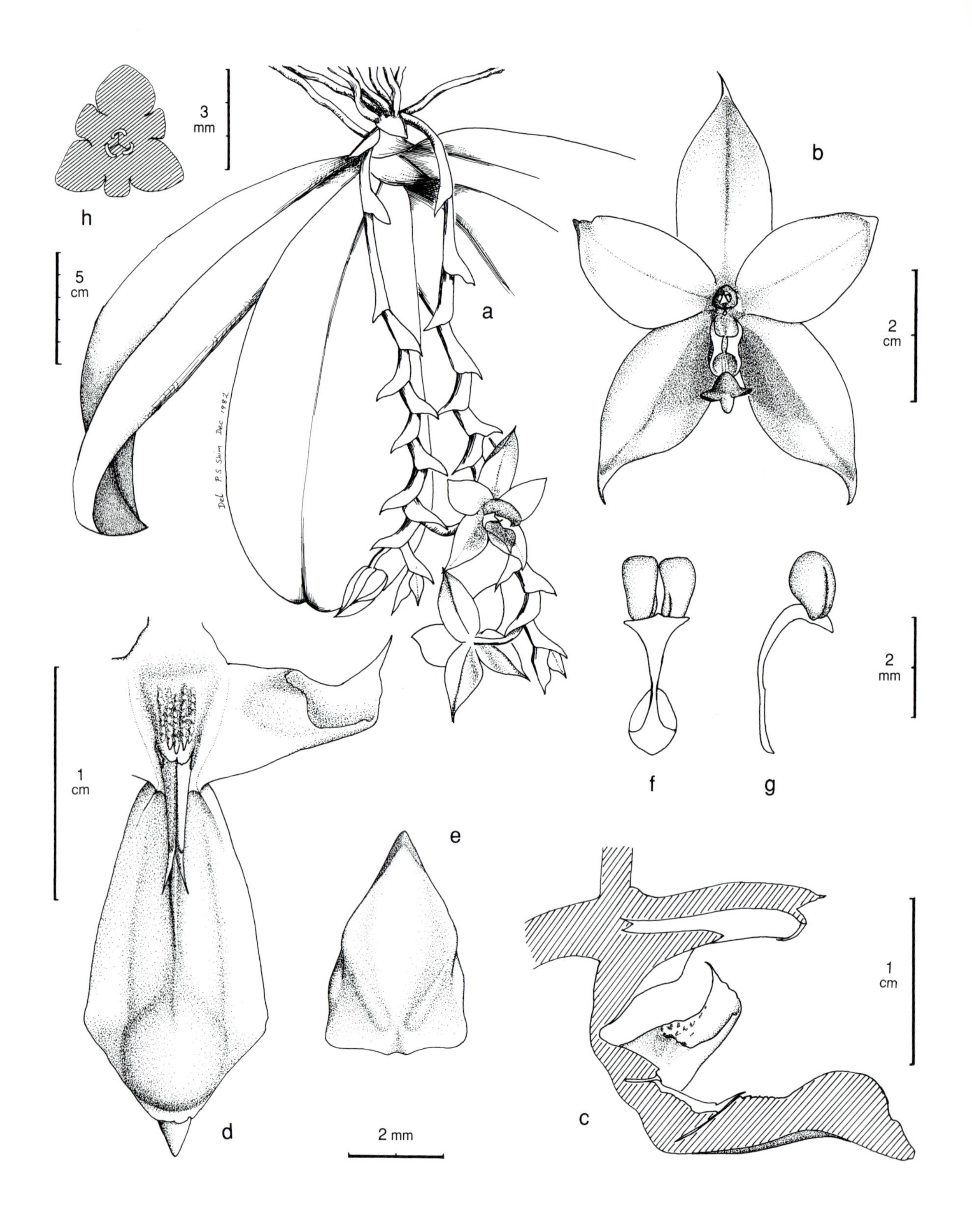

Figure 78. Phalaenopsis violacea Witte - A: plant. - B: flower, front view. - C: column and lip, longitudinal section. - D: lip, spread out. - E: anther. - F: pollinarium, front view. - G: pollinarium, lateral view. - H: ovary, transverse section. All drawn from cult. TOC by Shim Phyau Soon.

78. PHALAENOPSIS VIOLACEA Witte

Phalaenopsis violacea *Witte* in Ann. Hort. Bot. Leiden 4: 129 (1860). Type: Sumatra, Palembang, *Teysmann* s.n., cult. Bogor and Leiden, icon. *Witte* (holotype L).

Phalaenopsis violacea Witte var. *alba* Teijsm. & Binn. in Natuurk. Tijdschr. Ned.-Indië 24: 320 (1862). Type: Sumatra, Palembang, *Teysmann* s.n. (holotype ? L).

Stauritis violacea (Witte) Rchb.f. in Hamburger Garten-Blumenzeitung 18: 34 (1862).

Stauropsis violacea Rchb.f., Xenia Orch. 2: 7 (1862). Type: Sumatra, introduced and flowered by *Willink* (holotype W).

Polychilos violacea (Witte) Shim in Malayan Nat. J. 36: 27 (1982).

Epiphytic ***herb***. ***Roots*** heavy, flexuous, glabrous. ***Stem*** erect or ascending, strong, well-developed, completely enclosed by distichous imbricate leaf sheaths. ***Leaves*** 20-25 x 7-12 cm, fleshy or leathery, waxy and shining, elliptic, obovate to oblong-elliptic, acute or obtuse at apex, tapered toward conduplicate base, articulate with leaf sheaths. ***Inflorescence*** robust, suberect or arcuate with prominently fractiflex rachis, remotely few-flowered, flowers produced 1-2 at a time, eventually up to 20 or more flowers per scape; floral bracts ovate, cucullate, acute, to 7 mm long. ***Flowers*** fleshy, waxy, scented; sepals and petals white to creamy yellow or creamy green, lateral sepals almost covered by a large purple blotch; lip side lobes yellow with magenta tips and reddish purple dots on the inner surface, apex yellow to off-white, mid-lobe magenta to purple; column magenta to purple. ***Dorsal sepal*** (1.2-)2-3.6 x (0.5-)0.8-1.5 cm, oblong-elliptic or elliptic to ovate or ovate-lanceolate, somewhat concave, prominently carinate dorsally. ***Lateral sepals*** (1.2-)2.2-3.5 x (0.7-)1.1-1.7 cm, obliquely adnate to base of column, ovate-lanceolate, subfalcate above with a conduplicate apex, prominently carinate dorsally. ***Petals*** (1.1-)2-3 x (0.4-)0.7-1.7 cm, elliptic or ovate-elliptic with an asymmetrical base, acute to subobtuse at apex. ***Lip*** (1.1-)2-2.8 x (0.8-)1.6-2.3 cm, 3-lobed; side lobes 0.7 x 0.5 cm, erect, linear-oblong, truncate at apex with a semilunate thickening in middle; mid-lobe ovate, apiculate, fleshy, convex, with a median lamellate keel tapering into an ovoid thickening or callus just below apex, completely glabrous, very rarely with a few warts on apical callus; disc between side lobes of lip pulvinately incrassate, papillate in front, merging into 1 to 3 superimposed, often divided, fleshy lamellae; junction of mid-lobe and side lobes provided with an elevated bifid fleshy projection prominently grooved in centre. ***Column*** (0.7-)1.5 cm long, fleshy, slightly arcuate, cylindric, dilated toward base. Plate 16F & 17A.

NB. figures in brackets refer to *Clements* 3372 (K), from Sabah, which has smaller flowers than other material at Kew and measurements given by Sweet (1980).

HABITAT AND ECOLOGY: Trees beside rivers and streams in mixed primary lowland forest and swampy riverine forest. It generally grows low down on trees, or on branches overhanging streams as well as on lianas, requiring shade in a moist habitat with a uniform high rainfall of over 2500 mm per year. Alt. below 200 m. Flowering January-April, July-September.

DISTRIBUTION IN BORNEO: KALIMANTAN. SARAWAK: Lundu District.

GENERAL DISTRIBUTION: Peninsular Malaysia, Sumatra, Borneo, Mentawai & Simeuluë.

NOTES: *P. violacea* is placed in section Zebrinae subsection Lueddemannianae by Sweet (1980). This species is almost extinct in the wild in Borneo, but is cultivated worldwide. It is commonly called the "Lundu Orchid" and is the State flower of Sarawak.

The Bornean plants differ slightly from Peninsular Malaysian and Sumatran ones and may warrant separate taxonomic recognition.

The variety *alba* was recorded from Sarawak by Au Yong Nang Yip who cultivates plants from the Lundu District. It is distinguished from the typical plant by its waxy white sepals and petals with greenish tips, white lip, yellow callus and yellow base to the side lobes.

DERIVATION OF NAME: The specific epithet is derived from the Latin *violaceus*, violet, referring to the colour of the flowers.

Figure 79. Plocoglottis acuminata Blume - A: plant. - B: flower, front view. - C: flower, lateral view. - D: petal. - E: lateral sepal. - F: dorsal sepal. - G: lip, abaxially. - H: lip, adaxially. - I: anther, adaxially. - J: anther, abaxially. - K: pollinia. - L: ovary, transverse section. All from living plant collected from Sapong Waterfall, Tenom. Drawn by Chan Chew Lun and Jaap J. Vermeulen.

79. PLOCOGLOTTIS ACUMINATA Blume

Plocoglottis acuminata *Blume*, Mus. Bot. Ludg. Bat. 1: 46 (1849). Types: Java & Sumatra, *Blume* s.n. (syntypes L).

Medium-sized to large terrestrial ***herb*** with a stout creeping rhizome. ***Pseudobulbs*** 5-8 cm long, 0.3-0.7 cm in diameter, cylindrical, several-noded, unifoliate at apex. ***Leaf*** 18-40 x 5.5-10 cm, erect, elliptic to ovate, acuminate; petiole slender, 2-14 cm long. ***Inflorescence*** erect, lateral from the flower nodes of the pseudobulb, laxly to densely few- to many-flowered, up to 50 cm long; peduncle and rachis terete, densely shortly pubescent; floral bracts lanceolate, acuminate, 0.6-1 cm long. ***Flowers*** yellow with red spots on the sepals and petals. ***Pedicel*** with ***ovary*** 1.2-1.7 cm long, shortly pubescent. ***Dorsal sepal*** 1.3-2.2 x 0.3-0.4 cm, spreading, lanceolate, acuminate. ***Lateral sepals*** 2.5 x 0.9 cm, slightly falcate and positioned back-to-back. ***Petals*** 1.5-1.7 x 1.5-0.3 cm, slightly oblique, lanceolate, acuminate. ***Lip*** 0.7-1 cm long and wide, strongly deflexed, recurved, convex, subquadrate, shortly apiculate. ***Column*** 0.4-0.5 cm long, erect; pollinia 4. Plate 17B.

HABITAT AND ECOLOGY: Lowland dipterocarp to mixed hill-dipterocarp forest on alluvium and alluvial soils derived from sandstones, shales and mudstones, or limestone; sometimes in secondary forest. Found on well-drained soils on hill slopes above rivers to seasonally flooded alluvial flats. Prefers dense shade and moist, humid conditions. Alt. sea level to 1000 m. Flowering May-July, October-December.

DISTRIBUTION IN BORNEO: BRUNEI: Belait District. KALIMANTAN: Belajan River. SABAH: Tenom District; Sipitang District, Mt. Lumaku; Mt. Kinabalu. SARAWAK: Mt. Dulit; Bidi Cave; Mt. Mulu National Park.

GENERAL DISTRIBUTION: Sumatra, Java, Borneo and the Philippines.

One of the "mousetrap" orchids, though the insect pollinator is not recorded. The lip is strongly hinged and, after being "sprung" by a visiting insect, snaps up against the column (a human hand can reset it). It is distinguished from the closely related *P. javanica* Blume by its longer, narrower sepals and petals. *P. acuminata*, a common and widespread orchid in Borneo, is free flowering and showy. Some plants from Sabah and Sarawak have large round cream to yellow spots on the leaves.

DERIVATION OF NAME: The generic name is derived from the Greek *ploke*, binding or plaiting together, and *glotta*, tongue, alluding to the lip which is shortly joined with the base of the column, to which it is joined on each side by a membranous fold. The specific epithet is derived from the Latin *acuminatus*, tapering abruptly to a narrow point, referring to the narrow, pointed sepals and petals.

Figure 80. Porphyroglottis maxwelliae Ridl. - A: plant. - B: seed pod. - C: flower, fron view. - D: flower, lateral view. - E: column and lip, lateral view. - F: anther, lateral view - G: anther, abaxially. - H: pollinarium. - I: ovary, transverse section. A from cult. TOC. I - I from *Lamb* AL 121/83. Drawn by Chan Chew Lun.

80. PORPHYROGLOTTIS MAXWELLIAE Ridl.

Porphyroglottis maxwelliae *Ridl.* in J. Linn. Soc., Bot. 31: 290, pl. 15 (1896). Type: Borneo, Sarawak, *Maxwell* s.n. (holotype SING).

Large epiphytic ***herb***. ***Pseudobulbs*** up to 120 cm long, clustered, elongate, cane-like, leafy. ***Leaves*** up to 30 x 0.8-1 cm, distichous, linear-ligulate, acuminate. ***Inflorescence*** lateral, erect to porrect, simple or sometimes with a lateral branch, slender, up to 150 cm or more long, bearing a succession of many flowers, only one open at a time; floral bracts small, ovate, adpressed to rachis. ***Flowers*** non-resupinate, unscented; sepals and petals pale pinkish; lip purple-brown outside, with a large yellow apical patch; column whitish, spotted purple. ***Pedicel*** with ***ovary*** 3-4 cm long, slender. ***Sepals*** and ***petals*** reflexed, lying against pedicel with ovary. ***Sepals*** 2.4-2.8 x 0.8-1 cm, oblong-elliptic, obtuse. ***Petals*** a little smaller than sepals, oblong-elliptic, obtuse. ***Lip*** 1.6-1.8 cm long, 0.5-0.6 cm wide at base, 0.6-1 cm wide at middle and apex, attached to column-foot by a strap-like hinge, mobile, broadly obovate, rounded, entire, sides deflexed so that the whole is convex when viewed from above, scoop-shaped when viewed from below, hairy, bee-like. ***Column*** 1-1.8 cm long, curved, hollowed up the front, with 2 large curved, triangular acute, spreading arms at about the middle; stigma round, much broader than anther; anther-cap ovate, cucullate; pollinia 2, large, cleft, attached by flattened caudicles to a broad, round viscidium. ***Fruit*** 7-10 cm long. Plate 17C & D.

HABITAT AND ECOLOGY: Open, stunted, low crowned heath forest on podsolic soils, often growing in the forks of tree trunks, particularly of *Tristaniopsis* spp. with many short stiff roots pointing upwards into the air which act as traps for dead leaves and twigs falling from the canopy above. This root system is also home to fierce biting ants, which are associated with the roots of other orchids and ant plants in this nutrient-poor environment. Flowers are produced one at a time and appear to mimic a large solitary bee resting or waiting to attract a mate. Alt. sea level to 500 m. Flowering in August.

DISTRIBUTION IN BORNEO: KALIMANTAN BARAT: Pontianak area. SABAH: Nabawan area. SARAWAK: Kuching area.

GENERAL DISTRIBUTION: Peninsular Malaysia, Sumatra and Borneo.

NOTES: Non-flowering specimens of this strange orchid can easily be mistaken for young plants of *Grammatophyllum speciosum*, which occasionally occurs in the same habitat. The long arching inflorescence bear flowers which open one at a time along the rachis, each twisting so that they are above the rachis. On the first day the sepals and petals reflex back along the ovary and together with the purple-black and yellow hairy lip, look like a large solitary bee sitting on a twig, resting or waiting for a mate. Any insect alighting would be tipped by the delicately balanced lip onto the column, effecting pollination. The next day the sepals and petals open fully to a spreading position, and it may be that unpollinated flowers can be self-pollinated by the bending over of the apex of the column. On the third day the flower has closed and the next flower is open. The pollinator has not yet been recorded, but Holttum (1953) suggests carpenter bees (*Xylocopa* spp.). This is a very rare plant in Sabah and Sarawak where its habitat is endangered. However, in south, central and west Kalimantan, there are patches of similar podsol forest that have hardly been visited by botanists, which would provide a suitable habitat.

DERIVATION OF NAME: The generic name is derived from the Greek *porphyra*, purple, and *glotta*, tongue, describing the dark purple lip. The specific epithet honours Mrs Maxwell who collected the type.

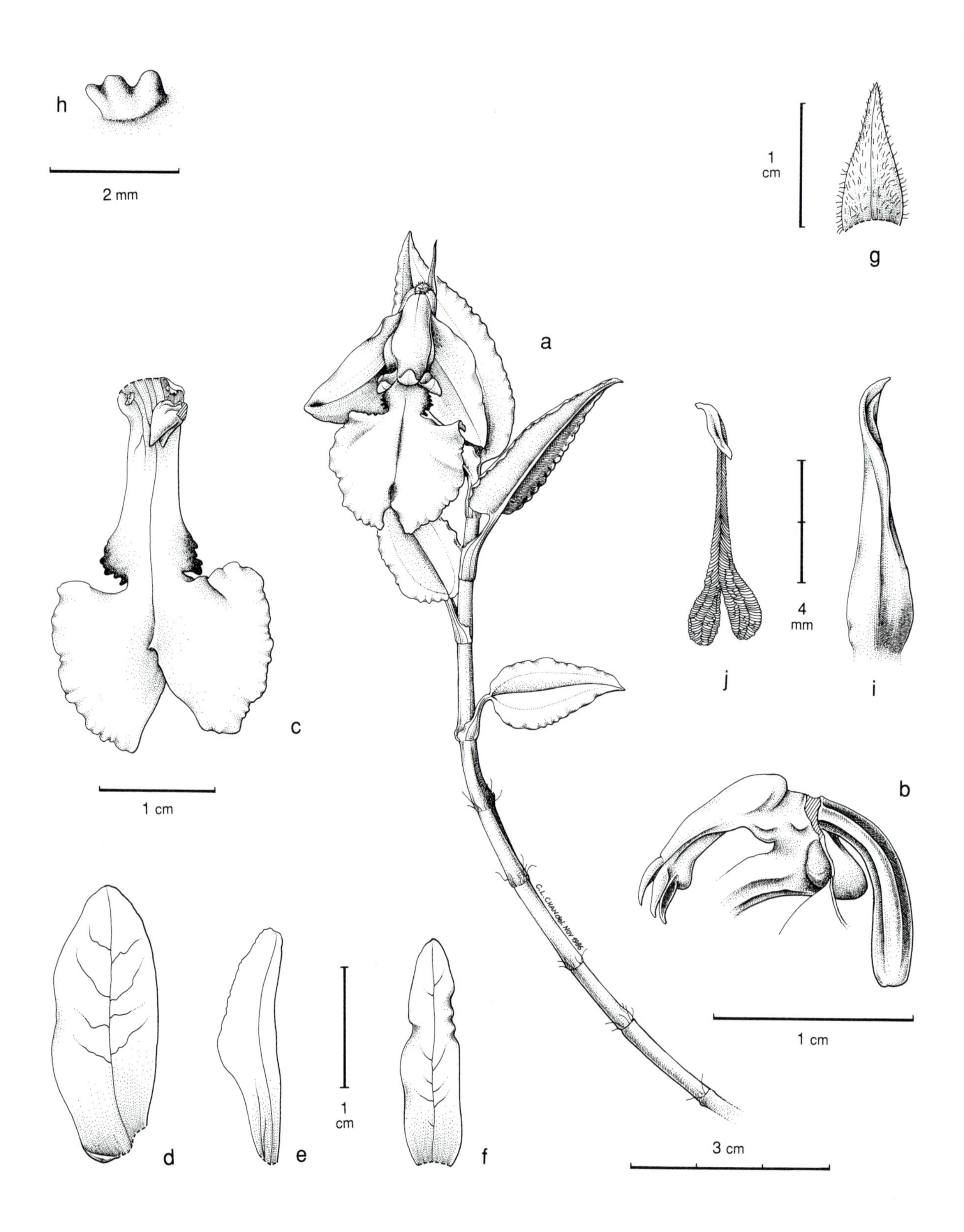

Figure 81. Pristiglottis hasseltii (Blume) Cretz. & J.J. Sm. - A: plant. - B: column, lip base and pedicel, lateral view. - C: lip. - D: lateral sepal. - E: petal. - F: dorsal sepal. - G: floral bract. - H: gland inside the lip. - I: anther. - J: pollinarium. All drawn from *Lamb* AL 633/86 by Chan Chew Lun.

81. PRISTIGLOTTIS HASSELTII (Blume) Cretz. & J.J.Sm.

Pristiglottis hasseltii (*Blume*) *Cretz. & J.J.Sm.* in Acta Fauna Fl. Universali, ser. 2, Bot., 1, no. 14: 4 (1934). Type: Java, Bantam Province, *van Hasselt* s.n. (holotype BO).

Cystopus hasseltii Blume, Fl. Javae, ser. 2, 1, Orch. (Coll. Orch. Arch. Ind.): 86, t. 30, fig. 4, t. 36B (1858).

Anoectochilus hasseltii (Blume) Miq., Fl. Ind. Bat. 3: 734 (1859).

Small terrestrial ***herb***. ***Rhizome*** subterranean, to 6 cm, producing erect, aerial stems at intervals. ***Stems*** 3-6 cm high, green, glabrous, slightly angular, leafy. ***Leaves*** 2-4.5 x 0.7-1.8 cm, ovate-oblong, acute, margin often undulate, sheathing petiole 0.5-1 cm long, dark green to greenish purple, often with a pink midnerve above, dark purple below. ***Inflorescence*** very short, to 2 cm, greenish white, pubescent, usually 2-flowered. ***Flowers*** white with a bright dark green area around base of lip. ***Sepals*** oblong-elliptic or linear-elliptic, obtuse, pubescent on outside. ***Dorsal sepal*** 1.3-2.3 x 0.4-0.6 cm, connivent with petals to form a hood. ***Lateral sepals*** 1.5-2.7 x 1.2 cm, spreading. ***Petals*** 1.2 x 0.3-0.5 cm, oblong to oblong-elliptic, obtuse. ***Lip*** 2.5-3 cm long, saccate base 0.4-0.5 cm long, claw 1 cm long, 0.5 cm wide at base, sides erect, with 3-5 short dark green crenations each side, expanding into a shallowly bilobed 2-3.5 x 1.5-2.4 cm blade, lobules with irregular outer margins. ***Column*** 0.5 cm long, with 2 parallel anterior lamellae; pollinia 2. Plate 17E.

HABITAT AND ECOLOGY: Lower montane cloud forest, either in leaf litter or in moss, or on mossy stream banks with rocks. Usually on sandstone formations. Alt. 1200 to 1600 m. Flowering observed in August and November.

DISTRIBUTION IN BORNEO: SABAH: Mt. Kinabalu; Kimanis Road. SARAWAK: Mt. Matang; Bukit Lambir; Mt. Batu Lawi area.

GENERAL DISTRIBUTION: Java and Borneo.

NOTES: This small, pretty terrestrial is striking for the contrasting leaves with deep purple undersides, and the one to two white flowers which are large for the size of the plant. The bright green base of the lip also contrasts vividly with the pure white of the remainder of the flower.

DERIVATION OF NAME: The generic name is derived from the Greek *pristes*, saw, and *glotta*, tongue, referring to the crenate margin of the lip. The specific epithet honours J.C. van Hasselt who collected the type.

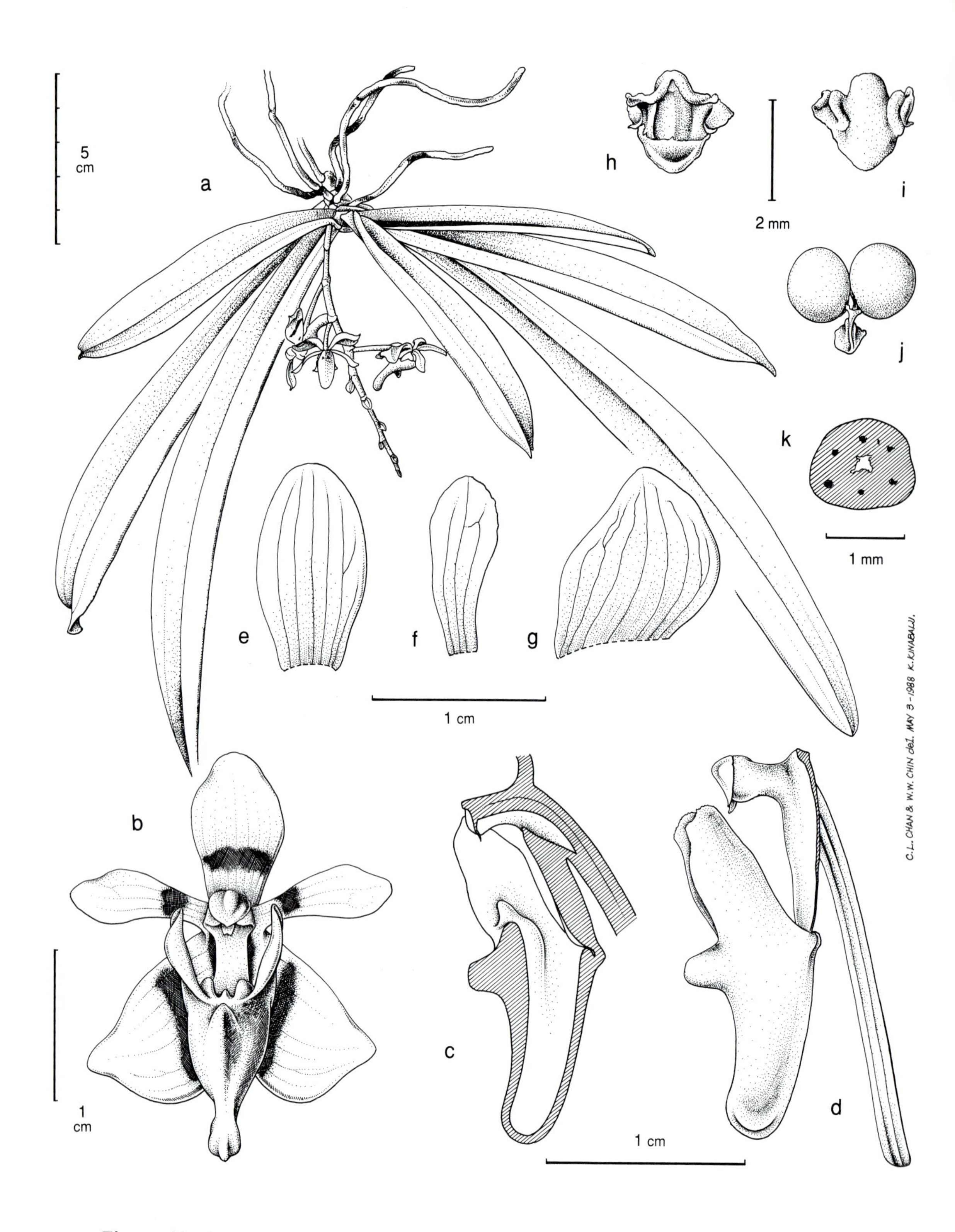

Figure 82. Pteroceras fragrans (Ridl.) Garay - A: plant. - B: flower, front view. - C: column and lip, longitudinal section. - D: flower with sepals and petals removed, lateral view. - E: dorsal sepal. - F: petal. - G: lateral sepal. - H: anther, adaxially. - I: anther, abaxially. - J: pollinarium. - K: ovary, transverse section. All drawn from cult. TOC by Chan Chew Lun and Chin Wan Wai.

82. PTEROCERAS FRAGRANS (Ridl.) Garay

Pteroceras fragrans (*Ridl.*) *Garay* in Bot. Mus. Leafl. 23(4): 193 (1972). Type: Borneo, Sarawak, Matang estate, *Ridley* s.n. (holotype SING).

Sarcochilus fragrans Ridl. in J. Straits Branch. Roy. Asiat. Soc. 49: 38 (1907).

Sarcochilus spathipetalus J.J.Sm. in Bull. Jard. Bot. Buitenzorg, ser. 3, 11: 151 (1931). Type: Borneo, Kalimantan Timur, Long Temelen, Sungai Mahakam, *Endert* 2851 (holotype L).

Pteroceras spathipetalum (J.J.Sm.) Garay in Bot. Mus. Leafl. 23(4): 194 (1972).

Epiphytic ***herb***. ***Stem*** 1.7-6.5 cm long. ***Leaves*** 5-9 per plant, 7.8-17 x 1.2-2.4 cm, linear-oblong to linear-lanceolate, somewhat oblique or falcate, apex unequally bilobed. ***Inflorescences*** up to 3 per shoot; peduncle 1.4-2 cm long; rachis 2.6-12.5 cm long, terete, at each flower node dilated into two parallel, semi-lanceolate, entire wings or keels; floral bracts 1.2-2.5 mm long, scale-like. ***Flowers*** (3-)4-16(-20) per rachis, 1-2 open at a time, resupinate, usually sweetly scented; sepals and petals white to pale yellowish green with 1-2 transverse ochre-yellow to brownish red bars on inner surface, markings rarely absent; lip with white side lobes, mid-lobe yellow, spur white to pale greenish yellow; column white to pale green, foot with a brownish red line on each side running up and meeting behind column. ***Dorsal sepal*** 0.7-0.95 x 0.3-0.7 cm, narrowly obovate to obovate-elliptic, obtuse, usually somewhat carinate. ***Lateral sepals*** adnate to column-foot for its entire length, 0.56-1 x 0.54-0.9 cm, broadly elliptic to broadly ovate, oblique, obtuse, carinate. ***Petals*** 0.6-0.9 x 0.2-0.4 cm, subspathulate, apex rounded, margins slightly wavy. ***Lip*** 0.75-1 cm long, sessile; side lobes 0.3-0.5 x 0.23-0.45 cm, broadly oblong to somewhat triangular; mid-lobe represented by a prominent, obliquely cone-shaped callus proximally on the upper spur wall, 0.2-0.25 x 0.15-0.18 cm; 3 small bulb-like calli present between bases of side lobes; spur (conical-)cylindric. ***Column*** 0.3-0.4 cm high, column-foot 0.4-0.6 cm long; anther-cap transversely elliptic with an obtuse, triangular apex, with a horn-like dorsal keel; pollinia 2, sulcate; stipes linear oblong to subspathulate. ***Fruit*** up to 21.5 cm long. Plate 17F.

HABITAT AND ECOLOGY: Lowland and hill-dipterocarp forest, and mixed hill-forest on sandstone-derived soils, alluvial soils and ultramafic substrate. Often collected on the drier rain shadow side of the mountain ranges with a rainfall of 1500-2000 mm. Usually epiphytic on small branches and understorey trees, in dense shade and humid conditions. Alt. 400 to 1000 m.

DISTRIBUTION IN BORNEO: KALIMANTAN TIMUR: Long Temelen area. SABAH: Mt. Kinabalu; Tambunan District; Tenom District. SARAWAK: Mt. Matang.

GENERAL DISTRIBUTION: Endemic to Borneo.

DERIVATION OF NAME: The generic name is derived from the Greek *pteron*, feather or quill, and *keras*, horn, which describes the two narrow wing-like appendages at the base of the lip. The specific epithet is derived from the Latin *fragrans*, fragrant, sweet-smelling, referring to the flowers.

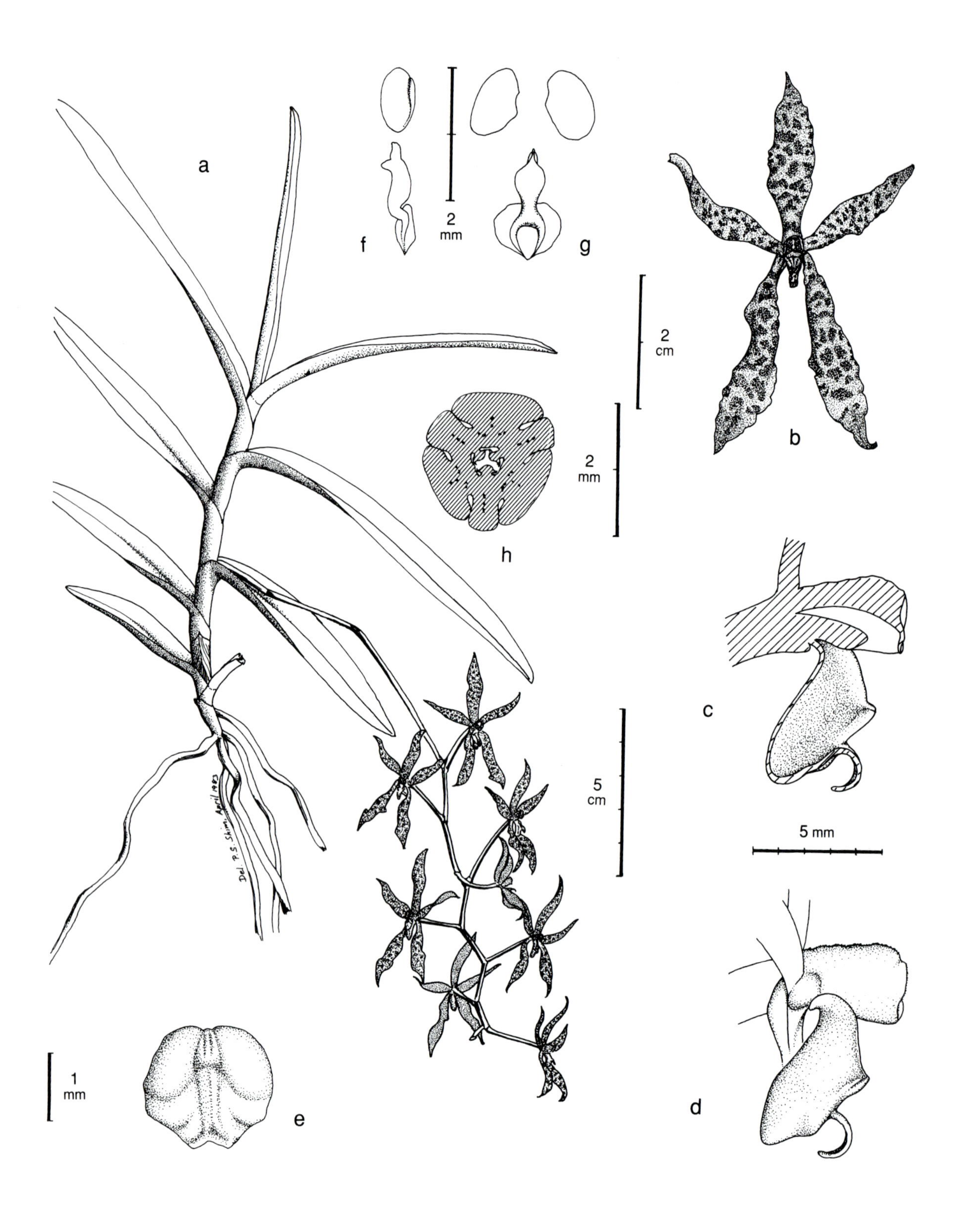

Figure 83. Renanthera bella J.J. Wood - A: plant. - B: flower, front view. - C: column and lip, longitudinal section. - D: column and lip, lateral view. - E: anther. - F: pollinarium, lateral view. - G: pollinarium, front view. - H: ovary, transverse section. All drawn from cult. TOC by Shim Phyau Soon.

83. RENANTHERA BELLA J.J.Wood

Renanthera bella *J.J.Wood* in Orchid Rev. 89: 116 (1981). Type: Borneo, Sabah, Mt. Kinabalu, Hempuen Hill, 15 July 1979, *Lamb* SAN 89640 (holotype K).

Epiphytic ***herb*** to 75 cm long including inflorescence. ***Aerial roots*** up to 38 x 0.3-0.4 cm long, produced from each node, verruculose. ***Stem*** simple, 20-50 cm long, enclosed in persistent leaf sheaths below, leafy above. ***Leaves*** 12-13 x 1-1.1 cm, ligulate, unequally bilobed, articulated to a sheathing base 1.5-2 cm long, coriaceous. ***Inflorescence*** 13- to 16(-25)-flowered, lax, simple, occasionally with 1 or 2 branches, ± horizontal or gently drooping; peduncle 17-28 cm long, with 3-5 small greyish-brown, ovate, obtuse sterile bracts; rachis 14-19 cm long; floral bracts 2-3 mm long, ovate, obtuse. ***Flowers*** with sepals and petals yellowish-cream below, with pink or mid crimson-red blotches above, pale yellow to apricot at the base; lip yellow with many dark red and a few purple-red blotches; column yellow suffused dark red or entirely purple-black, anther-cap dark red with a yellow median ridge. ***Pedicel*** with ***ovary*** 2-3.3 cm long. ***Dorsal sepal*** 3.2 x 1.1 cm, narrowly elliptic, undulate, cuneate at base, apex acute and strongly cucullate. ***Lateral sepals*** 2.6-3.7 x 1.1-1.2 cm, narrowly elliptic, clawed at base, apex carinate, mucronate, attenuate, deflexed, usually parallel and often overlapping distally, margins recurved, undulate. ***Petals*** 2.6-2.8 x 0.8-0.9 cm, narrowly elliptic, often rather oblique, sometimes carinate, apex strongly or weakly cucullate, slightly undulate. ***Lip*** 0.6-0.65 cm long, 0.6 cm broad at base, obscurely 3-lobed, fleshy, glabrous; side lobes 0.3 cm long, oblong, rounded, clasping column, slightly spreading at base; mid-lobe 0.13-0.2 cm long, oblong, obtuse, minutely mucronate, reflexed; spur 0.3-0.35 cm long, conical, obtuse; callus high, oblong, on either side at base of side lobes at junction of mid- and side lobes. ***Column*** 0.5 x 0.45 cm, oblong, apex papillose; pollinia 4, appearing as 2 pollen masses. Plate 18A.

HABITAT AND ECOLOGY: Confined to hill-forest on ultramafic substrate, eg. serpentine rock, commonly epiphytic on *Gymnostoma sumatrana*, in slightly shaded positions, low down on trunks, to 10 metres above ground. Alt. 800 to 1100 m. Flowering February and March, July-September.

DISTRIBUTION IN BORNEO: SABAH: Mt. Kinabalu; Lahad Datu District.

GENERAL DISTRIBUTION: Endemic to Borneo (Sabah only).

NOTES: This orchid has been widely collected and is now both rare and endangered in the wild. To protect it further it has been placed on CITES (Appendix I), for which no trade is allowed except for nursery-raised flasked seedlings. Because of its compact habit compared to most *Renanthera* species, and its long-lasting, medium-sized, deep red flowers it has great potential as an ornamental plant, and as a parent in hybridisation for the cut-flower trade. Several hybrids have been made but seedling growth has been found to be very slow. The flowers are long-lasting and if not pollinated a spray will last four weeks or more.

DERIVATION OF NAME: The generic name is derived from the Latin *renes*, kidney, and the Greek *anthera*, anther, describing the kidney-shaped pollinia of the type species of the genus. The specific epithet is derived from the Latin *bellus*, beautiful, referring to the attractive flowers.

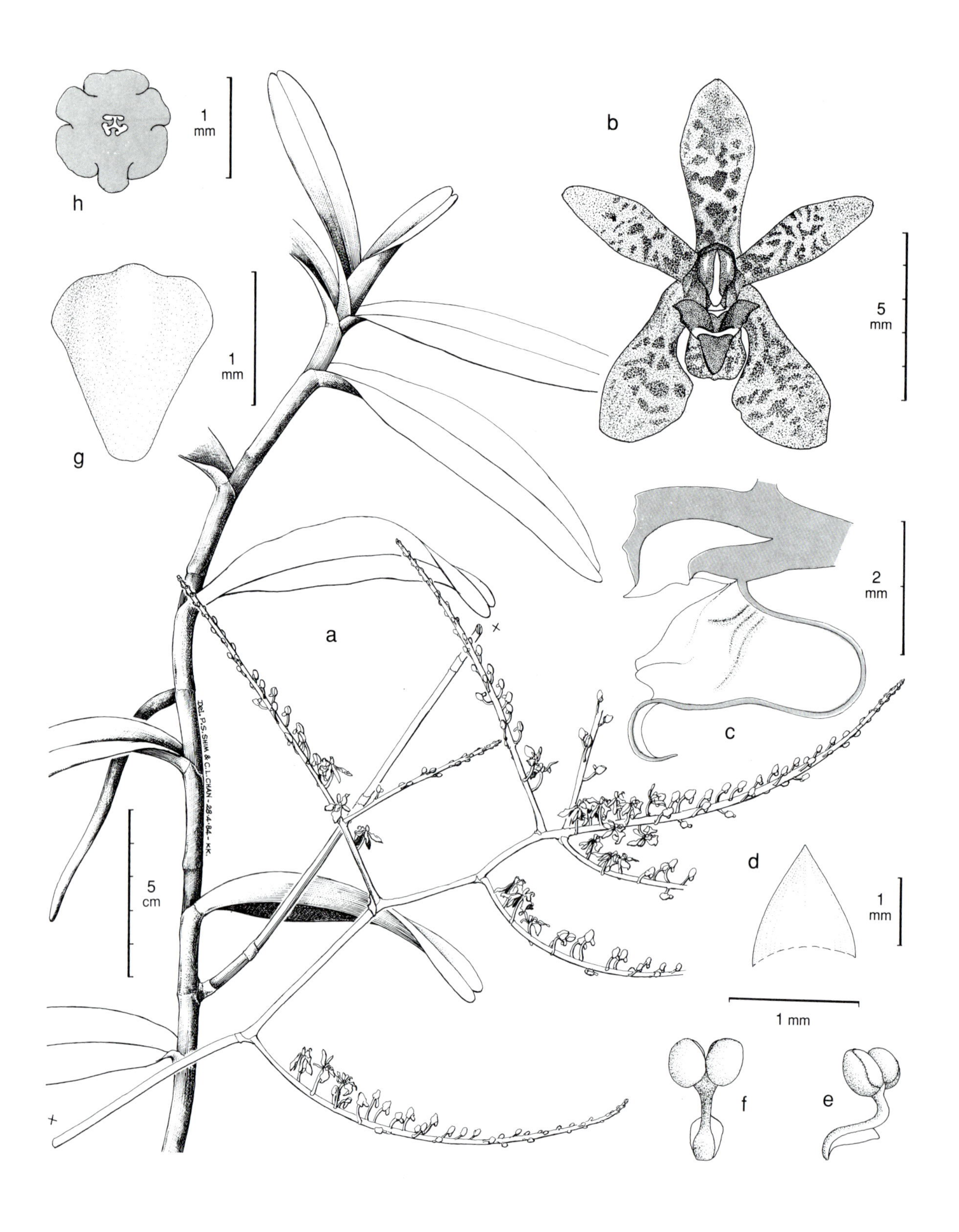

Figure 84. Renanthera elongata (Blume) Lindl. - A: plant. - B: flower, front view. - C: column and lip, longitudinal section. - D: floral bract. - E: pollinarium, lateral view. - F: pollinarium, front view. - G: anther. - H: ovary, transverse section. All drawn from *Lamb* AL 74/83 by Shim Phyau Soon and Chan Chew Lun.

84. RENANTHERA ELONGATA (Blume) Lindl.

Renanthera elongata (*Blume*) *Lindl.*, Gen. Sp. Orch. Pl.: 218 (1833). Type: Java, Kuripan, *Blume* s.n. (holotype L).

Aerides elongatum Blume, Bijdr.: 366 (1825).

Saccolabium reflexum Lindl., Gen. Sp. Orch. Pl.: 225 (1833). Type: Singapore, *Wallich* 7309 (holotype K).

Renanthera micrantha Blume, Rumphia 4: 54 (1849) nom. nud.; Mus. Bot. Lugd. 1: 60 (1849). Type: Philippines, *Cuming*, hort. *Duke of Devonshire* (holotype K).

Porphyrodesme micrantha (Blume) Garay in Bot. Mus. Leafl. 23(4): 191 (1974).

Large scrambling or climbing epiphytic or lithophytic ***herb*** with a long flexuous stem up to several metres long, 0.5-0.8 cm in diameter, leafy in apical half, covered by persistent leaf sheaths. ***Roots*** adventitious, elongate, seldom branching, grey. ***Leaves*** 5.5-11.5 x 1.5-2.5 cm, coriaceous, oblong, rounded and somewhat unequally bilobed at the apex, articulated to persistent leaf bases up to 2.7 cm long. ***Inflorescences*** emerging through the sheaths opposite a leaf, suberect or erect, pyramidal in outline, many-branched, subdensely many-flowered; peduncle stout, terete, 15-20 cm long, purple; branches up to 15 cm long, purple; floral bracts minute, triangular, 1-1.5 mm long. ***Flowers*** small, scarlet to orange with a yellow lip heavily marked with maroon. ***Pedicel*** with ***ovary*** 0.7-1.2 cm long. ***Dorsal sepal*** 0.6-0.8 x 0.1 cm, erect, oblanceolate, acute. ***Lateral sepals*** 0.8-0.9 x 0.2 cm, spreading widely, narrowly clawed, spathulate, obtuse. ***Petals*** 0.6-0.7 x 0.1 cm, oblanceolate, rounded at the apex. ***Lip*** 0.25 x 0.2 cm, fleshy, very small, 3-lobed, ecallose, spurred at the base; side lobes held at 45°, obliquely triangular, truncate; mid-lobe oblong-elliptic, obtuse, recurved, apex bilobed, minutely papillose, 0.2 x 0.21 cm; spur saccate, rounded at the apex, 0.15 cm long. ***Column*** 0.1 cm long, fleshy; pollinia 4, ovoid, in 2 closely adpressed pairs, attached by a slender oblanceolate stipes to a small round viscidium. Plate 18B.

HABITAT AND ECOLOGY: Swamp forest, mixed hill-dipterocarp forest, often in more open and secondary forest on poor soils. Often starting life as a terrestrial, becoming a scrambling and climbing epiphyte, climbing with the use of adventitious roots from the stem internodes to attach itself to trunks and branches, sometimes attaining many metres in length and reaching into the sunny canopy where flowering takes place. Rarely lithophytic on limestone and sandstone cliffs where it is more of a scrambler than a true lithophyte. Alt. sea level to 1000 m. Flowering January-April, July-October, sporadically at other times.

DISTRIBUTION IN BORNEO: KALIMANTAN SELATAN: Banjarmasin area. KALIMANTAN TIMUR: Balikpapan area. SABAH: Tenom District. SARAWAK: Bako National Park; Kuching area; Bintulu area.

GENERAL DISTRIBUTION: Thailand (Terutao Island), Peninsular Malaysia, Singapore, Sumatra, Java, Mentawai, Borneo and the Philippines.

NOTES: This species was formerly common and widespread in the lowland and lower hill forests, but much of its habitat has been logged for timber. The flowers remain open for two or three weeks.

DERIVATION OF NAME: The specific epithet is derived from the Latin *elongatus*, elongated, referring to the long climbing stems.

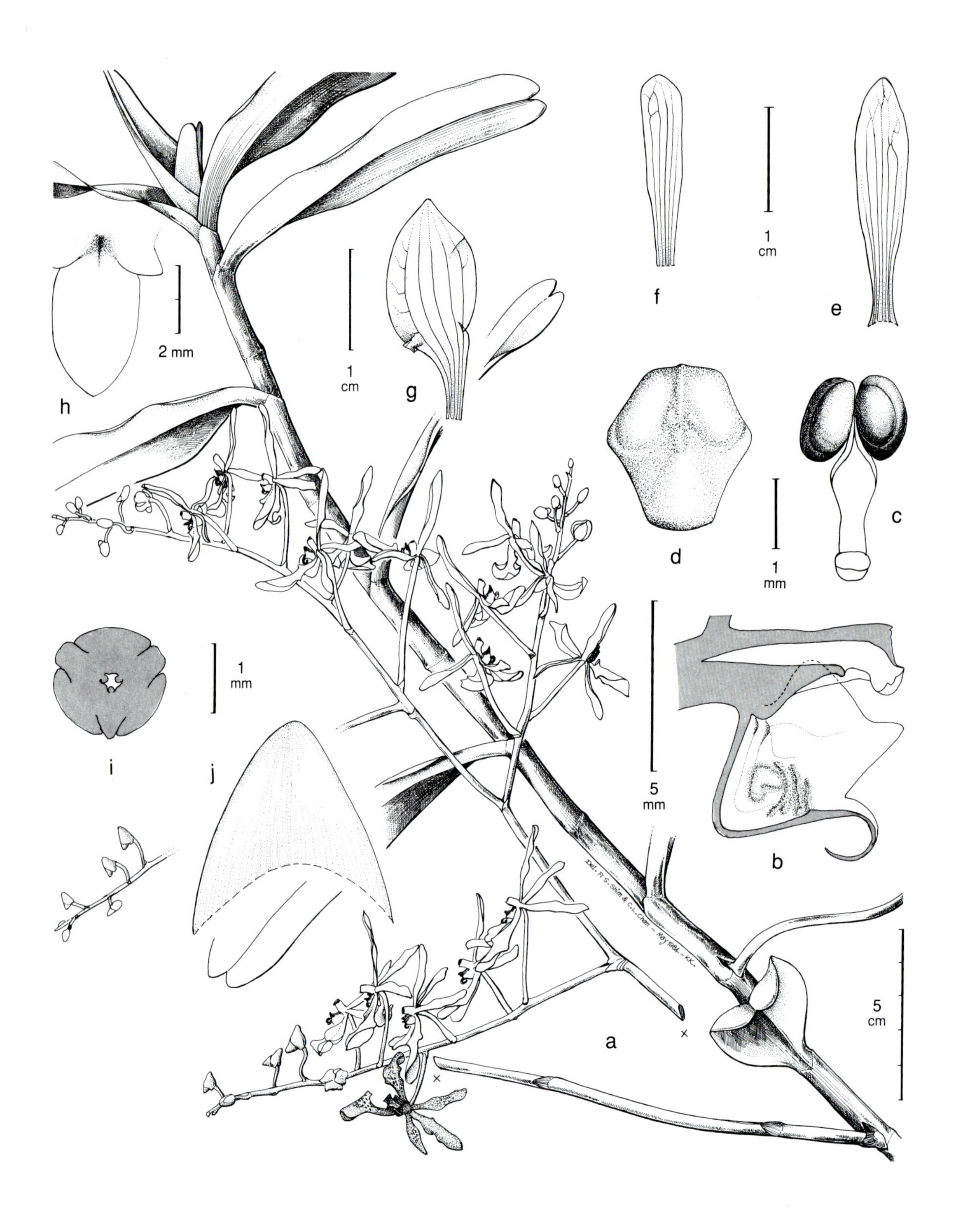

Figure 85. Renanthera isosepala Holttum - A: plant. - B: column and lip, longitudinal section. - C: pollinarium, adaxially. - D: anther. - E: dorsal sepal. - F: petal. - G: lateral sepal. - H: lip, spread out. - I: ovary, transverse section. - J: floral bract. All drawn from cult. TOC by Shim Phyau Soon and Chan Chew Lun.

85. RENANTHERA ISOSEPALA Holttum

Renanthera isosepala *Holttum* in Seidenf. & Smitinand, Orch. Thailand 4: 825 (1964). Type: Thailand, *Sagarik* 201 (holotype K).

Clambering epiphytic or terrestrial ***herb*** to several metres long. ***Aerial roots*** smooth, not verruculose. ***Stems*** simple, or something branching, internodes 3.5-5.5 cm long. ***Leaves*** 9-11.5 x 2.5-3.5 cm, ligulate, unequally bilobed, coriaceous, articulated to a purple-spotted sheathing base 2.5-2.8 cm long. ***Inflorescence*** many-flowered; peduncle 12-18(-22.5) cm long, with 3-5 ovate, adpressed sheaths 4-5 mm long; rachis 25-40 cm long, with 3 or 4 branches 11-14 cm long, each 20-35-flowered; floral bracts ovate, acute, those subtending branches larger, 4 mm long, those subtending flowers smaller, 1-2 mm long. ***Flowers*** with orange-red sepals and petals, sometimes with indistinct darker spotting; lip red, side lobes flushed yellow on the outside, calli yellow with a few red spots between; spur and column red, anther-cap reddish yellow. ***Pedicel*** with ***ovary*** 2.7-3.5 cm long. ***Dorsal sepal*** 2.2-2.4 x 0.5 cm, ligulate-spathulate, obtuse or subacute, erect. ***Lateral sepals*** 2-2.2 x 0.8 cm, spathulate, obtuse, undulate, unevenly revolute, drooping. ***Petals*** 1.7-1.9 x 0.3-0.4 cm, ligulate, obtuse, reflexed. ***Lip*** distinctly 3-lobed; side lobes 0.3-0.4 x 0.2-0.3 cm, rounded below, rather pointed above, slightly divergent, clasping column at base; mid-lobe 0.4 x 0.25 cm, oblong to oblong-elliptic, subacute, strongly recurved; spur 0.3-0.35 cm long, conical, obtuse; callus consisting of obscure thickened areas along base of side lobes at junction with mid-lobe. ***Column*** 0.5 x 0.3-0.4 cm, oblong; anther-cap 0.2 cm long, cucullate, truncate. Pollinia 4. Plate 18C.

HABITAT AND ECOLOGY: Swamp forest, coastal forest on peat soils to calcareous coastal sands. Alt. below 50 m. It has the same habit as *R. elongata* of climbing into the tops of trees that are usually stunted, in the swamp forest, where it can receive sufficient sunlight for flowering. Flowering sporadically once every two to three years.

DISTRIBUTION IN BORNEO: SABAH: Kuala Labuk; Dent Peninsula.

GENERAL DISTRIBUTION: Thailand and Borneo.

DERIVATION OF NAME: The specific epithet is derived from the Greek *iso*, equal, and *sepalum*, sepal, referring to the nearly uniform size of the dorsal and lateral sepals.

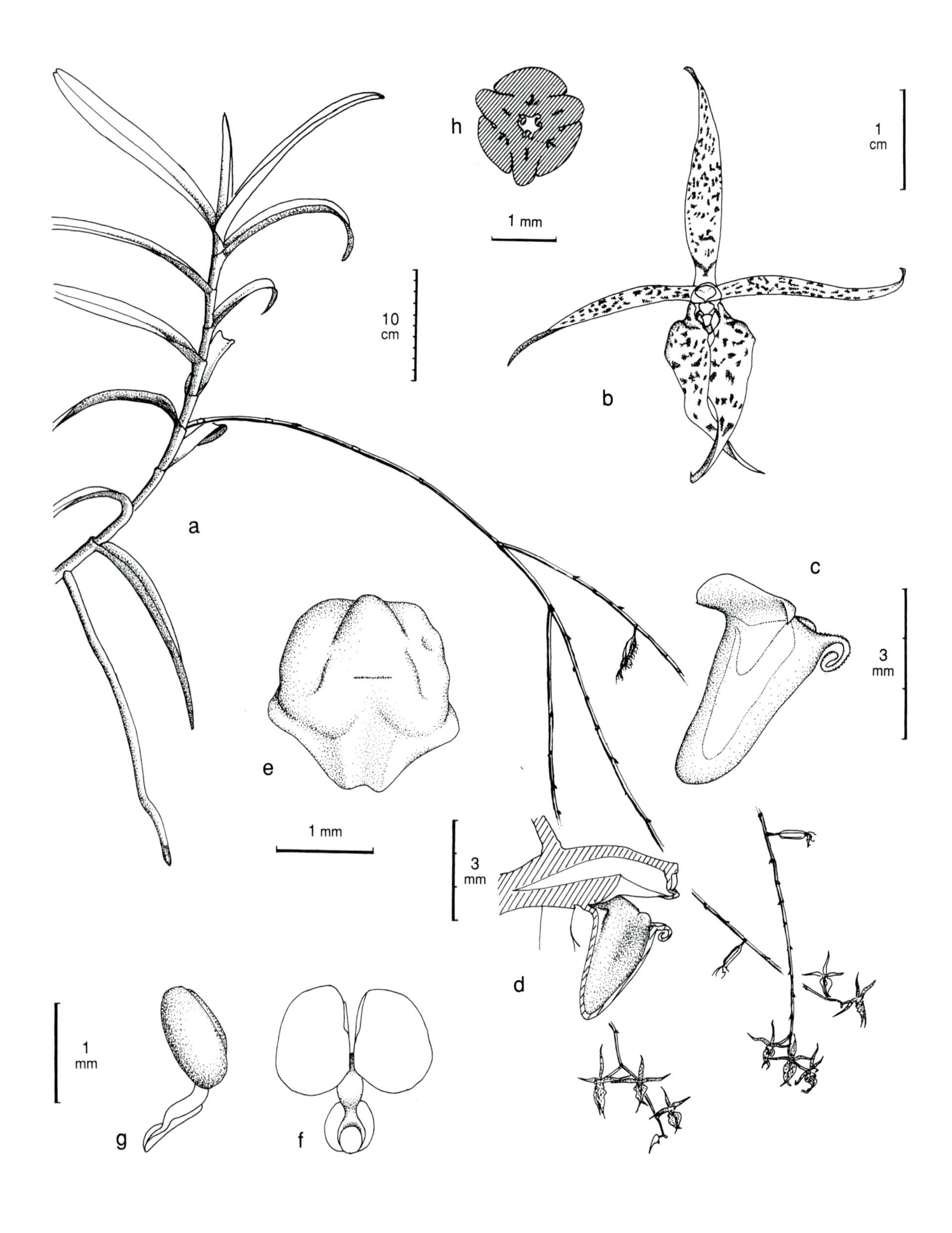

Figure 86. Renanthera matutina (Blume) Lindl. - A: plant. - B: flower, front view. - C: lip, lateral view. - D: column and lip, longitudinal section. - E: anther. - F: pollinarium, adaxially. - G: pollinarium, lateral view. - H: ovary, transverse section. All drawn from *Lamb* AL 80/83 by Shim Phyau Soon.

86. RENANTHERA MATUTINA (Blume) Lindl.

Renanthera matutina (*Blume*) *Lindl.*, Gen Sp. Orch. Pl.: 218 (1833). Type: Java, Mt. Salak, *Blume* s.n. (holotype BO).

Aerides matutina Blume, Bijdr.: 366 (1825).

Renanthera angustifolia Hook.f., Fl. Brit. Ind. 6: 49 (1890). Type: Java, Mt. Salak, *Blume* s.n. (holotype L).

Neprantha matutina (Blume) Hassk., Cat. Hort. Bog.: 44 (1844).

Climbing, hanging epiphytic ***herb*** to over 1 m long. ***Aerial roots*** smooth, not verruculose. ***Stems*** simple, internodes 2-3 cm long. Leaves 7-22.5 x 1.2-1.6(-2) cm, ligulate, unequally bilobed, thick, fleshy and rigid, dark grey-green, articulated to a darker sheath 2-3 cm long. ***Inflorescence*** horizontal to gently drooping, many-flowered, 40-90 cm long with 2 or 3 branches each up to 40 cm long and elongating for some time, flowers well spaced; peduncle with 3 or 4 adpressed ovate sheaths 5-8 mm long; floral bracts 4 x 4 mm, triangular-ovate, acute. ***Flowers*** opening in succession, minutely pubescent; sepals and petals light crimson or yellowish (normally yellow in Sabah form) with deeper fine red spots, particularly on lateral sepals; lip side lobes orange-yellow and white, spotted red, mid-lobe red brown, spur yellow with one or several crimson blotches; column yellow spotted deep crimson mostly at base. ***Pedicel*** with ***ovary*** 2-2.5 cm long, red. ***Dorsal sepal*** 2.2-2.5 x 0.35-0.4(-0.5) cm, ligulate, acuminate, erect. ***Lateral sepals*** 1.7-2.1 x 0.5 cm, narrowly elliptic, acuminate, clawed at base, parallel and drooping, margins undulate and connivent for a short distance near the base, apex carinate. ***Petals*** 2-2.2 x 0.25-0.3 cm, ligulate, acuminate, spreading horizontally. ***Lip*** 3-lobed; side lobes c. 0.2 cm long, rounded, with narrow, pointed and curving subulate tips; mid-lobe c. 0.15 cm long, ligulate, obtuse, strongly recurved, with thickened area either side at junction with side lobes; spur 0.4 cm long, cylindrical, obtuse; callus consisting of 2 thickened areas at base of side lobes. ***Column*** 0.4 cm long; anther-cap 0.2 cm long; pollinia 4. Plate 18D.

HABITAT AND ECOLOGY: Lowland and mixed hill-dipterocarp forest; podsolic heath forest, often epiphytic on *Dacrydium pectinatum*. Most often found in forests on sandstone-derived soils. It appears to prefer to grow on branches in the tree canopy with light to moderate shade. Alt. 100 to 600 m. Flowering sporadic, usually starting in the rainy season and extending into the dry period.

DISTRIBUTION IN BORNEO: SABAH: Tawau District; Nabawan area; Tenom District; Mt. Kinabalu.

GENERAL DISTRIBUTION: Peninsular Malaysia, Sumatra, Java, Borneo and the Philippines.

NOTES: This appears to be a widespread but rare species in Sabah. Lim & Saharan (1992) have noted differences between populations in Peninsular Malaysia and Sabah. The main differences are shorter, stiffer grey-green leaves and yellow flowers with more numerous and finer red spots in the Bornean specimens while those from Peninsular Malaysia have green leaves and red flowers. Further minor differences are found in the lip, mentum and anther-cap. Further study of the species is required throughout its range to ascertain the status of the Bornean plant.

DERIVATION OF NAME: The specific epithet is derived from the Latin *matutinus*, of or belonging to the morning, in reference to the flowering time.

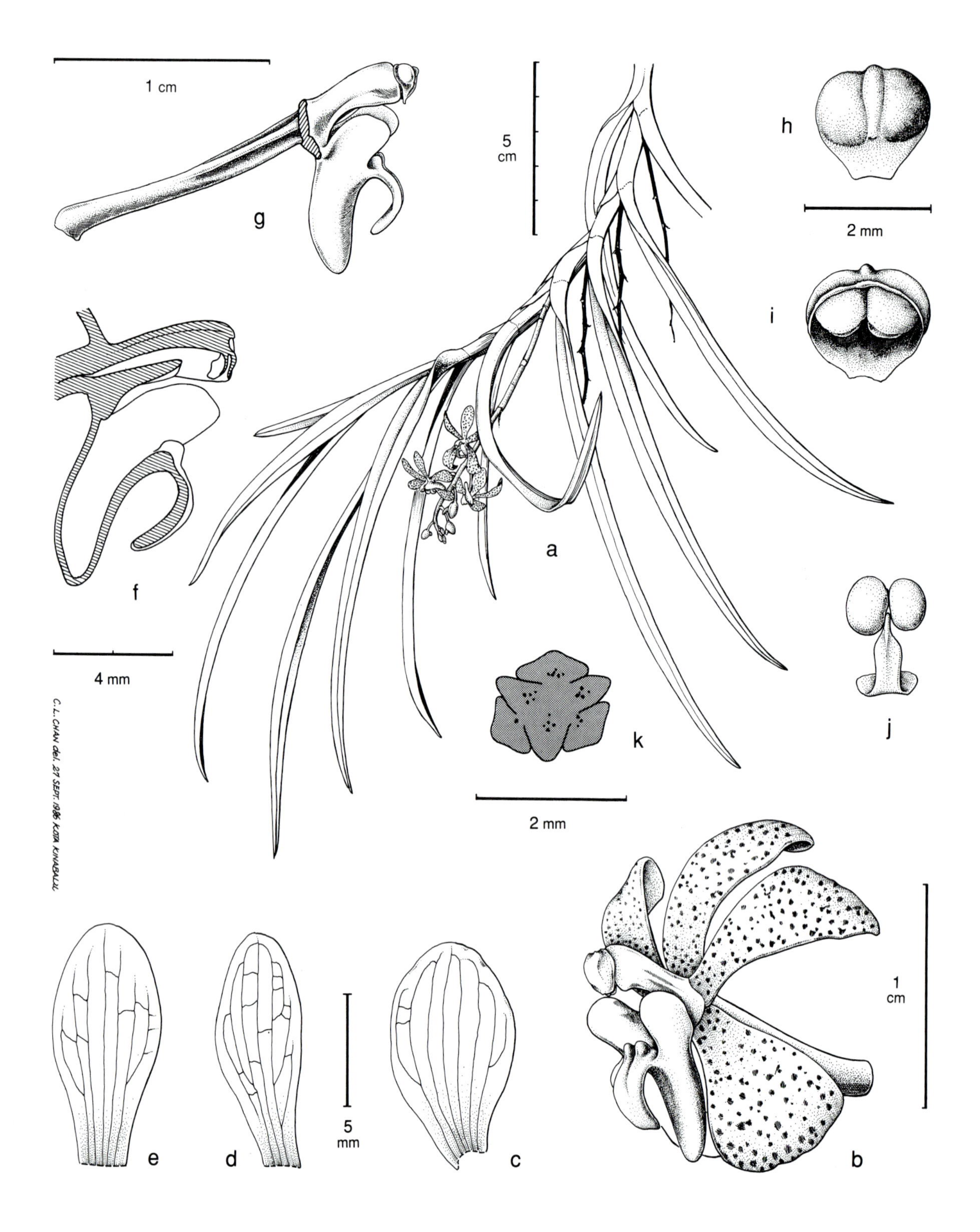

Figure 87. Renantherella histrionica subsp. auyongii (Christenson) Senghas - A: plant. - B: flower, lateral view. - C: lateral sepal. - D: petal. - E: dorsal sepal. - F: column and lip, longitudinal section. - G: flower with sepals and petals removed, lateral view. - H: anther, abaxially. - I: anther, adaxially. - J: pollinarium. - K: ovary, transverse section. All drawn from cult. Au Yong Nang Yip by Chan Chew Lun.

87. RENANTHERELLA HISTRIONICA (Rchb.f.) Ridl. subsp. AUYONGII (Christenson) Senghas

Renantherella histrionica *(Rchb.f.) Ridl.* subsp. **auyongii** *(Christenson) Senghas* in Schlechter, Die Orchideen ed. 3, 1(21): 1327 (1988). Type: Sarawak, near Mt. Santubong, *Au Yong Nang Yip* s.n., cult. *J. Levy* 1277 (holotype SEL).

Renanthera auyongii Christenson in Orchid Digest 50: 169 (1986).

Medium-sized or small epiphytic ***herb,*** often forming a dense mass. ***Roots*** adventitious, emerging through the leaf sheaths, unbranched, up to 2.5 mm in diameter, greyish. ***Stem*** erect to spreading or subpendent, branching towards base, up to 40 cm long, 0.3-0.6 cm in diameter, somewhat flattened, leafy along length, covered by persistent leaf sheaths. ***Leaves*** 10-18.5 x 0.6-1 cm, coriaceous, spreading widely, conduplicate, linear, tapering, minutely unequally obtusely bilobed at apex, articulated to a sheathing leaf base up to 2.2 cm long. ***Inflorescence*** 1 to several, axillary from nodes some 5 to 10 cm below the stem apex, 6-8 cm long, 6- to 12-flowered; peduncle terete, 1.5-3 cm long; floral bracts triangular, obtuse, 1-2 mm long. ***Flowers*** produced in succession, 0.7-0.8 cm apart on the rachis, non- resupinate, small but showy, 2-2.5 cm across, fleshy; sepals and petals orange-red with red spots; lip yellow-orange with red-spotted side lobes and a red mid-lobe edged with yellow. ***Pedicel*** with ***ovary*** 1-1.2 cm long, cream-coloured. ***Dorsal sepal*** 0.9-1 x 0.4 cm, erect and somewhat convex, obovate, rounded at apex. ***Lateral sepals*** 0.7-1 x 0.5 cm, somewhat recurved, obliquely obovate-spathulate, obtuse. ***Petals*** 1 x 0.3 cm, somewhat reflexed, spathulate, apex rounded. ***Lip*** 0.6-0.7 cm long, 0.6 cm wide when flattened, 3-lobed, spurred at the base; side lobes 0.25-0.3 cm long, erect, oblong-elliptic, apex rounded; mid-lobe 0.3-0.4 cm long, reflexed, ligulate, bearing two globose calli at the base; spur 0.5 cm long, cylindrical, slightly incurved. ***Column*** 0.4 cm long, slightly incurved, terete; pollinia 4, in two unequal adpressed pairs, attached by a short ovate stipes to a reniform viscidium. Plate 18E.

HABITAT AND ECOLOGY: On trees near the coast. Alt. sea level to 50 m. Flowering recorded in May.

DISTRIBUTION IN BORNEO: SARAWAK: Damai Beach.

GENERAL DISTRIBUTION: Endemic to Borneo (Sarawak only).

NOTES: This subspecies is only known from the type locality at Damai Beach near Kuching in Sarawak. The site has now been lost since the development of a hotel complex. It differs from subsp. *histrionica* from Thailand and Peninsular Malaysia in having non-resupinate orange, red-spotted flowers with a longer spur.

DERIVATION OF NAME: The generic name is a diminutive of *Renanthera.* The subspecific epithet honours Mr Au Yong Nang Yip, a keen and energetic orchid enthusiast living in Kuching, who collected the type.

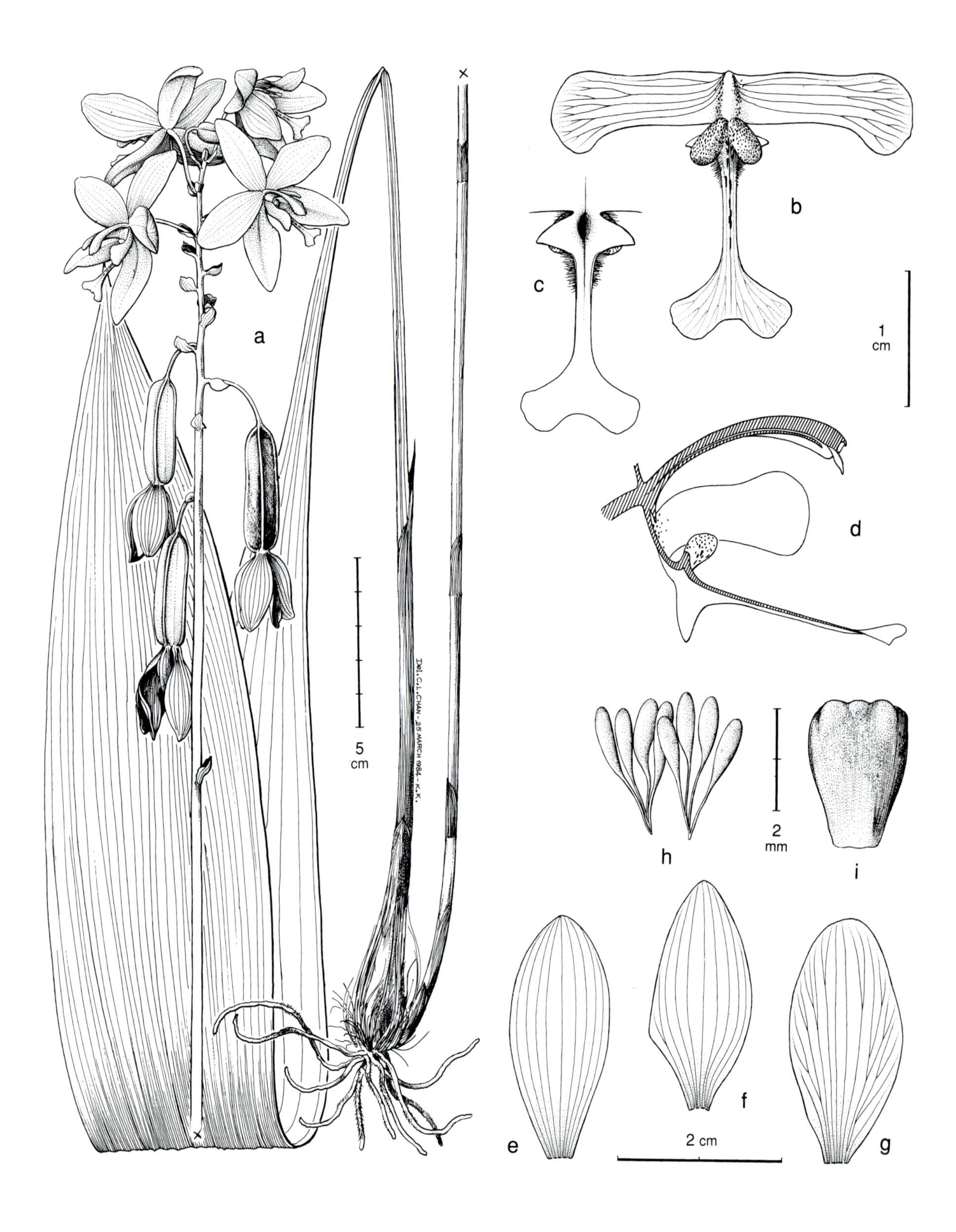

Figure 88. Spathoglottis gracilis Rolfe ex Hook. f. - A: plant. - B: lip spread out, front view. - C: mid-lobe, back view. - D: column and lip, longitudinal section. - E: dorsal sepal. - F: lateral sepal. - G: petal. - H: pollinia. - I: anther. All drawn from *Beaman* 9070a by Chan Chew Lun.

88. SPATHOGLOTTIS GRACILIS Rolfe ex Hook.f.

Spathoglottis gracilis *Rolfe ex Hook.f.* in Bot. Mag. 120: t. 7366 (1894). Type: Borneo, *Foerstermann* s.n., cult. *Sander & Co.* (holotype K).

Terrestrial ***herb***. ***Pseudobulbs*** 2 x 1.5 cm, ovoid, enclosed in sheaths. ***Leaves*** to 52 x 3-7 cm, linear-lanceolate, acuminate, plicate, long-petiolate; petiole enclosed at the base by a brown sheath up to 18 cm long. ***Inflorescence*** emerging from base of pseudobulb; peduncle 30-40 cm long, erect, bearing a few scattered sheaths 2-3 cm long; rachis 3-15 cm long; floral bracts 0.5-1 cm long, ovate, concave, cymbiform, obtuse. ***Flowers*** bright yellow; side lobes of lip, claw of mid-lobe and callus spotted with red. ***Pedicel*** with ***ovary*** 3-4 cm long, slender, curved. ***Sepals*** 2.5 x 1.5 cm, broadly oblong to oblong-elliptic, obtuse, spreading. ***Petals*** 2.5 x 1.7 cm, obovate-oblong or obovate-elliptic, obtuse. ***Lip*** 1.8-2 cm long; side lobes 1.5 x 0.6-0.8 cm, subspathulate-oblong, rounded, truncate, incurved, erect to spreading; mid-lobe with a narrow claw 1 mm wide, with 2-3 mm triangular-rounded, pilose auricles at base, apex expanded and transversely obtusely bilobed, 0.7-0.9 cm wide; callus bilobed, each lobe oblong, rounded, erect, divaricate, 2-3.5 mm long. ***Column*** 1.7 cm long, slender, curved; anther-cap 2.5 x 1-1.5 mm; pollinia 8, in 2 groups of 4. Plate 18F.

HABITAT AND ECOLOGY: Hill-forest and lower montane forest; *Gymnostoma* and *Tristaniopsis* forest; among rocks under bamboo, on ultramafic substrate. Alt. 800 to 1650 m. Flowering recorded in February, April, June, August and November.

DISTRIBUTION IN BORNEO: SABAH: Mt. Kinabalu; Mt. Nicola. SARAWAK: Lambir National Park.

GENERAL DISTRIBUTION: Peninsular Malaysia and Borneo.

NOTES: Closely allied to *S. kimballiana* Hook.f. which has larger flowers often flushed red on the outer surface of the sepals and a lip with a broader claw. The relationship between these species needs further investigation. However, *S. gracilis* is normally found in shaded forest, in leaf litter or in rocky and mossy habitats on ridges. *S. kimballiana* has much narrower leaves and often occurs in more sunny situations on river banks.

DERIVATION OF NAME: The generic name is derived from the Greek *spathe*, spathe, and *glotta*, tongue, alluding to the broad mid-lobe of the lip in some species. The Latin specific epithet *gracilis* means thin or slender, in reference to the habit.

SPONGIOLA J.J.Wood & A. Lamb

Spongiola *J.J.Wood et A.Lamb* **gen. nov.** (Vandeae-Aeridinae) *Malleolae* J.J.Sm. et Schltr. affinis. Herba epiphytica caulem brevem et foliatum gaudens, foliis apice obtusis, leviter asymmetricis; inflorescentia racemosa foliis multo longiore, bracteis minutis, floribus fugacibus, sepalis petalisque patentibus semipellucidis, labello affixo trilobato ecalloso, lobo medio labelli spongioso lobis lateralibus grandiore, lobis lateralibus parvis auriculatis, calcari brevi aseptato, columna brevi, rostello longo gracili, polliniis duobus, integris. Typus: *S. lohokii* J.J.Wood et A.Lamb.

Epiphytic ***herb***. ***Stems*** short and leafy. ***Leaves*** obtuse and slightly asymmetrical at apex. ***Inflorescence*** pendent, much longer than leaves, racemose; floral bracts minute. ***Flowers*** short-lived, several open together at regular intervals, with spreading semi-translucent sepals and petals. ***Lip*** immobile, 3-lobed, ecallose; mid-lobe longer than side lobes, resembling a small sponge-like pouch, hollow near the mouth of the spur, solid below, shallowly concave on the underside, papillose verrucose; side lobes small, auriculate; spur short, aseptate. ***Column*** short, foot absent; rostellum prominent; anther-cap terminal, operculate, acuminate; pollinia 2, entire; stipes spathulate, broadened below pollinia; viscidium cucullate. Type: *S. lohokii* J.J.Wood & A. Lamb.

Spongiola belongs to the group of genera within the subtribe Aeridinae of the tribe Vandeae which have two instead of four entire pollinia and no column-foot and include, among others, *Malleola, Pennilabium, Porrorhachis* and *Tuberolabium*. The vegetative habit of *Spongiola* is similar to *Malleola* and *Brachypeza*, particularly *B. archtyas* (Ridl.) Garay from Christmas Island near Java, and to certain *Pteroceras* and *Sarcochilus*. The curious lip mid-lobe, which resembles a small sponge-like pouch, readily distinguishes it from all these related genera. The lip structure is relatively simple compared with many genera of the Aeridinae, having neither calli nor a septate spur.

Spongiola is distinguished from *Malleola* by the large sponge-like lip mid-lobe and the long slender rostellar projection. In *Malleola* the mid-lobe is very small, conical or linear, never sponge-like and the rostellum is short. The lip of *Spongiola* is immobile and not jointed to a long column-foot as in *Grosourdya, Pteroceras* and *Sarcochilus*. *Grosourdya* may also be distinguished by the long, bent column and the lip mid-lobe being replicate on to the prominent spur. The mid-lobe of the lip of *Pteroceras* is much smaller than the side lobes and the two pollinia are sulcate into subequal halves. In *Sarcochilus* the flowers are longer lasting and have a callose lip and four unequal pollinia.

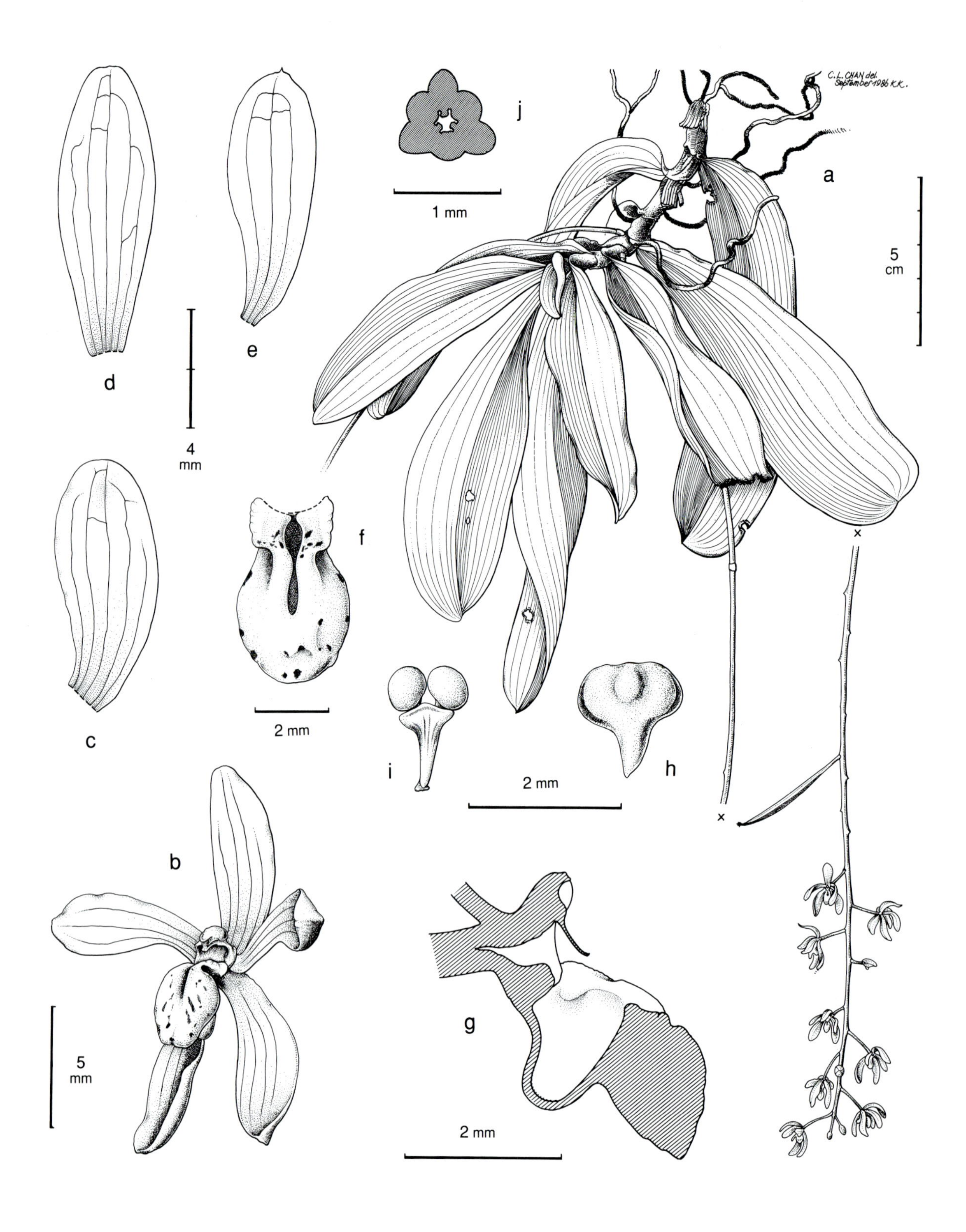

Figure 89. Spongiola lohokii J.J. Wood & A. Lamb - A: plant. - B: flower, front view. - C: lateral sepal. - D: dorsal sepal. - E: petal. - F: lip. - G: column and lip, longitudinal section. - H: anther. - I: pollinarium. - J: ovary, transverse section. All drawn from *Lohok* in *Lamb* AL 426/85 by Chan Chew Lun.

89. SPONGIOLA LOHOKII J.J.Wood & A.Lamb

Spongiola lohokii *J.J.Wood et A.Lamb* **sp. nov.** distincta. Herba epiphytica foliis obtusis, inflorescentia racemosa foliis longiore, bracteis minutis, floribus fugacibus, sepalis petalisque concavis, lobis lateralibus labelli auriculatis, lobo medio grandiore spongioso ecalloso, calcari brevi aseptato. Typus: Borneo, Sabah, Kinabatangan District, on border of Pensiangan District, Batu Urun, Batu Ponggol, February 1985, flowered in cultivation October 1985, *Lohok* in *Lamb* AL 426/85 (holotypus K; isotypus SAN, spirit material only).

Spreading epiphytic ***herb***. ***Stem*** 5-7 cm long, covered in persistent leaf sheaths. ***Leaves*** 3-6 per stem, 7.5-19.5 x 1.8-3(-4.2) cm, oblong or oblong-elliptic, apex obtuse and asymmetrical, coriaceous, spreading, sheaths 0.5-1.4 cm long. ***Inflorescence*** lateral, pendent, emerging from the base of the leaf sheaths and usually growing out from behind the leaves, racemose, occasionally branched; peduncle and rachis narrow and wiry, pale green; peduncle 12-15 cm long, with 3-4 remote, adpressed, ovate sterile bracts, the basal 5 mm long, the upper 2 mm long; rachis 18-20 cm long; floral bracts 1 mm long, ovate. ***Flowers*** up to 25, usually 5-6 open at a time progressing downwards in succession, sometimes opening in the middle of the raceme, ephemeral, each lasting one day only; sepals and petals semi-translucent pale yellow to yellowish green, edges and tips semi-translucent whitish; lip white with pale to dark purple spots, side lobes flushed purple on inner surface, spur pale yellow; column pale yellow with a purple flush at base, anther-cap yellow. ***Pedicel*** with ***ovary*** 0.9-1 cm long, straight. ***Sepals*** and ***petals*** spreading. ***Dorsal sepal*** 0.9-1 x 0.35 cm, oblong-elliptic, obtuse. ***Lateral sepals*** 0.9 x 0.35-0.4 cm, oblong-elliptic, obtuse and obscurely carinate at apex, concave, inner margins somewhat undulate. ***Petals*** 0.8 x 0.3-0.35 cm, oblong-elliptic, obtuse, concave, margins slightly undulate. ***Lip*** 3-lobed, ecallose; side lobes 0.1 cm long, auriculate, obtuse, inner surface gibbous at the base, mid-lobe 0.3 x 0.2-0.25 cm, resembling a small spongy pouch, hollow above, solid towards the apex, shallowly concave on the underside, papillose-verrucose; spur 2 mm long, conical, obtuse, minutely racemose around the entrance. ***Column*** 0.15 x 0.18-0.19 cm, oblong, foot absent; rostellar projection 0.1 cm long; anther-cap 0.15 x 0.18 cm, ovate, apex caudate acuminate, sometimes obscurely tridentate; pollinia 2, ovoid, entire; stipes c. 0.15 cm long, spathulate, widened below pollinia; viscidium cucullate. Plate 19 & 20A.

HABITAT AND ECOLOGY: Branches of small trees in damp, shaded and humid situations in riverine forest on limestone and sandstone ridges. Alt. 300 to 500 m. Flowering January, September and October.

DISTRIBUTION IN BORNEO: SABAH: Kinabatangan District.

GENERAL DISTRIBUTION: Endemic to Borneo (Sabah only).

DERIVATION OF NAME: The generic name is Latin for a little sponge and refers to the lip mid-lobe. The specific epithet honours Mr Harry Lohok, formerly of the Tenom Orchid Centre, now supervisor of Poring Orchid Centre, Sabah Parks, who collected the type.

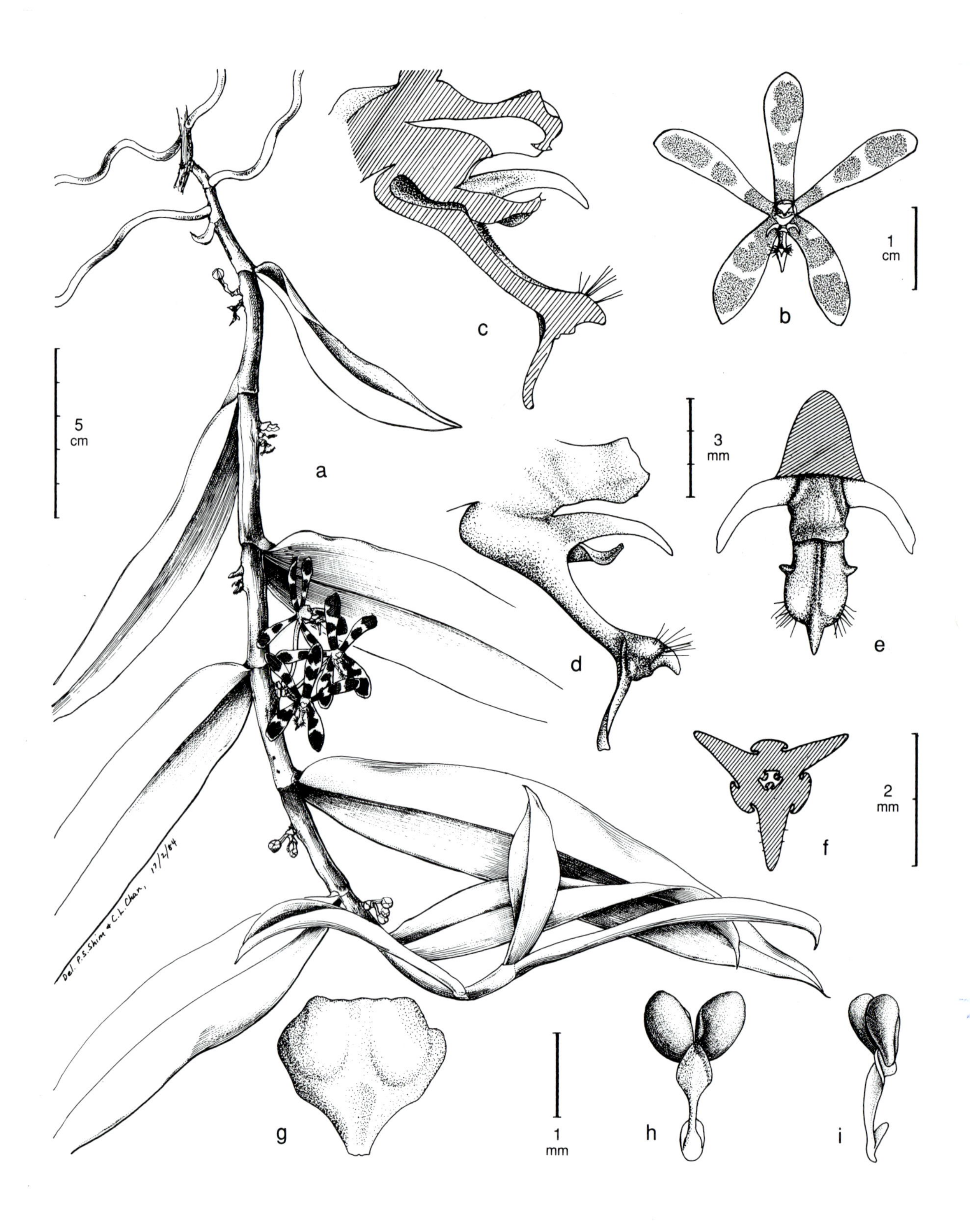

Figure 90. Trichoglottis bipenicillata J.J. Sm. - A: plant. - B: flower, front view. - C: column and lip, longitudinal section. - D: column and lip, lateral view. - E: lip. - F: ovary, transverse section. - G: anther. - H: pollinarium, front view. - I: pollinarium, lateral view. All drawn from *Beaman* 8602 by Shim Phyau Soon and Chan Chew Lun.

90. TRICHOGLOTTIS BIPENICILLATA J.J.Sm.

Trichoglottis bipenicillata *J.J.Smith* in Icon. Bogor. 2: 125, t. 125A (1903). Type: Borneo, Kalimantan, Sungai Semitau, *Hallier* s.n. (holotype BO).

Pendulous epiphytic ***herb***. ***Stems*** up to 70 cm long, internodes 1.5-3.5 cm long. ***Leaves*** 5-15.5 x 1.5 x 2.4 cm, oblong-ligulate to narrowly elliptic, cuneate at base, apex obtusely unequally bilobed or acutely asymmetric. ***Inflorescence*** 1- to 3-flowered, abbreviated; peduncle 2-3 mm long; floral bracts 2-3 mm long, triangular-ovate, acute. ***Flowers*** 3 cm across, sweetly scented; sepals and petals pale to dark yellow with large irregular cinnamon or orange brown blotches or in some plants, small spots; lip pale yellow with a cinnamon callus. ***Pedicel*** with ***ovary*** 1.6-2.3 cm long. ***Sepals*** and ***petals*** spreading. ***Dorsal sepal*** 1.5-1.7 x 0.4-0.5 cm, oblanceolate-spathulate, obtuse. ***Lateral sepals*** 1.3-1.5 x 0.4-0.5 cm, oblanceolate, obtuse. ***Petals*** 1.5-1.6 x 0.3-0.35 cm, narrowly oblanceolate, obtuse. ***Lip*** 0.7-0.9 cm long, 0.15-0.16 cm wide across centre of mid-lobe; side lobes 0.25-0.3 cm long, falcate, acute, erect; mid-lobe oblong, with a deflexed narrowly triangular-ligulate, acute apex, 0.3 cm long; callus central, cuspidate, 0.2 cm long, porrect, papillose at base, with tufts of long white hairs each side; back wall tongue ligulate, spathulate, apical margins ciliate. ***Column*** 0.25-0.4 cm long; pollinia 4, in 2 unequal masses. Plate 20B.

HABITAT AND ECOLOGY: Mixed ridge-top or hill-dipterocarp forest on ultramafic and sandstone ridge substrate. Alt. 400 to 600 m. Flowering observed in April, May and August.

DISTRIBUTION IN BORNEO: KALIMANTAN BARAT: Semitau area. SABAH: Mt. Kinabalu; Tenom District.

GENERAL DISTRIBUTION: Endemic to Borneo.

NOTES: This appears to be a rather rare species. The flowers of some plants are deep orange and there is quite a considerable degree of variation in markings, from light spotting to broad orange-cinnamon bars and blotches.

DERIVATION OF NAME: The generic name is derived from the Greek *tricho*, hair, and *glotta*, tongue, referring to the pubescent tongue on the back wall of the lip. The specific epithet is derived from the Latin *bi*, two and *penicillatus*, pencil-shaped, in reference to the two horn-like lip side lobes.

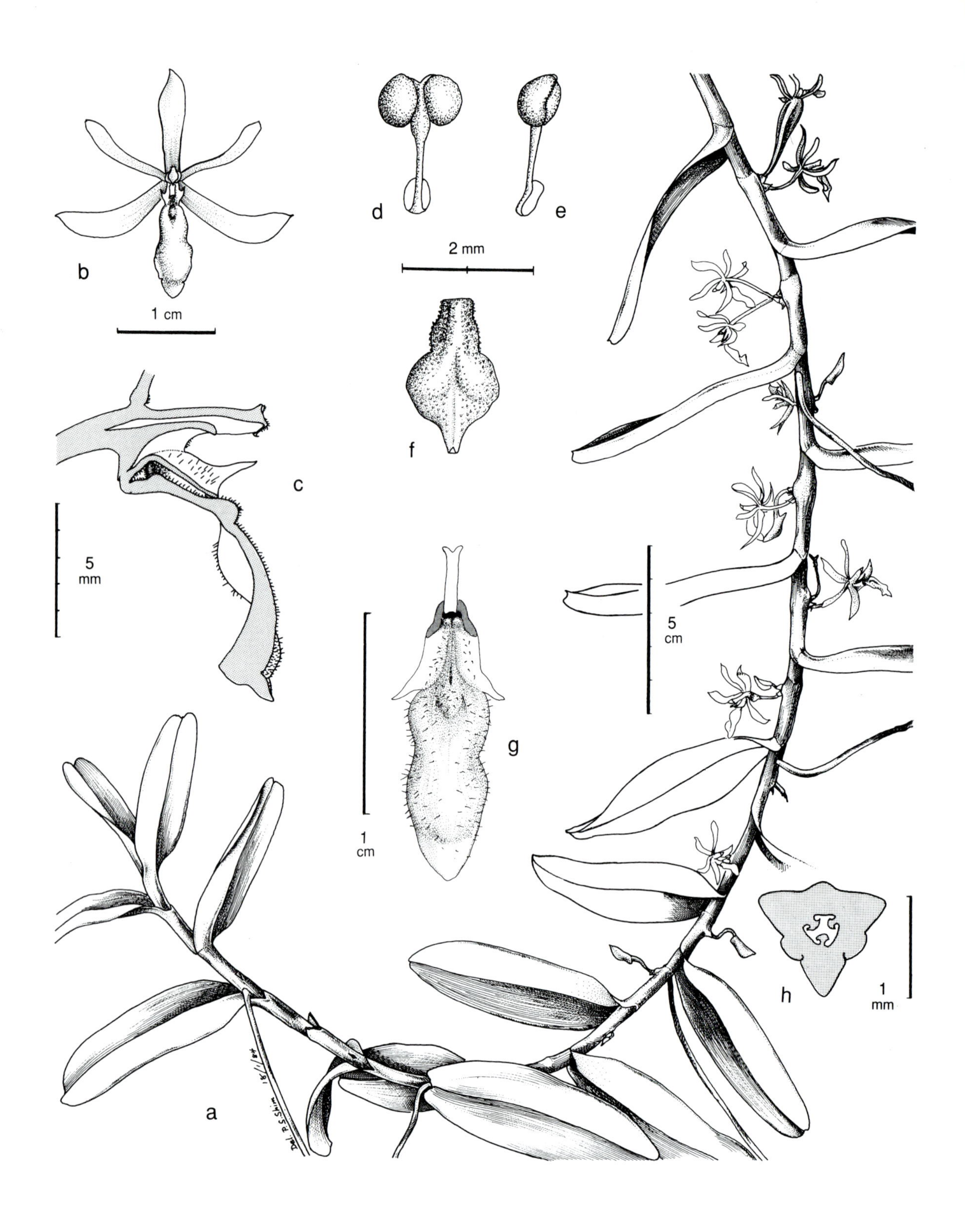

Figure 91. Trichoglottis retusa Blume - A: plant. - B: flower, front view. - C: column and lip, longitudinal section. - D: pollinarium, front view. - E: pollinarium, lateral view. - F: anther. - G: lip. - H: ovary, transverse section. All drawn from cult. TOC by Shim Phyau Soon.

91. TRICHOGLOTTIS RETUSA Blume

Trichoglottis retusa *Blume*, Bijdr.: 360, pl. 2, fig. 8 (1825). Type: Java, Buitenzorg & Bantam, *Blume* s.n. (holotype BO).

Scandent epiphytic ***herb***. ***Stems*** up to 1 m long, branching, internodes 1-3.5 cm long. ***Leaves*** 7-12 x 1.2-2 cm, oblong-ligulate, roundly unequally bilobed, leaf sheaths 2.5 cm, enclosing part of the stem. ***Inflorescence*** 1-flowered, 1-3 at a node; floral bracts 1-2 mm long. ***Flowers*** sweetly lemon-scented, lasting 1-2 weeks, with greenish yellow to lemon-yellow sepals and petals (without large reddish brown spots found in Javan form); lip white (sometimes with a red or purple blotch) and a white, sometimes yellow callus at the base, mid-lobe rarely with a few brown apical spots; column white at base, greenish yellow at apex. ***Pedicel*** with ***ovary*** 1.3-1.5 cm long. ***Dorsal sepal*** 1.1-1.3 x 0.3-0.5 cm, oblong-elliptic, acute. ***Lateral sepals*** 1.2-1.5 x 0.3-0.5 cm, narrowly elliptic, acute. ***Petals*** 1.1-1.3 x 0.2-0.25 cm, ligulate, subacute. ***Lip*** 1.2-1.4 cm; side lobes 0.4-0.5 x 0.1 cm, hairy inside, falcate, acute, erect to spreading; mid-lobe 0.9-1 x 0.4-0.5 cm, oblong-triangular, acute, margins deflexed, covered in white hairs except apex, with a fleshy tooth-like projection underneath just below apex; disc with a small hairy callus at base of side lobes; spur 0.15-0.3 cm long, obtuse; back wall tongue narrow, forked, shortly hairy. ***Column*** 0.45-0.6 cm long, apical projections hairy; pollinia 4, free, unequal. Plate 20C.

HABITAT AND ECOLOGY: Often found on the large lower branches and forks of large trees, especially dipterocarps in lowland mixed dipterocarp to hill-dipterocarp forest on alluvial soils and soils derived from sandstone, mudstones and shales. Often receiving dense early morning mist, then dappled sunlight in their partly shaded habitat. Also rarely found in leaf litter on ledges of sandstone cliffs. Alt. lowlands to 500 m. Flowering recorded in June, August and November.

DISTRIBUTION IN BORNEO: SABAH: Sandakan Peninsula; Sepilok Forest Reserve; Tawau District; Tenom district. SARAWAK: Kuching area.

GENERAL DISTRIBUTION: Thailand, Indochina, Peninsular Malaysia, Sumatra, Mentawai, Java, Sumbawa and Seram.

NOTES: Flowering is sporadic, twice to several times a year, but most often during the rainy seasons, with profuse flowering in November in Sabah.

This epiphyte is seldom collected, probably because of its inaccessibility high up in the forks and branches of tall dipterocarps and other trees.

DERIVATION OF NAME: The specific epithet is derived from the Latin *retusus*, retuse, with a rounded, shallowly notched end, referring to the broadening, diverging to gently forked apex of the tongue attached to the back wall of the spur of the lip.

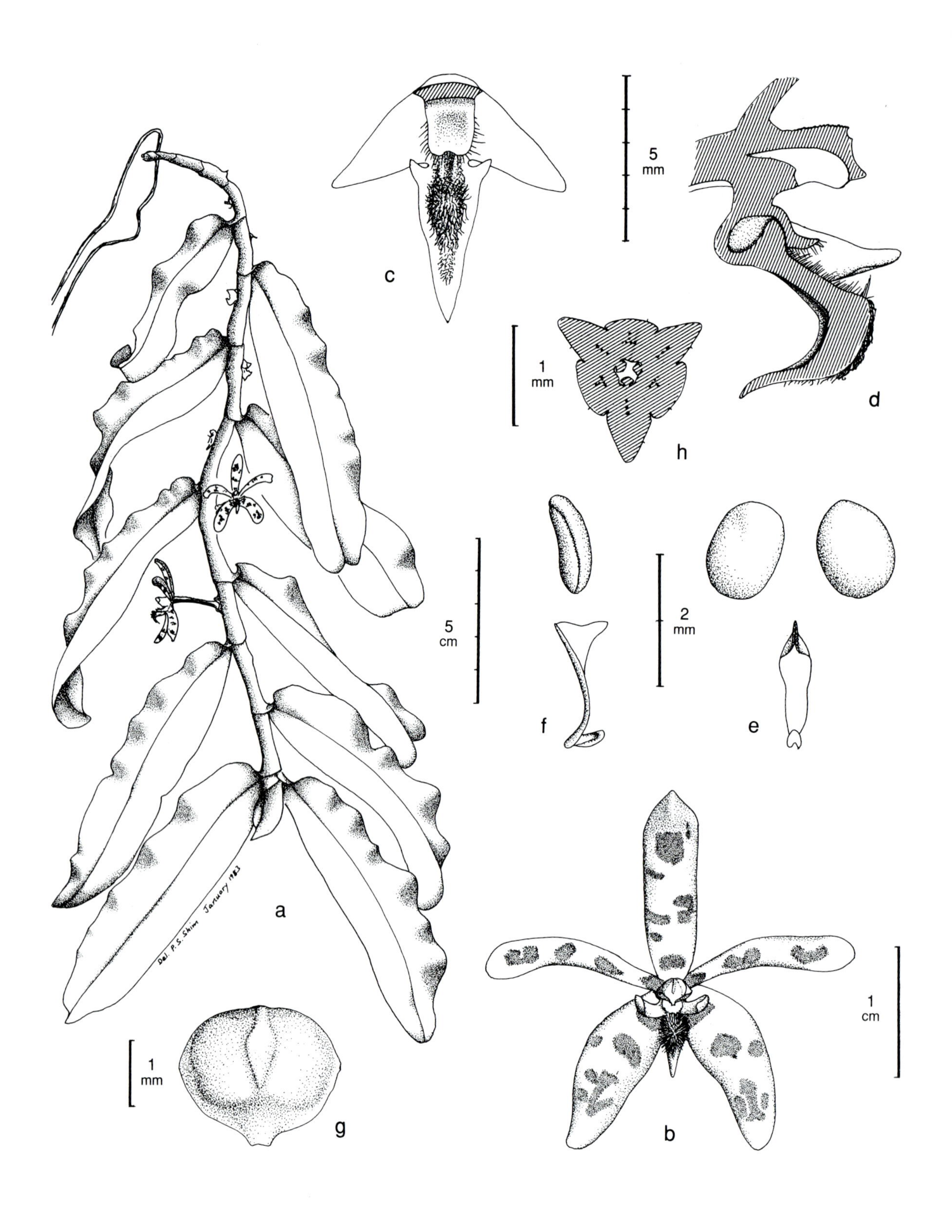

Figure 92. Trichoglottis scapigera Ridl. - A: plant. - B: flower, front view. - C: lip. - D: column and lip, longitudinal section. - E: pollinarium. - F: pollinarium, lateral view. - G: anther. - ovary, transverse section. All drawn from cult. TOC by Shim Phyau Soon.

92. TRICHOGLOTTIS SCAPIGERA Ridl.

Trichoglottis scapigera *Ridl.* in J. Linn. Soc., Bot. 32: 357 (1896). Type: Peninsular Malaysia, Penang, Government Hill, *Curtis* 1964 (holotype SING, isotype K).

Pendent epiphytic ***herb*** with stems up to 50 cm or more long, 4-5 mm in diameter, leafy throughout length. ***Roots*** climbing from basal half of stem. ***Leaves*** up to 10.5 x 3.5 cm, coriaceous, distichous, twisted at base to lie in one plane vertical to substrate, elliptic to oblong-elliptic, obliquely and minutely bilobed at obtuse apex, articulated at base to a tubular sheath 2-3.5 cm long. ***Inflorescences*** several emerging one at a time through base of leaf sheath opposite leaf below, one-flowered; peduncle very short; floral bract elliptic, very short. ***Flowers*** 3 cm across, faintly sweetly scented, fleshy; sepals and petals white marked with brown; lip white marked with brown and yellow on side lobes and with an orange-yellow hairy callus. ***Pedicel*** with ***ovary*** 1.3 cm long. ***Sepals*** 1.3-1.4 x 0.5-0.6 cm, subspathulate, rounded at apex. ***Petals*** similar but narrower, 1.3 x 0.3 cm, rounded to acute. ***Lip*** very fleshy, 3-lobed, 0.6-0.7 cm long, saccate at base; side lobes linear, erect, 3 mm long; mid-lobe recurved, triangular, acute, auriculate at base, with a hairy cushion on upper surface, apex hairless, acute; saccate spur almost closed at mouth by a dorsal and ventral fleshy callus. ***Column*** 0.25-0.3 cm long, with a very short foot; stipes obovate, hooked at apex, attached at base to a small horseshoe-shaped viscidium; pollinia 4, bilaterally compressed, in 2 pairs. Plate 20D.

HABITAT AND ECOLOGY: In the crowns of trees in primary mixed dipterocarp forest on yellow sandy clay soils and in coffee plantations. Alt. lowlands to 400 m. Flowering January and February, July-September.

DISTRIBUTION IN BORNEO: SABAH: Keningau District; Tenom District. SARAWAK: Mt. Matang; Bintulu area.

GENERAL DISTRIBUTION: Thailand, Peninsular Malaysia and Borneo.

NOTES: Flowering sometimes extends down the whole length of the stem for over 1 metre, with 2-3 flowers at each leaf node. Plants from Borneo have broader, more obtuse leaves and larger flowers than those from the mainland. They may, on further study, require taxonomic recognition. The description given above refers to Bornean plants only.

DERIVATION OF NAME: The specific epithet is derived from the Latin *scapiger*, bearing a scape, referring to the long pedicel with ovary.

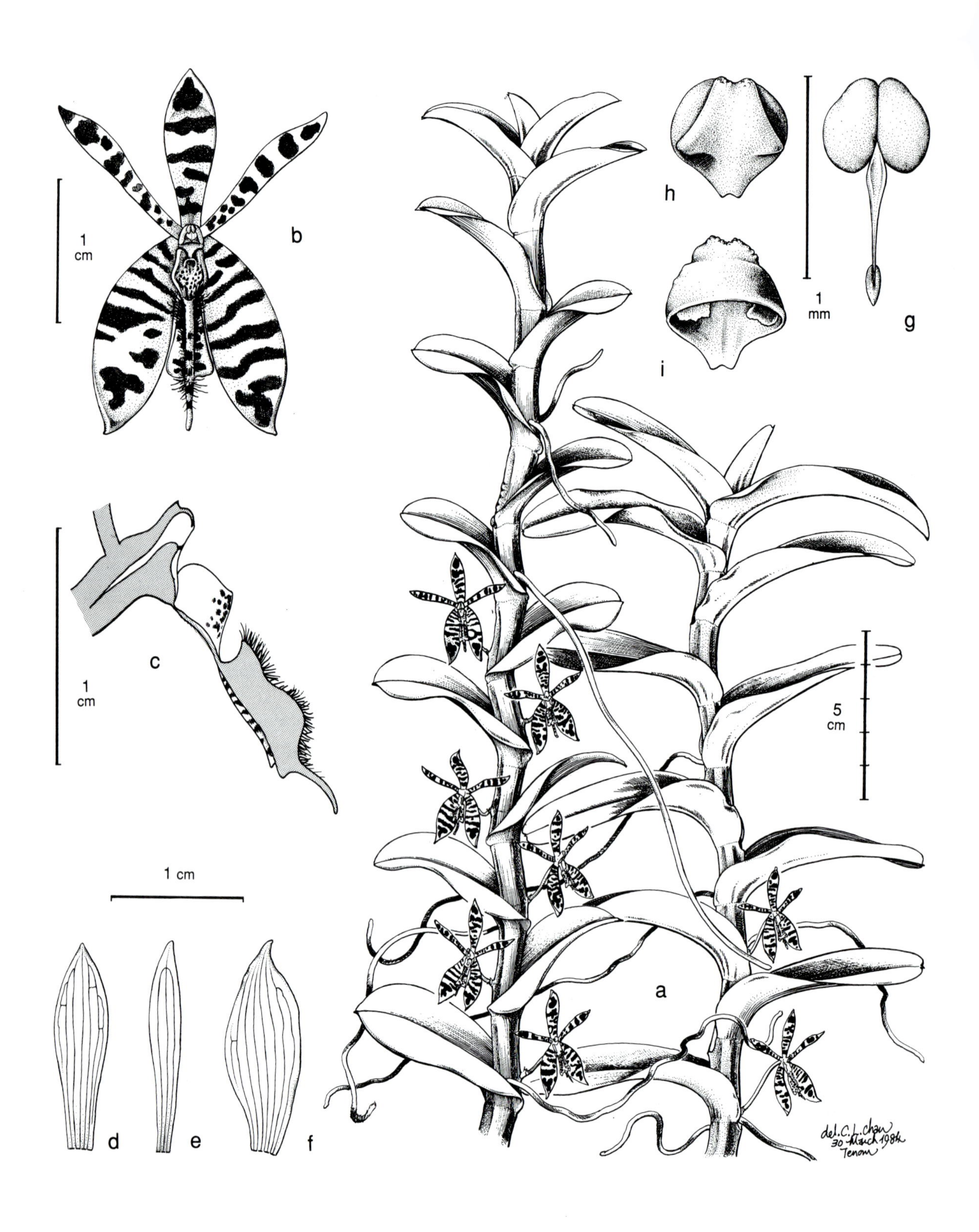

Figure 93. Trichoglottis smithii Carr - A: plant. - B: flower. - C: column and lip. - D: dorsal sepal. - E: petal. - F: lateral sepal. - G: pollinarium. - H: anther, abaxially. - I: anther, adaxially. All drawn from cult. TOC by Chan Chew Lun and Jaap J. Vermeulen.

93. TRICHOGLOTTIS SMITHII Carr

Trichoglottis smithii *Carr* in Gard. Bull. Straits Settlem. 8: 125 (1935). Type: based on same type as *T. quadricornuta* J.J.Sm.

T. quadricornuta J.J.Smith in Bull. Jard. Bot. Buitenzorg, ser. 3, 9: 188 (1927) non Kurz (1876). Type: Borneo, Kalimantan Timur, Koetai, cult. *van Gelder* 24 (holotype BO).

T. appendiculifera Holttum in Gard. Bull. Singapore 25: 108 (1969). Type: Borneo, Sabah, Sook Plain, *Alphonso* 1/110/66 (holotype K).

Scandent, epiphytic ***herb***. ***Stems*** 24-40 cm (up to 2 m) long, rooting at nodes, transversely elliptic in cross section, internodes 0.8-1.5 cm long. ***Leaves*** 4-5.5(-8) x (0.8-)1-1.6(-2) cm, oblong-ligulate or oblong-elliptic, obtusely unequally bilobed or retuse, midrib prominent, carinate on reverse, extended as a short apical mucro, sheaths dorsally carinate. ***Inflorescence*** 1-flowered, abbreviated; peduncle c. 3 mm long; floral bracts minute, triangular. ***Flowers*** 2.5 cm across, slightly scented; sepals and petals creamy white or pale greenish yellow, with irregular transverse reddish brown, reddish orange or crimson markings; lip white with pale violet or deep magenta spots and streaks, sometimes flushed yellow near base, hairs white and violet; column white. ***Pedicel*** with ***ovary*** 1.6-1.8 cm long. ***Sepals*** and ***petals*** spreading. ***Dorsal sepal*** 1.3 x 0.35-0.4 cm, narrowly oblanceolate, obtuse to acute. ***Lateral sepals*** 1.4 x 0.45-0.55 cm, obliquely narrowly elliptic, acute. ***Petals*** 1.3 x 0.2 cm, linear-elliptic, acute. ***Lip*** 0.9-1.1 cm long, 0.4 cm wide across side lobes (when flattened), 0.3-0.35 cm wide across mid-lobe; side lobes 0.25 cm long, 0.15 cm high, oblong, erect; mid-lobe 0.6-0.7 cm long, oblong-cuneate, fleshy, apex 3-toothed, outer teeth falcate-triangular or rounded, central tooth subulate-linear, acute, upturned, with an acute keel on the underside; central fleshy keel 0.15-0.2 cm high, with dense covering of pilose hairs; back wall tongue obtusely bilobed, margin ciliate. ***Column*** 0.3 cm long, foot 0.2 cm long; pollinia 4, in 2 unequal pairs. Plate 20E.

HABITAT AND ECOLOGY: This species is often found climbing up the trunks of trees in lightly shaded situations, usually in fairly open forest on poor sandstone ridge soils, podsolic to peaty, swampy forest, in predominantly *Gymnostoma* forest on ultramafic substrate or in lower montane forest. Alt. lowlands to 1300 m. Flowering April to June, October and November.

DISTRIBUTION IN BORNEO: KALIMANTAN TIMUR: Kutai. SABAH: Sook Plain; Tenom District; Mt. Kinabalu. SARAWAK: Dulit Range; Kuching area.

GENERAL DISTRIBUTION: Borneo and possibly Sumatra.

NOTES: This is a vigorous species that can be grown up weathered concrete pillars. It often flowers profusely along the whole stem, each flower contrasting in a startling way with the pale green foliage.

DERIVATION OF NAME: Named in honour of the eminent Dutch orchidologist Johannes Jacobus Smith (1867-1947).

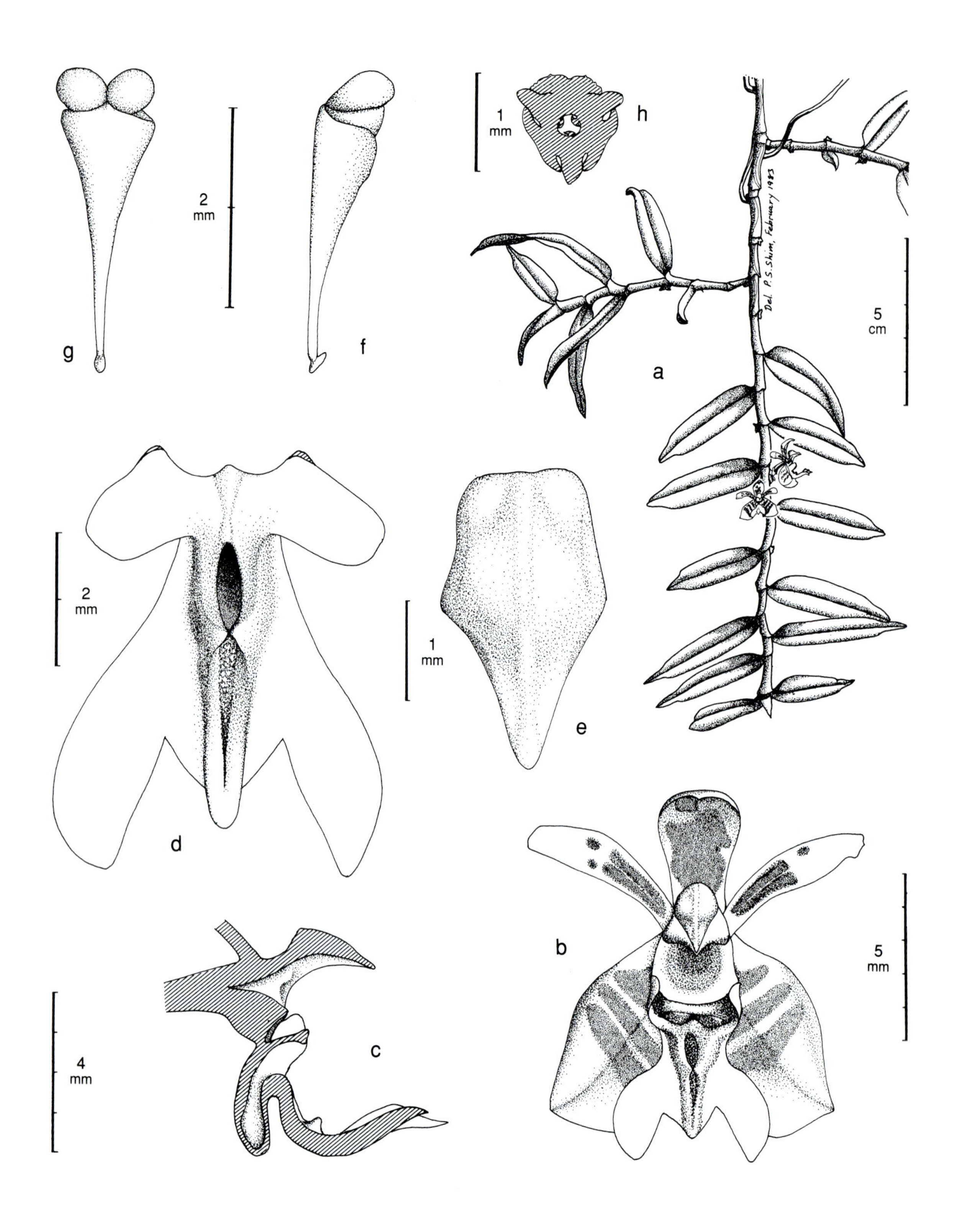

Figure 94. Trichoglottis winkleri J.J. Sm. var. **minor** J.J. Sm. - A: plant. - B: flower, front view. - C: column and lip, longitudinal section. - D: lip. - E: anther. - F: pollinarium, lateral view. - G: pollinarium, front view. - H: ovary. All drawn from living plant collected from Lohan River, Mt. Kinabalu by Shim Phyau Soon.

94. TRICHOGLOTTIS WINKLERI J.J.Sm. var. MINOR J.J.Sm.

Trichoglottis winkleri *J.J.Sm.* var. **minor** *J.J.Sm.* in Bull. Jard. Bot. Buitenzorg, ser. 2, 26: 102 (1918). Type: Java, Priangan, Tjisokan, Tjibeber, *Bakhuizen van den Brink* s.n., cult. *Winckel* (holotype BO).

Pendulous epiphytic ***herb***. ***Stems*** 20-70 cm long, internodes 1-2.5 cm long, 3 mm thick. ***Leaves*** 3-5.5 x 0.5-0.9 cm, oblong-elliptic, narrowed to an acute apex. ***Inflorescences*** 1-flowered, 1-3 at a node, sessile; floral bracts 1-2 mm long. ***Flowers*** unscented, white or pale yellowish with purple or pale reddish brown bars and spots on sepals and petals; lip white, backwall tongue often with pale yellow tips; column white to pinkish, mauve at base, anther-cap white. ***Pedicel*** with ***ovary*** 0.4-0.6 cm long. ***Dorsal sepal*** 0.6 x 0.25-0.3 cm, narrowly spathulate or oblong-elliptic, somewhat concave, apiculate. ***Lateral sepals*** 0.6-0.7 x 0.4-0.5 cm, ligulate, obtuse or subacute. ***Petals*** 0.4-0.45 x 0.15-0.18 cm, ligulate, obtuse or subacute. ***Lip*** 0.5-0.6 cm long; side lobes 0.1-0.2 cm long, oblong, truncate, erect, close to base of column, forward ends running on to 2 small, rounded, tooth-like calli at base of mid-lobe; mid-lobe 3-lobed, 0.35-0.4 cm wide, the outer lobes slightly longer, 0.35 cm long, somewhat falcate, obtuse or acute, median lobe shorter, oblong to triangular, obtuse, fleshy and shallowly sulcate; spur 0.2 x 0.2 cm, downward pointing, conical, apex rounded, transversely flattened, obtuse; back wall tongue bilobed. ***Column*** 0.3 cm long; anther-cap with medium ridge, oblong with long beak; pollinia 4 in 2 bodies on a triangular stipes. Plate 20F.

HABITAT AND ECOLOGY: Mixed hill-forest and dipterocarp forest on ultramafic and sandstone derived soils. Sometimes epiphytic on branches high up in the canopy, or lower down near the ground, often overhanging rivers and streams; tolerating both dense and light shade. Alt. 200 to 700 m. Flowering May and June, November and December.

DISTRIBUTION IN BORNEO: SABAH: Mt. Kinabalu; Tenom District.

GENERAL DISTRIBUTION: Peninsular Malaysia, Java and Borneo.

NOTES: Plants seen near the Lohan River on the lower slopes of Mt. Kinabalu had very purplish green leaves, probably due to the greater exposure to sunlight. *T. winkleri* is a variable species and intermediate forms linking var. *minor* with var. *winkleri* probably occur.

DERIVATION OF NAME: The typical variety is named after Hubert Winkler (1875-1941) who collected the type. Var. *minor* is so called because of the smaller dimensions of the flowers.

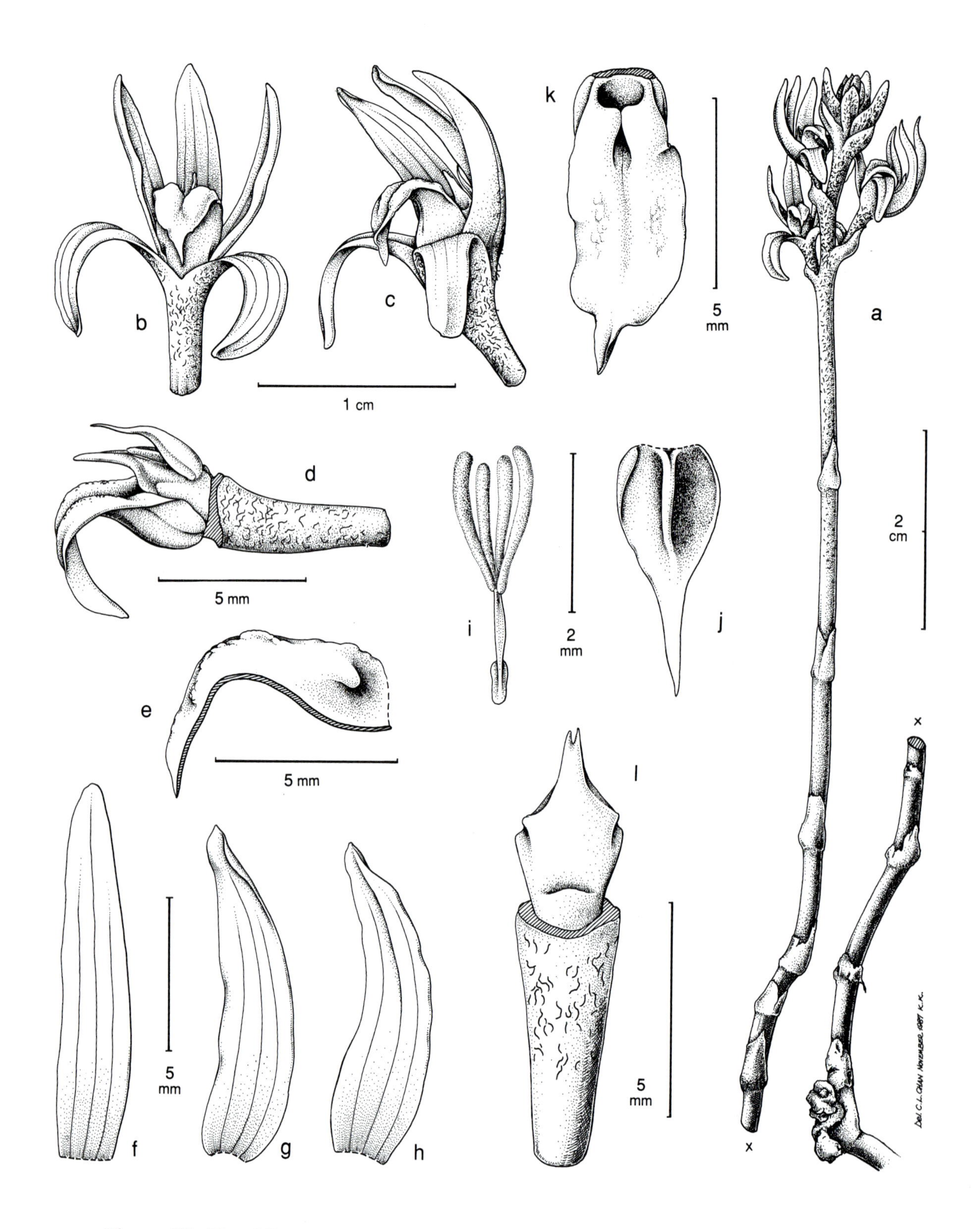

Figure 95. Tropidia saprophytica J.J. Sm. - A: plant. - B: flower, ventral view. - C: flower, lateral view. - D: flower with sepals and petals removed, lateral view. - E: lip, longitudinal section. - F: dorsal sepal. - G: petal. - H: lateral sepal. - I: pollinarium. - J: anther. - K: lip. - L: column and pedicel, dorsal view. All drawn from *Phillipps & Lamb* SNP 3019 by Chan Chew Lun.

95. TROPIDIA SAPROPHYTICA J.J.Sm.

Tropidia saprophytica *J.J.Sm.* in Mitt. Inst. Allg. Bot. Hamburg 7: 27, t. 3, fig. 16 (1927). Type: Borneo, Kalimantan, Sungai Raun, c. 350 m., 10 Feb. 1925, *Winkler* 1546 (holotype HBG).

Muluorchis ramosa J.J.Wood in Kew Bull. 39: 74, fig. 1 (1984). Type: Borneo, Sarawak, *Hansen* 437 (holotype C, isotype K).

Erect, leafless, saprophytic ***herb***. ***Stems*** 12-28 cm high, wiry, simple or repeatedly branching for much of their length, sparsely ramentaceous, chestnut-brown, bearing numerous remote ovate, amplexicaul, ramentaceous sheaths 3-5 mm long. ***Inflorescence*** 2.5-3.5 cm long, 6- to 12-flowered, lax or subdense; rachis and bracts ramentaceous; floral bracts 2-4 x 1-2 mm, stiff, ovate, acute, concave. ***Flowers*** non-resupinate, creamy-white or yellowish with a deep yellow area on lip, buds laterally compressed distally. ***Pedicel*** with ***ovary*** 0.25-0.3 cm long, ramentaceous. ***Sepals*** reflexed, ramentaceous. ***Dorsal sepal*** 0.6-0.9 x 0.2 cm, narrowly or oblong-elliptic, subacute. ***Lateral sepals*** 0.6-0.7 x 0.2-0.3 cm, narrowly elliptic, falcate, subacute, slightly carinate distally, connate to c. 0.1 cm from base when in bud. ***Petals*** 0.4-0.65 x 0.2-0.3 cm, oblong-ovate or narrowly elliptic, sometimes falcate, subacute, ramentaceous at base only. ***Lip*** 0.3-0.4 cm long, 0.2-0.34 cm wide across hypochile; hypochile concave, cymbiform, disc glabrous, with a short, low median ridge; epichile triangular, subacute to acute, strongly decurved, surface convex-inflated on either side at base, margin undulate at base. ***Column*** 0.2-0.36 x 0.2 cm, ovate-trullate; stigma entire; anther-cap ovate-elliptic, acuminate-rostrate; pollinia 2, sulcate. Plate 21A.

HABITAT AND ECOLOGY: In leaf litter on peaty top soil overlaying granitic or limestone colluvial soils. Usually in deep shade in humid hill-forest to mixed lower montane forest with chestnut and oak, often by streams. Alt. 300 to 1900 m. Flowering observed February to April and June.

DISTRIBUTION IN BORNEO: KALIMANTAN: Sungai Raun. SABAH: Mt. Kinabalu. SARAWAK: Mt. Mulu National Park.

GENERAL DISTRIBUTION: Endemic to Borneo.

NOTES: *Tropidia* is a pantropical genus containing about 20 species, only two of which are saprophytic. The remainder are rather tall, coarse plants with tough, plicate leaves and insignificant flowers that often open only partially. All have non-resupinate flowers with a spurred or saccate lip. The stellate flowers of *T. saprophytica* open widely, revealing the deep yellow epichile of the lip. A second saprophytic species, to be described shortly, was discovered recently on Mt. Lumaku in Sabah. This differs from *T. saprophytica* by its zig-zag rachis, connate lateral sepals forming a synsepal, and shortly spurred lip recalling *T. angulosa* (Lindl.) Blume.

In common with most saprophytes, *T. saprophytica* is rarely seen or collected and may be more widespread than extant collections indicate. It would be interesting to learn if the mycorrhizal fungi associated with this orchid is also associated with leaf litter or roots under specific tree species.

DERIVATION OF NAME: The generic name is derived from the Greek *tropideion*, keel, in reference to the boat-shaped lip. The specific epithet is derived from the Latin *saprophyticus*, a saprophyte, lacking chlorophyll and therefore unable to manufacture food through photosynthesis, but obtaining food materials by absorption of complex organic chemicals from the soil.

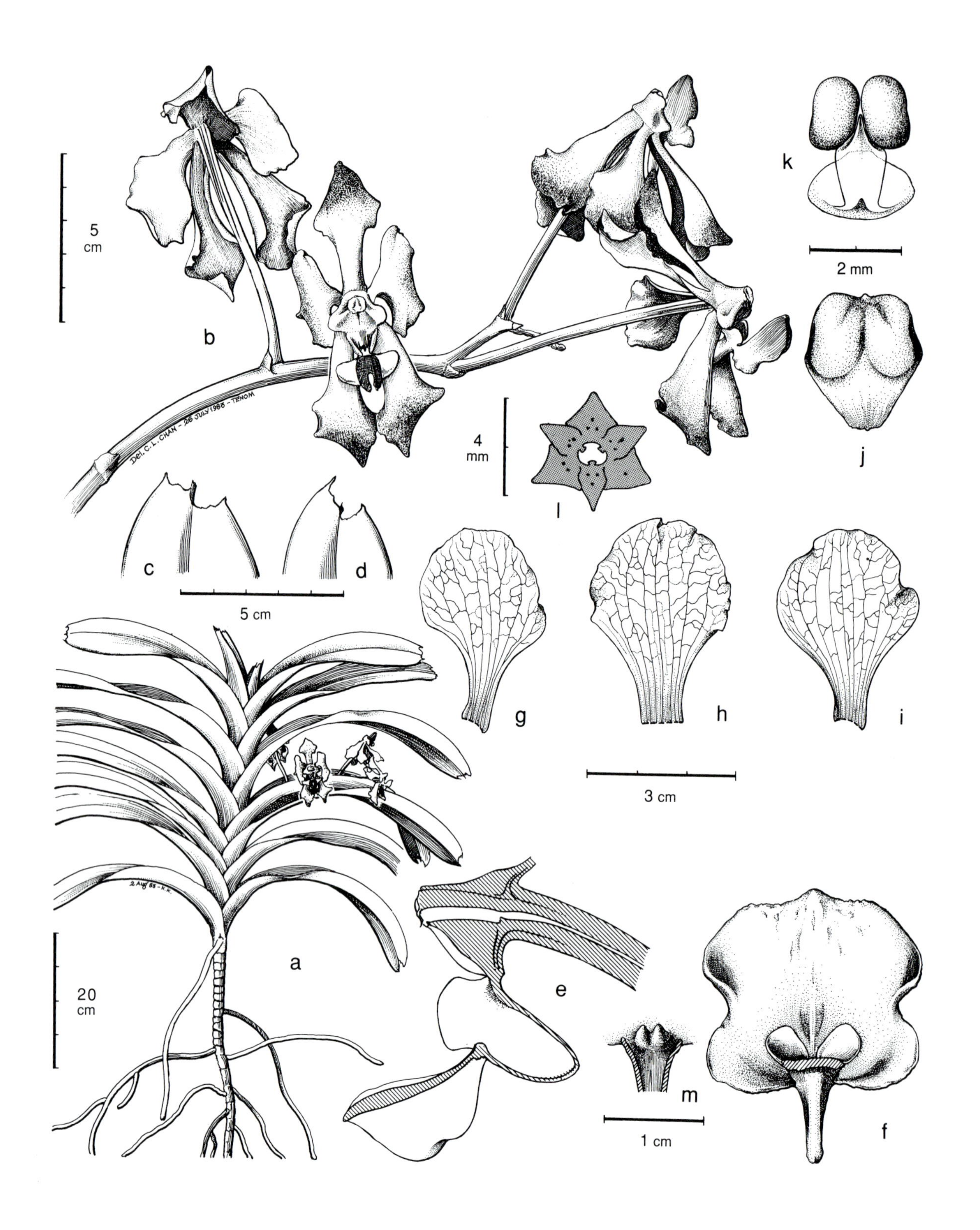

Figure 96. Vanda dearei Rchb. f. - A: plant. - B: inflorescense. - C & D: leaf apex. - E: column and lip. - F: lip. - G: petal. - H: dorsal sepal. - I: lateral sepal. - J: anther. - K: pollinarium. - L: ovary, transverse section. - M: lip base. A from cult. P.S. Shim. B - M from *Lamb* AL 106/83. Drawn by Chan Chew Lun.

96. VANDA DEAREI Rchb.f.

Vanda dearei *Rchb.f.* in Gard. Chron., n.s. 26: 648 (1886). Type: Borneo, hort. *Schroeder* (holotype W).

Epiphytic ***herb*** with thick, robust roots. ***Stems*** up to 100 cm or more long, erect to semi-pendulous, stout, the older ones having one or more new growths superposed on the basal leafless part, eventually forming a cluster of new plants, almost wholly enclosed in leaf sheaths. ***Leaves*** 35-45 x 3-3.8 cm, crowded, ligulate, apex bilobed and irregularly toothed, thick, coriaceous. ***Inflorescence*** usually 1 or 2 produced at a time at regular intervals during the year, up to 15 cm long, laxly 4- to 6-flowered; floral bracts 3-5 cm long. ***Flowers*** 7-9 cm across, fleshy, strongly scented, remaining open for one week; sepals and petals overlapping, in most plants becoming strongly reflexed after the first 2 days, pale cream, yellow (particularly in Sarawak populations) or pale yellow, flushed salmon or dull brown towards the tips and edges, rarely yellow suffused with brown all over; lip with white side lobes, mid-lobe white at base, with red lines at the junction with the side lobes, dark yellow at apex. ***Pedicel*** with ***ovary*** 6-9 cm long. ***Dorsal sepal*** 3.5-4 x 2-3 cm, broadly spathulate, with a slightly clawed narrow base, margins slightly undulate. ***Lateral sepals*** 3-4 x 2-3 cm, similar in shape. ***Petals*** 3-3.6 x 1.8-2.5 cm, spathulate, margins slightly undulate. ***Lip*** 3-lobed, 2.2-2.5 x 1.5-1.7 cm, attached to a very short column-foot, not flexible; side lobes 0.7 x 0.6 cm, inflexed; mid-lobe broad at base, with sides turned down towards apex; spur 0.7-1 cm long, with 2 small calli at the entrance, interior ecallose and aseptate. ***Column*** 1 cm long, with a very short, indistinct foot; pollinia 2, large, cleft, on a short, broad stipes; viscidium large. Plate 21B & C.

HABITAT AND ECOLOGY: Riverine forest. Based on herbarium collections and communications from other collectors, it is apparent that this species was once widespread on riverside trees, particularly in Sarawak, where it was formerly common. Alt. sea level to 300 m in the interior. Flowering March-May.

DISTRIBUTION IN BORNEO: KALIMANTAN BARAT: Sekayan River area. KALIMANTAN TIMUR: Kutai. SABAH: Kinabatangan District; Sapulot area; Tenom District, Tomani area; SARAWAK: Kuching area.

GENERAL DISTRIBUTION: Endemic to Borneo.

NOTES: This species was of great horticultural interest to breeders before and after the war. Its large, very fragrant flowers have been used in several well-known early hybrids including *Vanda* Tan Chay Yan, produced in Singapore. Cultivated plants produce flowers frequently throughout the year.

DERIVATION OF NAME: *Vanda* comes from a Sanskrit word referring to *Vanda roxburghii* R.Br. of India, the type of the genus, which is now known as *V. tessellata* (Roxb.) D.Don. The specific epithet honours Colonel Deare who sent plants to Baron Schroeder of Egham in England where it first flowered in 1886.

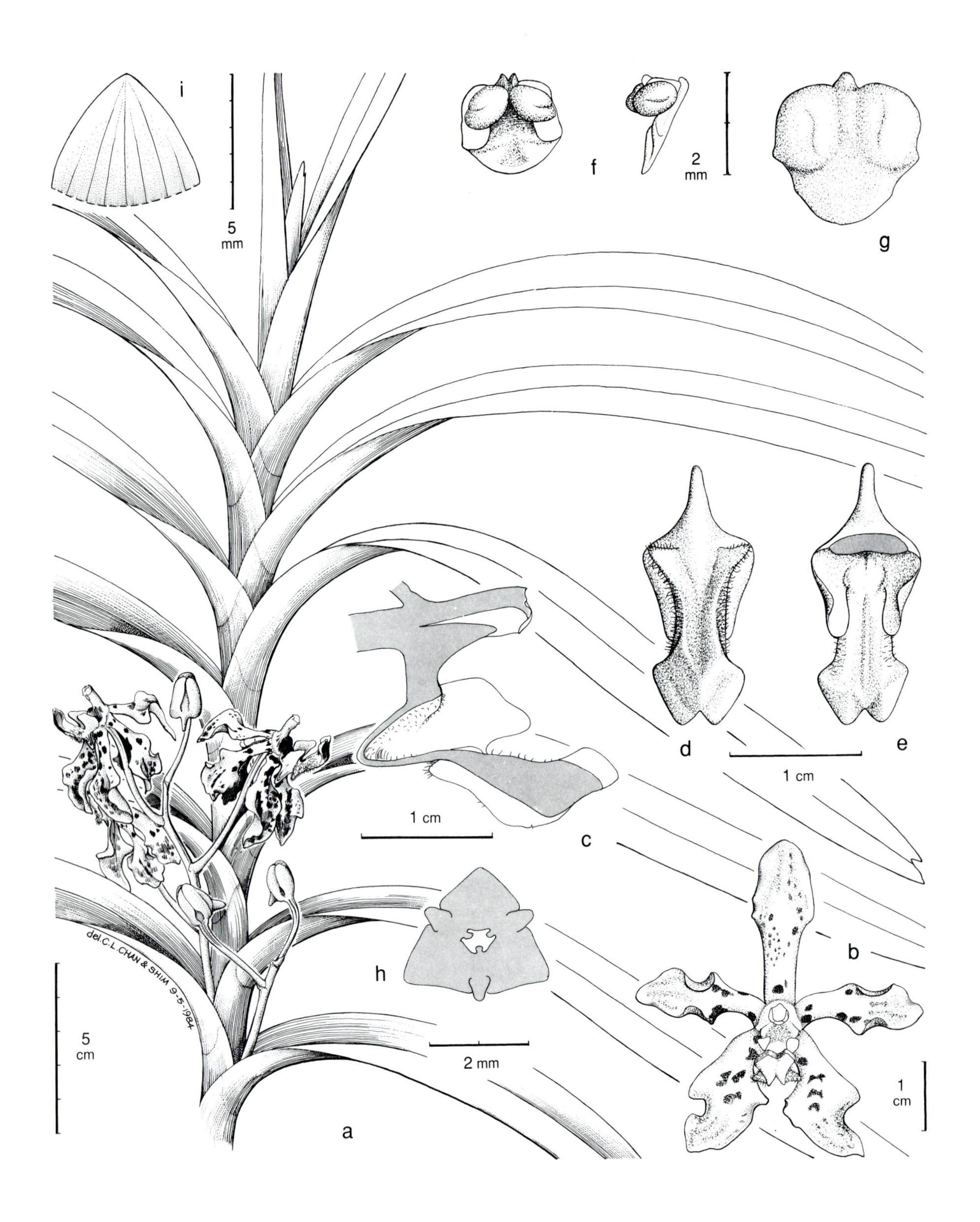

Figure 97. Vanda hastifera Rchb. f. var. **hastifera** - A: plant. - B: flower, front view. - C: colunm and lip, longitudinal section. - D: lip, back view. - E: lip, front view. - F: pollinarium, front and lateral view. - G: anther. - ovary, transverse section. - I: floral bract. All drawn from cult. Au Yong Nang Yip by Chan Chew Lun and Shim Phyau Soon.

97. VANDA HASTIFERA Rchb.f. var. HASTIFERA

Vanda hastifera *Rchb.f.* in Linnaea 41: 30 (1887). Type: Borneo, hort. *Linden* (holotype W).

Renanthera trichoglottis Ridl. in J. Linn. Soc., Bot. 31: 293 (1896). Type: Borneo, Sarawak, *Haviland* s.n. (holotype K).

var. **hastifera**

Epiphytic ***herb*** with thick, robust roots. ***Stems*** up to 100 cm or more long, the older ones having one or more new growths superposed on the basal leafless part, forming a small cluster of plants. ***Leaves*** 15-20 x 2-2.5 cm, crowded, ligulate, apex ± tridentate when young, becoming unequally bidentate, curved, articulated to a sheathing base, thin-textured. ***Inflorescence*** usually 1 or 2 produced at a time at regular intervals during the year, up to 8 cm long, laxly 4- to 6-flowered; floral bracts 2-3 mm long. ***Flowers*** 4.5-5 cm across, fleshy, scented, remaining open for up to 10 days; sepals and petals pale cream to pale yellow, unevenly and non-uniformly blotched brownish red; lip with white side lobes, sides and top of apex of mid-lobe shiny, waxy cream with underside of the apex purple to brownish purple and 2 reddish lines near the base, hairs on auricles and sides of mid-lobe white. ***Pedicel*** with ***ovary*** 4-5 cm long, white to cream. ***Sepals*** and ***petals*** reflexed. ***Dorsal sepal*** 2.2-2.5 x 0.9 cm, spathulate, edges undulate, incurved, basal claw 4 mm long. ***Lateral sepals*** 2 x 1 cm, broadly spathulate, basal claw 4 mm long. ***Petals*** 2 x 0.8 cm, spathulate, very undulate, claw narrow, 3 mm long. ***Lip*** 3-lobed, c. 2 cm long (including spur), attached to a short column-foot, not flexible; side lobes 0.6-0.8 x 0.3-0.4 cm, obliquely oblong, with 2 mm long, hirsute auricles at the base; mid-lobe 0.6 x 0.6-0.7 cm, narrowed at base, curved up to the ovate, emarginate, hoof-like apex, lower margins hirsute; spur 0.6 cm long, slightly flattened laterally. ***Column*** 0.7 cm long, with a very short, indistinct foot; pollinia 2, on a large, short, broad stipes; viscidium large. Plate 21D.

HABITAT AND ECOLOGY: Lowland, coastal and hill-forest; also recorded as a mangrove epiphyte on offshore islands. Alt. sea level. Flowering observed in December.

DISTRIBUTION IN BORNEO: KALIMANTAN BARAT: Pontianak area. SABAH: Kinabatangan District. SARAWAK: Kuching area.

GENERAL DISTRIBUTION: Endemic to Borneo.

NOTES: The typical variety differs from var. *gibbsiae* in having slightly smaller flowers with reflexed sepals and petals, and very hairy auricles and sides to the lip.

DERIVATION OF NAME: The specific epithet is derived from the Latin *hastifer*, spear-bearing, and refers to the leaf apex.

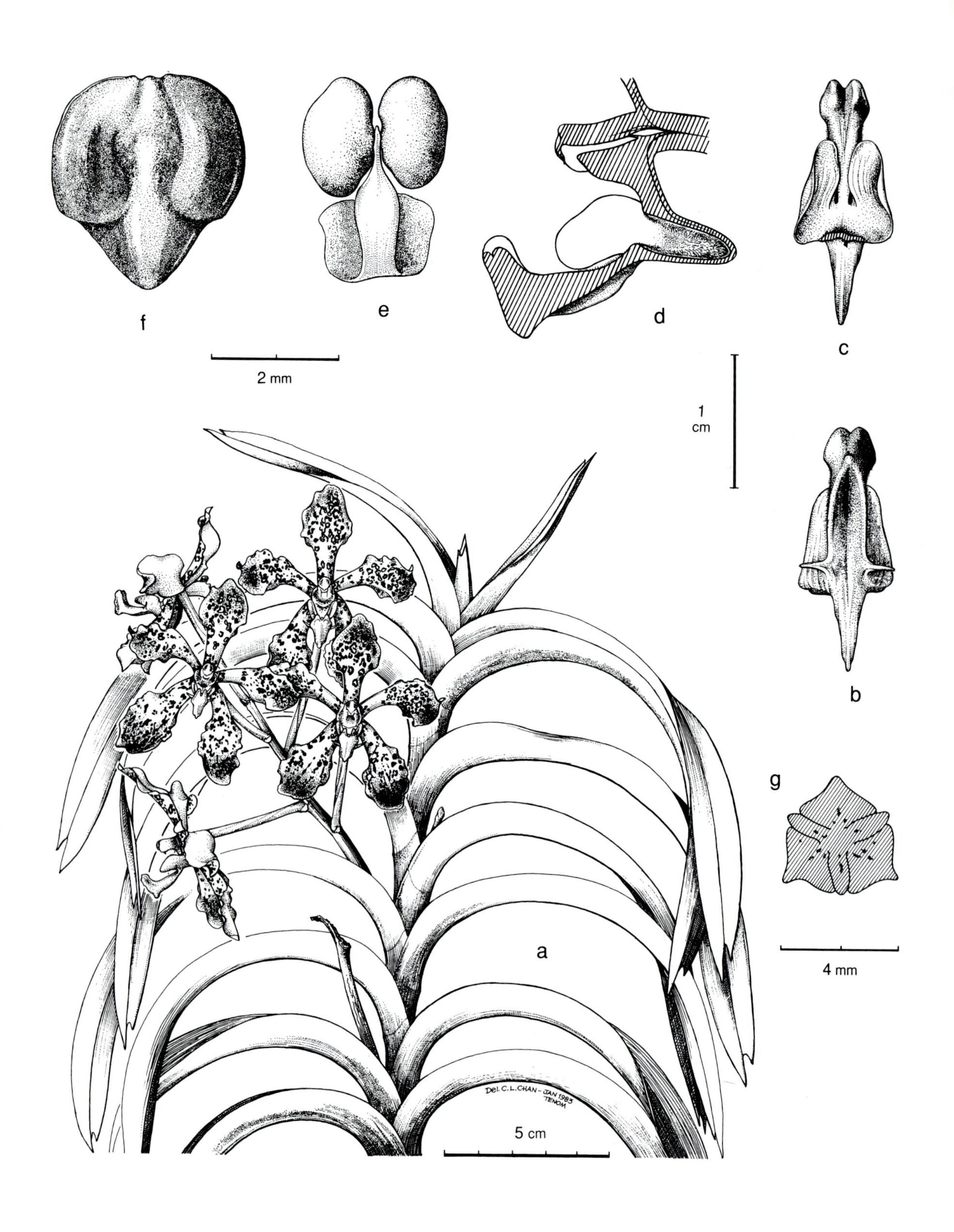

Figure 97a. Vanda hastifera Rchb. f. var. **gibbsiae** (Rolfe) Cribb - A: plant. - B: lip, ventral view. - C. lip. - D: column and lip, longitudinal section. - E: pollinarium. - F: anther. - G: ovary. All drawn from *Lamb* AL 63/83 by Chan Chew Lun.

97a. VANDA HASTIFERA Rchb.f. var. GIBBSIAE (Rolfe) P.J.Cribb

Vanda hastifera *Rchb. f.* var. **gibbsiae** *(Rolfe) P.J.Cribb* in Wood, Beaman & Beaman, Plants of Mount Kinabalu 2. Orchids: 339 (1993). Type: Borneo, Sabah, Mt. Kinabalu, Kiau/Kaung, *Gibbs* 3970 (holotype K, isotype BM).

Vanda gibbsiae Rolfe in Gibbs, J. Linn. Soc., Bot. 42: 158 (1914).

Epiphytic ***herb***. ***Stems*** up to 100 cm or more long, erect to semi-pendulous. ***Leaves*** 15-20 x 1.5-2.5 cm, ligulate, apex unequally bidentate, the teeth to 0.6 cm long. ***Flowers*** 3.5-6 cm across, sweetly scented; sepals and petals white and cream, uniformly marked with smaller, more regular reddish brown blotching; lip paler mauve-pink. ***Sepals*** and ***petals*** less reflexed than in var. *hastifera*. ***Dorsal sepal*** 2.5-2.8 x 1.3 cm. ***Lateral sepals*** 3 x 1.5 cm. ***Petals*** 2.8 x 1.6 cm. ***Lip*** with much less hirsute auricles and lower margins of the mid-lobe. Plate 21E.

HABITAT AND ECOLOGY: Hill-dipterocarp forest; lower montane oak/chestnut forest, usually epiphytic on very large trees 30 or 40 metres above ground level in light shade, preferring trunks or large branches. Alt. 800 to 1300 m. Flowering recorded in July and August.

DISTRIBUTION IN BORNEO: SABAH: Mt. Kinabalu; Tambunan District.

GENERAL DISTRIBUTION: Endemic to Borneo (Sabah only).

NOTES: This fragrant-flowered variety has far better colour and form that var. *hastifera*. Some crosses have recently been attempted with other vandas. Its original forest habitat around the lower southern and western slopes of Mt. Kinabalu has been largely cleared for vegetable cultivation.

DERIVATION OF NAME: Named after Miss Lillian S. Gibbs, the first woman botanist to climb Mt. Kinabalu, who collected the type.

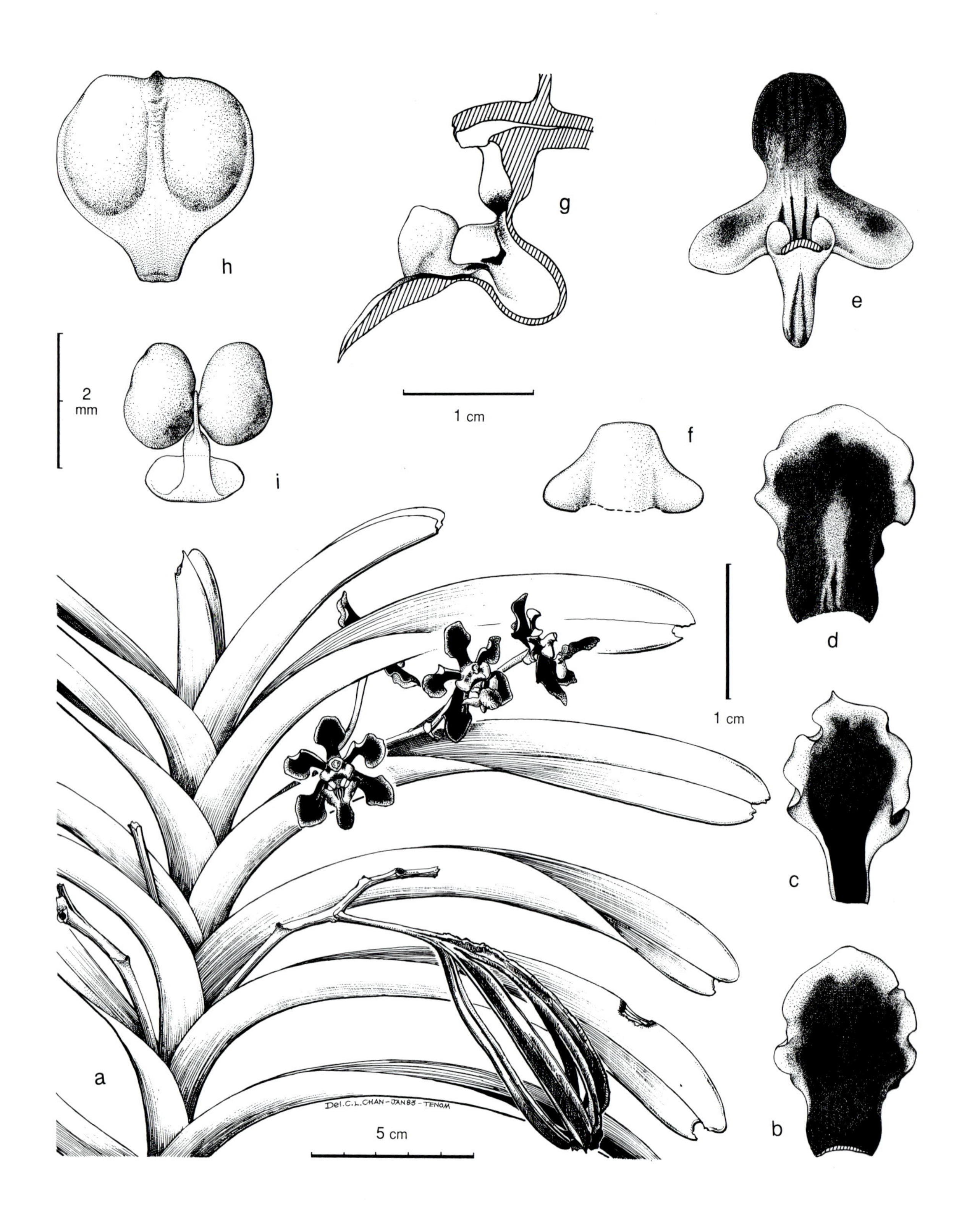

Figure 98. Vanda helvola Blume - A: plant. - B: dorsal sepal. - C: petal. - D: lateral sepal. - E: lip. - F: column, dorsal view. - G: column and lip, longitudinal section. - H: anther. - I: pollinarium. All drawn after *Jukian* in *Lamb* AL 33/82 by Chan Chew Lun.

98. VANDA HELVOLA Blume

Vanda helvola *Blume*, Rumphia 4: 49 (1849). Type: Java, *Blume* s.n. (holotype L).

Epiphytic ***herb*** with thick robust roots. ***Stems*** elongating to over 100 cm, erect but more commonly semi-pendulous to pendulous, lower portion of stem partly or wholly covered by old leaf-sheaths, older stems having one or more new growths superposed on the basal part, forming a small cluster of plants. ***Leaves*** 15-20 x 3.3-3.5 cm, close together, 2-3 cm apart, bases concealing stem, curving, rigid, leathery, ligulate, apex obtusely unequally bilobed and toothed. ***Inflorescence*** 8-11 cm long, usually 1 or 2 produced at a time at regular intervals during the year, laxly 4- to 6-flowered; floral bracts 3-4 mm long, broadly triangular. ***Flowers*** 4-5 cm across, wide-opening, fleshy, waxy, resupinate, lasting up to one week, scented, colour very variable; sepals from dull pale yellow-brown to coppery red with dull yellow edges; lip with purplish side lobes, base of mid-lobe yellow to ochre with purple lines, apex pale reddish brown or purple-brown, often with a waxy sheen. ***Pedicel*** with ***ovary*** 4 cm long, greenish white. ***Dorsal sepal*** 1.3-1.5 x 1.2 cm, spathulate, edges undulate, recurved at basal half. ***Lateral sepals*** similar but larger, 1.6-1.8 x 1.3-1.5 cm. ***Petals*** smaller, 1.3-1.5 x 0.8-1.1 cm, broadly falcate to spathulate, edges undulate and recurved at base. ***Lip*** 3-lobed, attached to a short column-foot, slightly hinged and movable; side lobes 0.3-0.4 x 0.4-0.6 cm, erect, curving towards apex, quadrate to oblong; mid-lobe 1.4-1.7 cm long, hastate, wide at base constricting to 0.5-0.6 cm before widening to 0.7-0.8 cm near apex, apex rounded, base with 4 longitudinal purple grooves; spur 0.7 x 0.25 cm, almost at right angles with blade of lip pointing obliquely downwards, laterally compressed, slightly narrowed vertically near entrance, hairy inside, apex rounded, with 2 hairy calli separated by a groove at entrance of spur. ***Column*** 0.7-0.8 cm long, thick, conical, with a short, indistinct foot; pollinia 2, cleft, large; stipes oblong; disc large; anther-cap 0.3 x 0.3 cm, with a truncate beak. ***Fruit*** 1.3 cm long. Plate 21F & 22A.

HABITAT AND ECOLOGY: Mixed hill-forest; riverine forest; mixed lower montane forest, generally on soils derived from sandstones, shales and mudstones. It is often found on trees overhanging rivers as well as on ridges. Alt. 400 to 1500 m.

DISTRIBUTION IN BORNEO: SABAH: Mt. Kinabalu; Tambunan area.

GENERAL DISTRIBUTION: Peninsular Malaysia, Sumatra, Java and Borneo. Recent collections have now been recorded from Papua New Guinea (N. Howcroft, pers. comm.) and the Philippines.

NOTES: Flowers are produced at regular intervals throughout the year, the coppery red mountain form being apparently more strongly scented. The form collected at Kundasang on the slopes of Mt. Kinabalu has beautiful cinnamon brown to coppery red coloured sepals and petals, similar to forms found in Sumatra. Paler yellowish to pinkish forms occur elsewhere in the Crocker Range.

DERIVATION OF NAME: The specific epithet is derived from the Latin *helvus*, light bay or pale red, the dingy red colour of cattle, referring to the colour of the flowers.

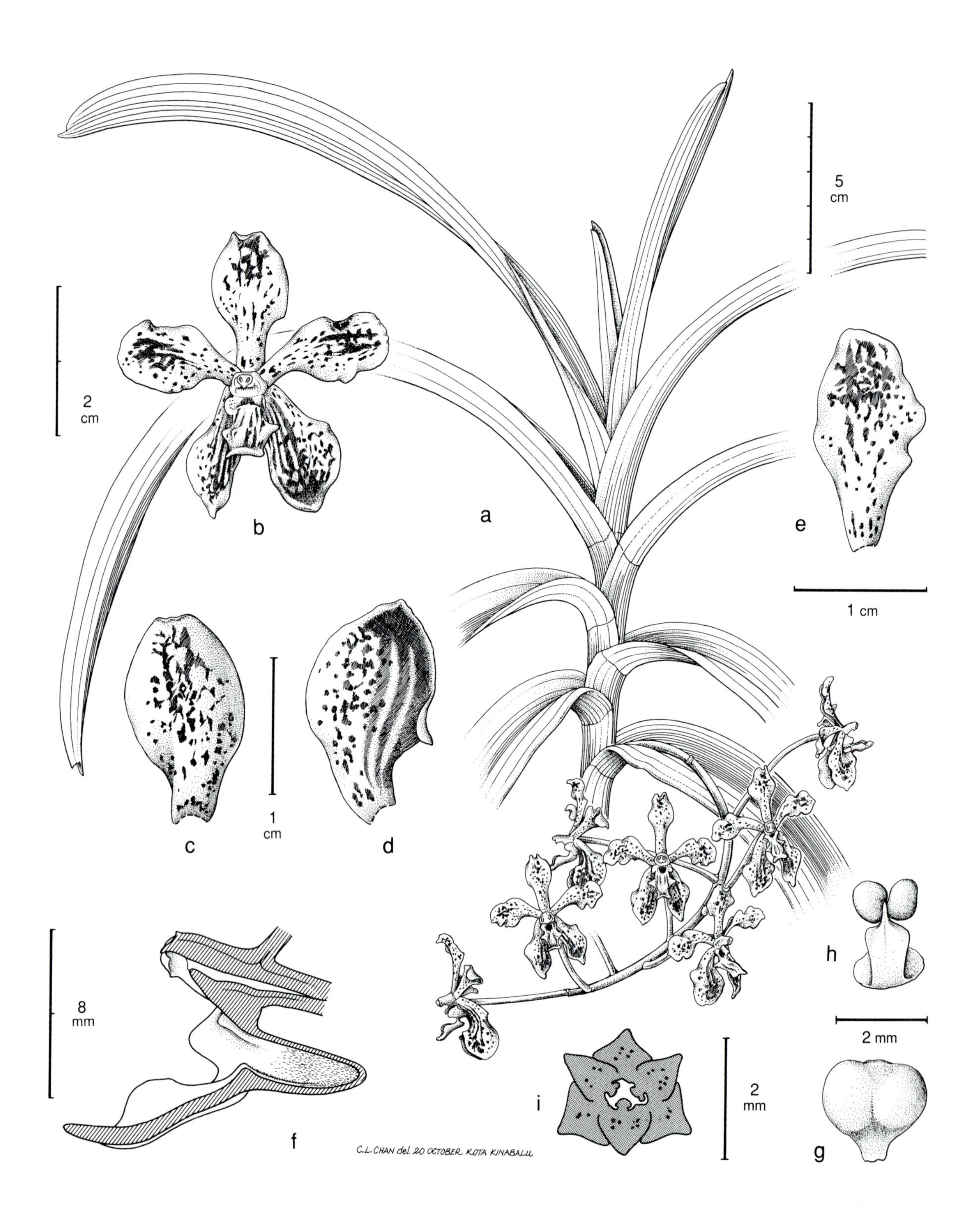

Figure 99. Vanda lamellata Lindl. - A: plant. - B: flower, front view. - C: dorsal sepal. - D: lateral sepal. - E: petal. - F: column and lip, longitudinal section. - G: anther. - H: pollinarium. - I: ovary, transverse section. All drawn from cult. A. Lamb by Chan Chew Lun.

99. VANDA LAMELLATA Lindl.

Vanda lamellata *Lindl.* in Bot. Reg. 24: misc. 66 (1838). Type: Philippines, Manila, cult. *Loddiges* s.n. (holotype K).

Vanda cumingii Lindl. ex Paxton in Loud., Hort. Brit. Suppl. 3: 654 (1850), nomen.

A medium-sized epiphytic ***herb***, often growing in clusters. ***Roots*** adventitious, unbranched, elongate, stout, up to 4 mm in diameter. ***Stem*** 6-30 cm long, 0.5-0.8 cm in diameter, covered by sharp sheathing leaf bases. ***Leaves*** 9-33 x 0.9-1.1 cm, coriaceous, conduplicate, linear, unequally roundly or obtusely bilobed at apex, articulated to sheathing leaf bases 1.3-1.5 cm long. ***Inflorescence*** erect or suberect, usually longer than the leaves, 20-30 cm long; floral bracts ovate-triangular, acute, 1.5-3 mm long. ***Flowers*** about 3 cm across, sweetly scented, spreading; sepals and petals brownish yellow to pale yellow; lip yellow and white, striped and tinged with reddish purple. ***Pedicel*** with ***ovary***, 4-5 cm long, slender, spreading, prominently six-ridged on ovary. ***Dorsal sepal*** 1.5-1.7 x 0.6 cm, reflexed, spathulate, obtuse, the sides reflexed and slightly undulate. ***Lateral sepals***, 1.4-1.6 x 0.8-0.9 cm, twisted at base to lie back to back, obliquely obovate, rounded at apex, with recurved sides towards the base and undulate sides above. ***Petals*** 1.5-1.7 x 0.6 cm, falcate-spathulate, rounded or obtuse. ***Lip*** 3-lobed, 0.9 x 0.6 cm; side lobes auriculate, erect, 0.2-0.3 cm long; mid-lobe obovate-subpandurate, truncate, 0.8 x 0.6 cm, bearing two raised fleshy lunate lamellae along length; spur 0.4-0.5 cm long, pointing backwards, subcylindric, narrowing at apex, laterally compressed, with two small raised calli in mouth and hairy within. ***Column*** 5 mm long, terete. Plate 22B.

HABITAT AND ECOLOGY: On branches and trunks of trees in forest on limestone, in coastal beach forest or in sea-cliff scrub. Often found in full sunlight and occasionally exposed to salt spray during storms. Alt. sea level to 100 m. Flowering March and April.

DISTRIBUTION IN BORNEO: SABAH: Mantanani Besar Island; Banggi Island; Tigabu Island; Kudat District.

GENERAL DISTRIBUTION: Borneo and the Philippines.

NOTES: *V. lamellata* was once plentiful on several islands off the Kudat Peninsula, but populations have been decimated in recent years. Shim (pers. comm.) says it was once a common sight on trees along estuaries on Banggi Island.

This species has been used in the Philippines as a parent to produce hybrids that make compact pot-plants with a wide range of colours. A few plants are well established in cultivation at Tenom Orchid Centre where it flowers at regular intervals throughout the year.

DERIVATION OF NAME: The specific epithet is derived from the Latin *lamellatus*, in layers or thin plates, in reference to the keels on the lip.

Figure 100. Vanda scandens Holttum - A: plant. - B: lip, ventral view. - C: lip, front view. - D: column and lip, longitudinal section. - E: dorsal sepal. - F: petal. - G: lateral sepal. - H: ovary. - I: anther. - J: pollinarium. All drawn from cult. TOC by Chan Chew Lun.

100. VANDA SCANDENS Holttum

Vanda scandens *Holttum* in Sarawak Mus. J. 5(2): 389 (1950). Type: Borneo, Sarawak, Kuching, cult. Singapore Bot. Gard. (holotype K).

Climbing, clump-forming epiphytic ***herb***. ***Roots*** adventitious, unbranched, up to 0.5 cm in diameter. ***Stems*** erect, long scandent, internodes c. 1.5 cm long, 0.8 cm in diameter, leafy in the upper part. ***Leaves*** up to 12 x 0.5-2 cm, spreading to suberect, or arcuate, linear, unequally bilobed at the apex, sometimes slightly twisted along length, articulated at the base to a sheathing leaf-base up to 1 cm long. ***Inflorescences*** lateral from upper nodes of stem, shorter than the leaves, 2-3 cm long, 2- to 5-flowered; floral bracts triangular, 2-3 mm long. ***Flowers*** 3.5-4 cm across, slightly sweetly scented; sepals and petals mustard-yellow with light brown spotting and speckling; lip pale cream to pale yellow marked with brown streaks on the mid-lobe and purple spots on the side lobes. ***Pedicel*** with ***ovary*** 4-5.5 cm long. ***Dorsal sepal*** 1.5-2 x 0.5-0.7 cm, erect, oblanceolate-subspathulate, obtuse, curving forward towards the apex and with the sides somewhat reflexed. ***Lateral sepals*** 1.7-2 x 0.8-0.9 cm, obliquely spathulate, rounded or obtuse at the apex, the sides somewhat undulate and recurved. ***Petals*** 1.7-2 x 0.5-0.7 cm, falcate-spathulate, obtuse, with somewhat undulate and recurved sides. ***Lip*** 3-lobed, 0.7-0.8 cm long; side lobes 0.3 cm long, erect, obliquely elliptic, curving forward; mid-lobe 0.5-0.6 cm long, very fleshy, clavate, laterally compressed, with a small spreading hairy tooth at the base on each side; spur 0.3-0.4 cm long, slightly tapering and laterally compressed, densely hairy within. ***Column*** 0.5 cm long, fleshy, pale yellow with brown spots on front at base; pollinia 2, very deeply cleft, attached by a short broad stipes to a large subcircular viscidium. Plate 22C.

HABITAT AND ECOLOGY: Lowland to hill-dipterocarp forest, generally on sandstone soils or limestone; lower montane forest. Preferring lightly shaded, humid positions in the canopy. Alt. sea level to 1000 m. Flowering regularly throughout the year.

DISTRIBUTION IN BORNEO: SABAH: Witti Range; Nabawan area; Tenom District. SARAWAK: Bau area.

GENERAL DISTRIBUTION: Endemic to Borneo.

NOTES: The short scapes and small, rather dull mustard-yellow-coloured flowers make this one of the least beautiful of vandas. However, it should provide good yellow colour in hybridisation. Much of its habitat has been logged or cleared for agriculture. Although endangered in the wild, it is well established in cultivation.

DERIVATION OF NAME: The specific epithet *scandens* is a Latin word meaning climbing and refers to the habit.

REFERENCES

Introduction

Ames, O. (1905-1922). Orchidaceae. Illustrations and Studies of the Family Orchidaceae. Fascicles 1-3, 5-7. The Merrymount Press, Boston.

— (1920). Orchidaceae Fascicle 6. The Orchids of Mount Kinabalu, British North Borneo. The Merrymount Press, Boston.

— (1925). Enumeration of Philippine Apostasiaceae and Orchidaceae. In Enumeration of Philippine Flowering Plants. Manila, Philippines.

Carr, C.E. (1935). Two Collections of Orchids from British North Borneo. Gard. Bull. Straits Settlem. 8(3): 165-240.

Comber, J.B. (1990). Orchids of Java. Bentham-Moxon Trust, The Royal Botanic Gardens, Kew.

Cribb, P.J. (1987). The genus Paphiopedilum. The Royal Botanic Gardens, Kew in association with Collingridge, London.

Dransfield, J., Comber, J.B., and Smith G. (1986). A Synopsis of *Corybas* (Orchidaceae) in West Malesia and Asia. Kew Bull. 41 (3):575-613.

DuPuy, D. and Cribb, P.J. (1988). The Genus Cymbidium. Christopher Helm, London.

Holttum, R.E. (1964). A Revised Flora of Malaya. Vol.1. Orchids of Malaya. 3rd Ed. 1964. Government Printing Office, Singapore.

Millar, A. (1978). Orchids of Papua New Guinea. Australia National Univ. Press, Canberra.

Minderhoud, M.E. and de Vogel, E.F. (1986). A taxonomic revision of the genus *Acriopsis*. Orchid Monographs 1:1-16. E.J. Brill and Rijksherbarium, Leiden.

Quisumbing, E.A. (1981). Philippine Orchids - The complete writings of Dr. E.A. Quisumbing. H. Valmayor (ed.) Vol. 1 & 2. Eugenio Lopez Foundation Inc., Manila.

Ridley, H.N. (1896). An enumeration of all Orchidaceae hitherto recorded from Borneo. J. Linn. Soc., Bot. 31: 261-305.

Schlechter, R. (1911). Die Orchideen von Deutsch-Neu-Guinea. Feddes Repert., Beih. 1: 1-1079. (1982). The Orchidaceae of German New Guinea (English translation of above). The Australian Orchid Foundation, Melbourne.

Seidenfaden, G. and Wood, J.J. (1992). The Orchids of Peninsular Malaysia and Singapore. Olsen & Olsen in association with The Royal Botanic Gardens Kew and Botanic Gardens Singapore.

Smith, J.J. (1905). Die Orchideen von Java. E.J. Brill, Leiden.

— (1908-1934). Die Orchideen von Niederlandisch-Neu Guinea. In Resultats de l'Expedition Scientifique Neerlandaises a la Nouvelle Guinee. E.J. Brill, Leiden.

— (1931). On a collection of Orchidaceae from Central Borneo. Bull. Jard. Bot. Buitenzorg, ser. 3, 11: 83-160.

— (1933). Enumeration of the Orchidaceae of Sumatra and Neighbouring Islands. Feddes Repert. 32: 129-386.

Sweet, H.R. (1980). The Genus *Phalaenopsis*. Orchid Digest Inc.

Turner, H. (1992). A revision of the Orchid Genera *Ania, Hancockia, Mischobulbum, Tainia*. Orchid Monographs 6. Rijksherbarium and Hortus Botanicus, Leiden.

Valmayor, H.L. (1984). Orchidiana Philippiniana. Vol 1 & 2. Eugenio Lopez Foundation Inc., Manila.

Vogel, E.F. de (1969). Monograph of the tribe Apostasieae (Orchidaceae). Blumea 17, 2:311-355.

— (1986). Revisions in Coelogyninae (Orchidaceae) II. The Genera *Bracisepalum, Chelonistele, Entomophobia, Geesinkorchis* and *Nabaluia*. Orchid Monographs 1: 17-81. E.J. Brill and Rijksherbarium, Leiden.

— (1988). Revisions in Coelogyninae (Orchidaceae) III. The Genus *Pholidota*. Orchid Monographs 3:1-118. Leiden.

— (1992). Revisions in Coelogyninae (Orchidaceae) IV. *Coelogyne* section *Tomentosae*. Orchid Monographs 6:1-42. Leiden.

Borneo, The Orchid Island

Anderson, J.A.R. (1963). The flora of the peat-swamp forests of Sarawak and Brunei including a catalogue of all recorded species of flowering plants, ferns and fern allies. Gard. Bull. Singapore 20:131-228.

Comber, J.B. (1990). Orchids of Java. Bentham-Moxon Trust, The Royal Botanic Gardens, Kew.

Flenley, J.R. (1979). The Equatorial Rain Forest: a geological history. Butterworths.

Kiew, R. (1991). The Limestone Flora, in: Kiew, R. (ed.). The State of Nature Conservation in Malaysia. Malayan Nature Soc. & IDRC, Canada.

Lamb, A. (1991). Orchids of Sabah and Sarawak, in: Kiew, R. (ed.) The State of Nature Conservation in Malaysia. pp. 78-88. Malayan Nature Soc. & IDRC, Canada.

Steenis, C.G.G.J. van (1971). Plant Conservation in Malaysia. Bull. Jard. Bot. Nat. Belg. 41: 189-202.

Whitmore, T.C. (1984). Tropical rainforests of the Far East. - Second edition. Oxford University Press.

Wong, K.M. (1990). In Brunei Forests - An Introduction to the Plant Life of Brunei Darussalam. Forestry Department, Ministry of Industry and Primary Resources, Brunei Darussalam.

Wood, J.J., Beaman, R.S. & Beaman, J.H. (1993). The Plants of Mount Kinabalu. 2: Orchids. Royal Botanic Gardens, Kew.

Wood, J.J. and Cribb, P.J. (in press). A checklist of the Orchids of Borneo. Royal Botanic Gardens, Kew.

The Life History of Orchids and Classification

Atwood, J. (1985). Pollination of *Paphiopedilum rothschildianum:* brood-site deception. National Geograph. Research, Spring 1985: 247-254.

Clements, M. (1982). Developments in the symbiotic germination of Australian terrestrial orchids. In Stewart, J. & C.N. van der Merwe, (eds.). Proc. 10th World Orchid Conference: 269-273.

— (1987). Orchid-fungus: host associations of epiphytic orchids. Proc. 12th World Orchid Conference: 80-83.

— (1988). Orchid mycorrhizal associations. Lindleyana 3: 73-83.

Dressler, R. (1981). The Orchids. Natural history and classification. Harvard University Press.

— (1983). Classification of the Orchidaceae and their probable origin. Telopea 2(4): 413- 424.

— (1990a). The Orchids. Natural history and classification, ed. 2. Harvard University Press.

— (1990b). The Neottieae in orchid classification. Lindleyana 5(2): 102-109.

— (1990c). The Spiranthoideae: grade or subfamily? Lindleyana 5(2): 110-116.

— (1990d). The major clades of the Orchidaceae-Epidendroideae. Lindleyana 5(2): 117- 125.

Warcup, J. (1985). *Rhizanthella gardneri* (Orchidaceae), its *Rhizoctonia* endophyte and close association with *Melaleuca uncinata* (Myrtaceae) in Western Australia. New Phytologist 99: 273-280.

Descriptions and Figures

Bechtel, H., Cribb, P.J. & Launert, E. (1992). The Manual of Cultivated Orchid Species, ed. 3. Blandford Press.

International Orchid Commission (1993). The Handbook on Orchid Nomenclature and Registration, ed. 4. I.O.C. London.

Jackson, B. Daydon (1971). A glossary of botanic terms, ed. 4. - Duckworth, London.

Schultes, R.E. & Pease, A.S. (1963). Generic Names of Orchids. Their Origin and Meaning. Academic Press.

Stearn, W.T. (1992). Botanical Latin, ed. 4. David & Charles.

IDENTIFICATION LIST

The following is an identification list based on a selection of specimens. Each species is arranged according to the descriptive part of the volume. Collections are cited, under each locality, in alphabetical order using the collector's name. Precise locality details for some species are withheld for reasons of conservation.

1. **Acanthephippium lilacinum:** SABAH: Crocker Range, Sinsuron Road: *Lamb* AL 329/85 (K) and *Lamb* AL 1407/92 (K). Ranau: *Collenette* 33 (BM). Keningau: *Amin* SAN 95320 (SAN); *Lamb* AL 1377 (TOC). Tenom, Paling Paling Hills: *Lamb* s.n. (TOC).

2. **Anoectochilus longicalcaratus:** SABAH: Mt. Kinabalu: Kiau, 900 m: *J. & M.S. Clemens* 164 (AMES); Kiau View Trail, 1500 m: *Lamb* s.n. (K); Lumu Lumu, 1500 m: *J. & M.S. Clemens* s.n. (BM); Penibukan, 1200 m: *Carr* 3084 (BM, K, SING), 1200 m, 1500 m: *J. & M.S.Clemens* s.n. (BM), 3051-7 (BM) & 40513 (K). Crocker Range: Mt. Alab, 1500 m: *Lamb* AL 1394/91 & 1531/92 (K, TOC). Sipitang District: 900 m: *Comber* 123 (K).

3. **Arachnis breviscapa:** SABAH: Tambunan: *Lamb* SAN 88541 (K). Tenom: *Lamb* s.n. (K). SARAWAK: Bidi Cave: *Brook* s.n. (SING).

4. **Arachnis flosaeris:** KALIMANTAN: Banjarmasin: *Korthals* s.n. (L) & *Motley* 1159 (K). SABAH: Button Island, Labuan: *unknown coll.* (W). Mt. Kinabalu: Dallas, 900 m: *Carr* 3015 (SING) & *J. & M.S. Clemens* 26019 (BM, K). Tenom District: Kallang Waterfall & Lagud Seberang, along the road to the Agricultural Research Station: *Lamb* s.n. (TOC). Tambunan District, Tambunan Road: *Mikil* SAN 31853 (SAN).

5. **Arachnis hookeriana:** BRUNEI: Tutong District: between Danan and Tutong: *van Niel* 3922 (L). SABAH: near Papar: *Lamb* C7 (K). Labuan: *Motley* s.n. (K). SARAWAK: Without exact locality: *Burton* 2335 (K) & *Hewitt* 1158 (SAR).

6. **Arundina graminifolia:** KALIMANTAN: Pontianak, Andjongan: *Mondi* 274 (K). Mampurah: *Enoh* 360 (K). Bukit Raya: *Nooteboom* 4454 (BO). Sinar Baru: *Kato et al.* 10638 (L). Sintung: *Teysmann* 8444 (BO). Oloe Tjihan: *Amdjah* 325 (BO). West Kalimantan: *Winkler* 1272 (BO). SABAH: Ranau District, Mt. Kinabalu, near Fellowship hostel, c. 1700 m: *Chan* SAN 86913 (SAN); Dallas, 900 m: *J. & M.S. Clemens* 27445 (K); Tenompok, 1500 m: *J. & M.S. Clemens* 30149 (K); Mt. Kinabalu, 600 m: *Puasa* 1540 (K); Mempait Rendagong: *Amin & Donggop* SAN 56010 (SAN), Bukit Hempuen: *Amin et al* SAN 117202 (SAN), Bayan: *Amin & Jarius* SAN 116349 (SAN), Tinosupok: *Amin & Jarius* SAN 116164 (SAN). Kota Belud District, Kinosopian: *Amin & Jarius* SAN 114342 (SAN). Tambunan District, Ranau to Tambunan Road: *Tiong & Dewol* SAN 85742 (K, SAN). Tambunan: *Amin, Tuyok & Suali* SAN 60835 (SAN). Rafflesia Forest Reserve, KM 68, Tambunan Road: *Madani & Ismail* SAN 11278 (SAN). Mt. Trus madi: *Sumbinig* SAN 125473 (SAN). Sook Plain: *Lamb* AL 77/83 (K). Mt. Alab: *Lamb* SAN 87141 (K, SAN). Penampang District, Togudon/Tungol KM 48, Tambunan/Penampang Road: *Fidilis*

SAN 127710 (SAN), Path to kampung Longkogungan, Sunsuran: *Madani & George* SAN 121773 (SAN). KM 45 Tunggol, Penampang/Tambunan: *Mantor* SAN 131446 (SAN) & Tunggol Forest Reserve, KM45, Kota Kinabalu Road: *Sumbing* SAN 114602 (SAN). Sipitang District, Jalan Syarikat Fajar, Masapol: *Amin* SAN 115074 (SAN). SARAWAK: Ulu Tiau, above N. Sekroh, Mujong, Balleh: *Asah & Unyong* S.21202 (K, SAN). Ulu Sg. Gaat, Kapit, 7th Division: *Lee & Dyg. Awa* S. 50003 (SAN). Bidi Cave: *J. & M.S. Clemens* 20714 (K) & *Curtis* s.n. (K).

7. Bulbophyllum mandibulare: SABAH: locality unknown, *Burbidge* s.n. (W). Mt. Kinabalu, Kiau, 900m: *J. & M.S. Clemens* 40502; Kaung, 400 m: *Carr* SFN 27921 (K, SING). Ranau District, kampung Marakau: *Tiong* SAN 88656 (SAN), kampung Nalumad: *Sigin et al.* SAN 110630 (SAN). Near Tambunan: *Lamb* SAN 88587 (K). Tenom/Keningau boundary, Tambunan Valley: *Lamb* AL 181/84 (K). Kahung: *Collenette* 1 (BM). Ranau, Lohan: *Lohok* 26 (K).

8. Bulbophyllum microglossum: SABAH: Mt. Kinabalu, Mesilau Cave, 1900-2200 m: *Beaman* 9572 (K); Penibukan, 1200 m: *J. & M.S. Clemens* 30604 (BM) & 50341 (BM); Tenompok, 1500 m: *J. & M.S.Clemens* 28553 (BM), 29633 (BM, K) & 30106 (K); West Mesilau River: 1600-1700 m, *Beaman* 8701 (K); Mt. Nungkek, 1200 m: *J. & M.S. Clemens* 32785 (BM). Mt. Lumaku, 900 m: *Lamb* SAN 93451 (SAN). Maliau Basin, 1500 m: *Lamb* AL 1414/92 (K). SARAWAK: 4th Division, Mt. Mulu National Park: *Lamb* SAN 88907 (SAN).

9. Bulbophyllum nabawanense: SABAH. Only known from the type.

10. Bulbophyllum pugilanthum: SABAH: Mt. Kinabalu: Park Headquarters: *Price* 151 (K); Kiau View Trail: *Chan* SAN 86914 (SAN). Kemburongoh, 2400 m: *J. & M.S. Clemens* s.n. (BM); Lumu-Lumu, 1800 m: *J. & M.S. Clemens* s.n. (BM); Numeruk Ridge, 1400 m: *J. & M.S. Clemens* 40055 (BM); Tenompok, 1500 m: *J. & M.S. Clemens* s.n. (BM); Tenompok/Tomis, 1700 m: *J. & M.S. Clemens* 29440 (BM). Sipitang District, Ulu Long Pa Sia, near confluence of Maga and Pa Sia Rivers: *Wood* 669 (K); Ulu Long Pa Sia, beside Pa Sia River: *Wood* 712 (K).

11. Calanthe crenulata: KALIMANTAN: Long Poehoes: *Endert* 2412 (BOG, K). Djaro Dam, c. 10 km NE. of Muara Uja: *de Vogel* 730 (BOG, K, L). SABAH: Crocker Range: *Lamb* SAN 93457 & *Lamb* AL 616/86 (K). Mt Kinabalu: Kinateki River/Marai Parai, 1200 m: *Collenette* A 134 (BM).

12. Calanthe sylvatica: KALIMANTAN: Mt. Medadem, N. of Sangkulirang: *Kostermans* 13380 (BO, K, L). SABAH: Tenom District, Ulu Tomani, Ulu Padas River Basin: *Lamb* AL 85/83 (K). Crocker Range, Tambunan District, Sinsuron Road: *Beaman* s.n. (K). Widespread in the Mt. Lumaku area of Sipitang District. SARAWAK: 4th Division: Bukit Mersing, Anap: *Sibat ak Luang* S.21996 (K, SAR). Bario District, Mt. Batu Lawi: *Lamb* AL 218 (K). Mt. Murud: *Yii* S.44667 (SAR). Ulu Baram: *Sarawak Museum* No. 782 (SAR).

13. Calanthe truncicola: SABAH: Mt. Kinabalu: Silau Silau Trail, 1500 m: *Chan, Kiat Tan & Phillipps* s.n. (SNP); Little Mamut River, 1400 m: *Collenette* 1013 (K); Marai Parai Spur, 1500 m: *Lamb* AL 44/83 (K). Lahad Datu District, Mt. Nicola, c. 800 m: *Lamb* AL 1425/92 (K).

14. Calanthe vestita: KALIMANTAN: Depok: Cult. *von Mueller* s.n. (BO). SABAH: *Low* (fide Rolfe); Mt. Kinabalu: Liwagu River: *Bacon* s.n. (K) & *Lamb* AL 169/84 (TOC); Lohan River, 800-900 m: *Beaman* 9482 (K); Bukit Hempuen: *Lamb* s.n. 1984 (TOC). SARAWAK: first collected in the 1st Division by *Beccari* (1865-1866).

15. Ceratochilus jiewhoei: SABAH: Mt. Kinabalu: Golf Course Site, 1700-1800 m: *Beaman* 10673 (K); Penibukan, 1200 m: *J. & M.S. Clemens* 30106 (BM, K); Sediken River/Marai Parai, 1500 m: *J. & M.S. Clemens* 35173 (BM); Tenompok, 1500 m: *J. & M.S. Clemens* 26874 (BM) & 30150 (K); Bundu Tuhan, 900 m: *Darnton* 520 (BM); Lumu-Lumu (Kiau View Gap), 1700 m: *Lamb* AL 1516/92 (K). Crocker Range, Mt. Alab, 1590 m: *Lamb* AL 611/86 (K) & 1200 m: *Vermeulen & Lamb* 431 (K, L).

16. Chrysoglossum reticulatum: SABAH: Mt. Kinabalu: Kiau View Trail, 1700 m: *Lamb* SAN 87136 (SAN). Crocker Range, Mt. Alab, 1700 m: *Lamb* AL 1204/89 (K) & 1500m. *Nooteboom* 968 (SAN). Kimanis Road, 1200 m: *Lamb* SAN 89700 (K). SARAWAK: 4th Division, Bario District, 1800 m: *Lamb* AL SAR 222/85 (K).

17. Cleisostoma discolor: KALIMANTAN SELATAN: Banjarmasin: *Nauen* A956 (SING). SABAH: Mt. Kinabalu: Lohan River: *J. & M.S. Clemens* 3393 (K) & 500 m: *Lamb* AL 75/83 (K). SARAWAK: Bau, Gunung Batu: *George* 205 (K).

18. Collabium bicameratum: SABAH: Tenom District, Kallang Waterfall: *Lamb* s.n.(K). Keningau District, Kimanis Road: sight record by *Lamb*. SARAWAK: Batu Buli near Bario: sight record by *Lamb*.

19. Collabium simplex: SABAH: Crocker Range, Sinsuron Road, slopes of Mt. Alab, Moyog area: *Lamb* AL 59/83 (K) & AL 307/85 (K).

20. Corybas pictus: SABAH: Mt. Kinabalu: Minitinduk/Kinateki Divide, 1100 m: *Carr* 3205 (SING); Mt. Nungkek, 1200 m: *J. & M.S. Clemens* 32540 (K); Penibukan, 1500 m: *J. & M.S. Clemens* 51723 (BM); Eastern Shoulder, 1100 m: *Collenette* 821 (K); Mamut Copper Mine, 1300 m: *Collenette* 1046 (K); Bukit Hempuen, 1000 m: *Cribb* 89/32 (K); Liwagu River Trail, 1400 m: *Dransfield* JD 5709 (K); Power Station, 1700 m: *Dransfield* JD 5558 (K); Park Headquarters, 1500 m: *Wood* 616 (K). Sandakan Zone, Mt. Tavai, 1000 m: *Lamb* AL 646/86 (K). Crocker Range, Mt. Alab, 1400 m: *Lamb* AL 1339/91 (K).

21. Cymbidium elongatum: SABAH: Mt. Kinabalu: Marai Parai Spur: *J. & M.S. Clemens* s.n. (BM); Penataran Basin, ridge north of river: *J. & M.S. Clemens* 34331 (BM); Marai Parai Spur: *Collenette* A33 (BM), A46 (BM) & *Lamb* AL 43/83 (K). Mt. Tembuyuken: *Lamb* (sight record). Mt. Monkobo, 1600 m: *Lamb* AL 807/87 (K). SARAWAK: Mt. Murud, north side: *Burtt & Martin* 5460 (E). Mt. Mulu National Park, summit: *Martin* 538773 (SAR).

22. Cymbidium finlaysonianum: SABAH: Mt. Kinabalu: Bundu Tuhan, 1200 m: *Carr* SFN 27414 (SING). Tenom, Kallang, 330 m: *Lamb* AL 148/83 (TOC). Papar, Papar Beach: *Lamb* AL 239/84 (TOC). SARAWAK: Miri River: *Hose* 565 (K); 1st Division, Sebuaran Bau, below 500 m: *Jacobs* 5478 (K, L). Bako National

Park, Telok Asam, sea level: *Purseglove* P.5577 (K, L, SAR, SING). 4th Division, Marudi, sea level: *Synge* S.40 (K).

23. Cymbidium rectum: SABAH: Sook Plain: *Bacon* s.n. (K) & *Lamb* AL 76/83 (K). Nabawan, 450 m: *Lamb* AL 144/83 (TOC). Keningau District, Mile 8-10 Benaut Road, off Sook-Nabawan Road: *Lamb* SAN 87115 (SAN).

24. Cystorchis aphylla: SABAH: Mt. Kinabalu: Bukit Hempuen, 600 m: *Beaman* 7409a (K) & *Sinclair* s.n. (E). Lahad Datu, hill near Sungai Bole, 850 m: *Lamb* AL 1438/92 (K). SARAWAK: Mt. Mulu: *Argent et al.* 642 (E) & *Hansen* 335 (C, K). Niah: *Synge* S.600 (K).

25. Dendrobium maraiparense: SABAH: Mt. Kinabalu: Marai Parai Spur, 1600: *Lamb* SAN 93368 (K); Marai Parai Spur, 1400-1600 m: *Bailes & Cribb* 825 (K); Marai Parai Spur, 1500 m: *J. & M.S. Clemens* 32242 (BM) & 1400 m: *J. & M.S. Clemens* 32358 (BM); Marai Parai Spur, 1200 m: *Collenette* A3 (BM) & 1500 m: A81 (BM); Penibukan, 1400 m: *Carr* 3069 (SING); Penibukan, 1200 m: *J. & M.S. Clemens* 30776 (BM), 1200 m: 30971 (BM) & 1200-1500 m: 31255 (BM); Pig Hill, 1700 - 2000 m: *Sutton* 4 (K).

26. Dendrobium microglaphys: KALIMANTAN TENGAH: 16 km east of Sampit: *Alston* 13053 (BM). KALIMANTAN TIMUR: Balikpapan: *Spoel* s.n. (BO). SABAH: Bungai primary forest, Pamol: *Brentnall* 36 (K). Keningau District, near Nabawan: *Lamb* AL 61/83, SAN 92275 (K). Labuk & Sugut District, Mt. Tawai, plateau south of summit: *Vermeulen* 785 (K). Mt. Kinabalu: Melangkap Kapa, 800 m: *Lamb* AL 972/88 (K). SARAWAK: Baram: *Hewitt* 2 (K). Kuching: *Hewitt* 21 (K).

27. Dendrobium olivaceum: SABAH: Crocker Range: Mt. Alab, 1100 m: *Lamb* AL 298/84 (K) & 1000 m: *Lamb* AL 154/92 (TOC). Kimanis Road: *Lamb* AL 354/85 (K). Mt. Kinabalu: Tenompok, 1500 m: *Carr* SFN 28021 (SING).

28. Dendrobium piranha: SABAH: Mt. Kinabalu, Marai Parai Spur, 1400-1700 m: *Lamb* SAN 89671(K); Marai Parai: *J. & M.S. Clemens* 32261 (BM) & *J. & M.S. Clemens* 33135 (BM).

29. Dendrobium sandsii: SABAH: Only known from the type.

30. Dendrobium sculptum: SABAH: Kinabatangan District, Maliau Basin, Mt. Lotung, 1500 m: *Lamb* SAN 93491, LMC 2305 (K) & 1400 m: *Lamb* AL 1490/92 (TOC). SARAWAK: Dulit Range: *Richards & Synge* 421 (K).

31. Dendrobium singkawangense: SABAH: Ranau District, Crocker Range, Limbang: *Collenette* 537 (K). Penampang District, Moyog, Sinsuron Road, Mt. Alab: *Lamb* AL 250/84 (K) & 1700 m: *Kiki, Intang & Lamb* in *Lamb* AL 1559/92 (TOC). Mt. Kinabalu: Gurulau Spur, 1400: *Carr* 3144 (L, SING).

32. Dendrobium spectatissimum: SABAH: Mt. Kinabalu: Marai Parai Spur, 1600 m: *Collenette* s.n. (K), 1600 m: *Lamb* AL 42/83 (K) & 1700 m: *Bailes & Cribb* 818 (K).

33. Dimorphorchis lowii: KALIMANTAN BARAT: sight record by *Fowlie*. KALIMANTAN TENGAH: Bukit Raya and Ulu Katungan, upper Samba River,

60-80 m, and NNW of Tumbang Samba, 200 m: *Mogea & de Wilde* 3735 (BO). SABAH: Mt. Kinabalu: recorded as being collected from the Liwagu River and sold along the roadside in the 1950's (Cult. Peradenya Botanic Gardens, Sri Lanka); Poring: *Lohok* POC 91/0078 (SNP). Ranau District, Mt. Tembuyuken, 2000 m *Aban* SAN 55403 (SAN). Lamag District, Lake below Mt. Lotung, Inarat: *Cockburn* SAN 83412 (SAN). Crocker Range, Sinsuron Road, Ulu Moyog: *Lamb* T2 (K). Tambunan District, Mt. Trus Madi, Kaingaran River: *Wood* 859 (K). SARAWAK: *Beccari* 28 (FI,K), 180 (FI,K) & 2823 (FI,K). Mt. Mulu National Park, near Deer Cave, c. 200 m: sight record by *Lamb*.

34. Dimorphorchis rossii var. rossii: SABAH: Mt. Kinabalu: headwaters of the Lohan River, 500 m: *Bacon* s.n. (K); Lohan River, 600-1200 m: *Lamb* s.n. (K) & *Lohok* POC 87/0121 (SNP).

34a. Dimorphorchis rossii var. graciliscapa: SABAH: Only known from the type.

34b. Dimorphorchis rossii var. tenomensis: SABAH: Only known from the type.

35. Dyakia hendersoniana: BORNEO: locality unknown: *Curtis* s.n. (K); KALIMANTAN BARAT: Pontianak: *Coomens de Ruiter* 30 (BO); SABAH: Sandakan Zone, near Telupid, Tavai Plateau, c. 600 m: *Vermeulen* s.n. (L). Locality unknown: *Lewis* s.n. (K). SARAWAK: *Sarawak Museum* s.n. (SAR). 1st Division, Mt. Matang, 100-150 m: cult. *Au Yong Nang Yip* in *Lamb* AL SAR 270/86 (K). Cult, ex Sarawak (TOC).

36. Epigeneium kinabaluense: SABAH: Mt. Kinabalu: Gurulau Spur, 2400-2700 m: *J. & M.S. Clemens* 50659 (BM); Kemburongoh, 2400 m: *J. & M.S. Clemens* s.n. (BM) & 197 (AMES, BM); Lumu-Lumu, 2100 m: *J & M.S.* Clemens s.n. (BM, BM, BM); Marai Parai, 1500 m: *J. & M.S. Clemens* 32208 (BM) & 1500 m: 32746 (BM), 2400-2700 m: 33133 (BM,K); 1500 m: *Collenette* A 17 (BM); Marai Parai Spur, 1700 m: *Bailes & Cribb* 816 (K); *J. & M.S. Clemens* 249 (AMES), 370A (AMES) & 1600 m: *Lamb* AL 40/83 (K); Mesilau Cave, 1900-2200 m: *Beaman* 9567 (K); Mesilau Spur, 2500 m: *Lamb* LKC 3175 (K); Minirinteg Cave, 2300 m: *J. & M.S. Clemens* 29214 (BM); Panar Laban, 3400 m: *Sidek* S.50 (K, SING); Penibukan, 1500 m: *J. & M.S. Clemens* 30943 (BM) & 1800 m: 50292 (BM); Pig Hill, 1700-2000 m: *Sutton* 5 (K), 1700-2000 m: 7 (K) & 9 (K); Summit Area, 3200-3400 m: *Sands* 3881 (K); Tenompok/Lumu-Lumu, 1800 m: *J. & M.S. Clemens* 27368 (BM). SARAWAK: Mt. Mulu, NW ridge of Mt. Tamacu, 1600 m: *Argent & Coppins* 1201 (E,K). Mt. Mulu, 1700, *Hansen* s.n. (C); 1800 m: *Lamb* MAL16 (K); 1930 m: *Martin* S.37106 (K,L,SAR). Mt. Rumput: *Anderson* 172 (holotype of *Sarcopodium suberectum*, K).

37. Epigeneium longirepens: SABAH: Mt. Kinabalu: Gurulau Spur, 1500 m: *J. & M.S. Clemens* 50384 (BM, G,K,) & 1700 m: 50409 (BM,G,K); Kiau 900 m: *J. & M.S. Clemens* 68A (AMES); Lubang: *J. & M.S. Clemens* 116A (AMES); Marai Parai Spur: *J. & M.S. Clemens* 372 (AMES); Penibukan, 1200 m: *J. & M.S. Clemens* 30639 (BM,K), 1200 m: 32077 (BM), 1200 m: 40659 (BM), 1500 m: 50203 (BM, K), 1500 m: 50393 (BM,G,K); Pinosuk Plateau, 1400 m: *Beaman* 10723 (K), 1500 m: *Lamb* AL 20/82 (K); Tenompok, 1600 m: *Carr* SFN 27196 (SING).

38. Epigeneium treacherianum: KALIMANTAN: Locality unknown: cult. *Gravenhorst* (BO); KALIMANTAN BARAT: cult. *J.J. Smith* (BO); SABAH: Sandakan zone, Telupid: *Lamb* SAN 86882 (SAN); Mt. Trus Madi, *Meijer*, cult. Tenom Orchid Centre (TOC). SARAWAK: Mt. Santubong: *Ridley* s.n. (K).

39. Eria ignea: SABAH: Nabawan, c. 450 m: *Lamb* AL 314/85 (K, TOC). Cult. Tenom Orchid Centre (TOC). Sandakan Zone, Telupid, Sungai Maliau, c. 200 m: *Lamb* s.n. (SAN).

40. Eria ornata: KALIMANTAN SELATAN: Banjarmasin: *Motley* s.n. (K). Martapura: *Korthals* s.n. (BO,L) & *Teysmann* s.n. (BO). Sandak Nyaband, *Teysmann* s.n. (BO). SABAH: Mt. Kinabalu: Dallas, 900 m: *J. & M.S. Clemens* 26502 (BM, K) & 30122 (K); Dallas/Tenompok: *J. & M.S. Clemens* 29676 (BM,K); Kadamaian River, 700 m: *Carr* SFN 27038 (SING); Kaung, 300 m: *Collenette* 3 (BM); Kiau, *J. & M.S. Clemens* 29008 (BM); Tenompok, 1500 m: *J. & M.S. Clemens* s.n. (K); Poring: *Lohok* 91/0241 (K, SNP); above kampong Lohan, 600 m: *Lamb* SAN 91520 (K). Membakut: *Keith* 9370 (K). SARAWAK: Bau, Mt. Bidi: *Lee* S.45859 (SAR). Mt. Meraja, south of Bidi: *Burtt* 8151 (SAR). Bukit Krian: *Anderson* S.31971 (K, SAR). Serian District, Bukit Selabor, Lobang Mawang: *Paie* S.28078 (K, SAR).

41. Eulophia graminea: BRUNEI: Tutong District, Bukit Beruang, *Forman* 836 (K). KALIMANTAN TIMUR: Kutai, Pedikan River, near Talang: *Kostermans* 10607 (BO); KALIMANTAN BARAT: Singkawang & Pontianak: *Polak* 260 (BO). SABAH: Kota Kinabalu: *Lamb* SAN 89646 (K). Likas Bay: *Shim* 8 (K). Mt. Kinabalu: Lohan River: *Lamb* AL 310/85 (K). Beluran, Bidu-Bidu Hill, *Maikin & Francis* SAN 130667 (SAN). SARAWAK: Simangang: *Alphonso* 264 (SING).

42. Eulophia spectabilis: KALIMANTAN TIMUR: Tabang, Belajan River: *Forman* 511 (K). SABAH: Labuan: *Motley* 297 (K). Sandakan: *Creagh* s.n. (K); *Elmer* 20289 (K); *Kloss* 18972 (K). Sandakan Zone, Ulu Dusun, *Lamb* AL 1190/89 (K). Tawau: *Elmer* 20512 & 21900 (K). Mt. Kinabalu: Dallas, 900 m: *J. & M.S. Clemens* 27629 (K). Sapagaya Forest Reserve: *Kadir & Enggoh* 10346 (K). Mamut Road, 900 m: *Lamb* SAN 92339 (K). Kinabatangan District, Maliu river valley, SW of Bukit Tawai: *Sands* 3851 (K). Sipitang District, Long Pa Sia: *Wood* 730 (K). Lamag District, Sg. Mongonsom, opposite Gunung Tavai, Karamuak: *Madani* SAN 81590 (SAN). Tambunan District, Ranau to Tambunan Road 2 km west of Ranau, 800 m: *Beaman* 10512 (K). SARAWAK: locality unknown: *Beccari* 3508 (FI,K).

43. Eulophia zollingeri: SABAH: Mt. Kinabalu: Tenompok, 1500 m: *J. & M.S. Clemens* 29782 (K); Bundu Tuhan, 900 m: *Lamb* AL 13/82 (K).

44. Gastrochilus patinatus: SABAH: Lahad Datu District, Danum Valley: *Argent* in *Lamb* AL 331/85 (K). Tenom District, Ulu Tomani, near Padas River: *Lamb* AL 102/83 (FI,K).

45. Habenaria setifolia: SABAH: Mt. Kinabalu, Haye-Haye River, 1000 m: *Lamb* AL 50/83 (K); near Dahobang River, 1300 m: *Phillipps & Lamb* SNP 3022 (K, SNP); Kundasang, 1200 m: *J. & M.S. Clemens* 51470 (BM); Penibukan, 1200-1500 m: *J. & M.S. Clemens* 31625 (BM).

46. Kingidium deliciosum: KALIMANTAN SELATAN: Banjarmasin: *Motley* s.n. (K). SABAH: Tenom District, Crocker Range, Melalap: *Lamb* SAN 87453 (SAN). Labuk Valley, Sungai Manta Langis: *Collenette* 595 (K).

47. Liparis latifolia: KALIMANTAN: Mt. Pamattin: *Korthals* s.n. (L). SABAH: Mt. Kinabalu: Bambangan River, 1500 m: RSNB 4529 (K); Kaung, 500 m: *Carr* SFN 27388 (SING) & 400 m: *Darnton* 385 (BM); Kiau, 900 m: *J & M.S. Clemens* 180 (AMES); Penibukan, 1200 m: *J. & M.S. Clemens* 40863 (BM); Tahubang River, 1200 m: *J. & M.S. Clemens* 40376 (BM); Tenompok, 1500 m: *J & M.S. Clemens* 26123 (BM, K), & 29677 (BM). Tenom District: Senagang River, 200 m: *Lamb* s.n. (TOC). Tambunan District, foothills of Mt. Trus Madi, Kaingaran village, 690 m: *Wood* 913 (K). SARAWAK: 5th Division, Bukit Tebunan, Ulu Trusan, *Lee* S.52336 (K, SAR).

48. Luisia curtisii: SABAH: Mt. Kinabalu: Bambangan River, 1500: *RSNB* 4529 (K); Kaung, 500 m: *Carr* SFN 27388 (SING); Tenompok, 1500 m: *J. & M.S. Clemens* 26123 (BM, K); Bukit Hempuen, c. 700 m: *Surat* s.n. (TOC). Sandakan Peninsula, Ulu Dusun, 50 m: *Lamb* AL 1168/89 (K, TOC). Tambunan District, Sinsuron Road: *Lamb* C17 (K).

49. Malaxis lowii: SABAH: Lahad Datu District, Ulu Sungai Danum: *Stone et al.* SAN 85238 (K, SAN). Mt. Kinabalu: Dallas, 900 m: *J. & M.S. Clemens* s.n. (BM), 26130 (K), 26602 & 26690 (BM); Bukit Hempuen, 800-1000 m: *Cribb* 89/28 (K); Mesilau River, 2100 m: *J. & M.S. Clemens* 51148 (BM); Penataran River, 500 m: *Beaman* 8858 (K); Tenompok, 1500 m: *J. & M.S. Clemens* 27980 (BM). SARAWAK: 4th Division, Mt. Mulu: *Nielson* 660 (AAU) & 578 (AAU).

50. Mischobulbum scapigerum: SABAH: Mt. Kinabalu: Dallas, 900 m: *J. & M.S. Clemens* s.n. (BM); Kaung/Kelawat, 800 m: *Carr* 3420 (SING). Tenom District, Kallang Falls, 350- 400 m: *Lamb* AL 197/84 (K). SARAWAK: Mt. Matang, *Ridley* s.n. (K).

51. Nephelaphyllum aureum: SABAH: Crocker Range, Mt. Alab, Sinsuron Road, 1500 m: *Lamb* AL 1332/91 (K, TOC). SARAWAK: locality uncertain, *Burtt* 5189, cult. Edinburgh 1968 (E).

52. Nephelaphyllum flabellatum: SABAH: Sipitang District, Long Pa Sia to Meligan trail: *Lamb* AL 487/85 (K). Ulu Long Pa Sia, *Wood* 746 (K).

53. Ornithochilus difformis var. difformis: SABAH: Crocker Range, Sinsuron Road, c. 900 m: *Lamb* AL 265/84 (K) & 438/85 (K).

53a. Ornithochilus difformis var. kinabaluensis: SABAH: only known from the type.

54. Paphiopedilum bullenianum: KALIMANTAN: Bukit Mulu: *Winkler* 514 (L). KALIMANTAN BARAT: Mt. Klamm, Kapuas River: *Hallier* s.n. (BO). SARAWAK: 10 miles upstream from Kuching, *Sheridan-Lea* s.n. (holotype of *P. linii*, TUB). Bako National Park: *Dransfield* s.n. (Photographs). 1st Division, *Au Yong Nang Yip* in *Lamb* AL 317/85 (K).

55. Paphiopedilum hookerae var. volonteanum: SABAH: Mt. Kinabalu: *Beaman* 8621 (K), 8990 (K) & 9393 (K); *Collenette* 540 (K); *RSNB* 4958 (K).

56. Paphiopedilum javanicum var. virens: SABAH: Mt. Kinabalu: *Bailes & Cribb* 677 (K); *Carr* 3163 (SING); *J. & M.S. Clemens* 27618 (BM, K); *Collenette* s.n. (AMES, BM, K). SARAWAK: unconfirmed reports from Kelabit Highlands.

57. Paphiopedilum lawrenceanum: SABAH: unconfirmed reports. SARAWAK: locality unknown.

58. Paphiopedilum philippinense: SABAH: offshore islands: *Hepburn & Lamb* in *Lamb* SAN 92348 (K), *Phillipps* in *Lamb* AL 1223/90 (K, TOC), *Carson* SAN 62079 & *Lamb* & *Cockburn* SAN 85346 (SAN).

59. Paphiopedilum rothschildianum: SABAH: Mt. Kinabalu: *Beaman* 9066 (K); *J. & M.S. Clemens* s.n. (BM, BO); *Collenette* s.n. (K).

60. Paphiopedilum sanderianum: SARAWAK: Mt. Mulu National Park: *Argent, Coppins & Jermy* 906 (SAR); *Lai & Jugah* S.44156 (SAR); *Lamb* AL 318/85 (K).

61. Paphiopedilum supardii: KALIMANTAN: only known from the type.

62. Papilionanthe hookeriana: KALIMANTAN BARAT: Pontianak: *Mondi* 287 (BO, K, L). KALIMANTAN SELATAN: Banjarmasin: *Motley* 922 (K). SABAH: Near Weston: *Lamb* s.n. (TOC). SARAWAK: Baram District: *Hose* 366 (AMES, K). 3rd Division, Btg. Lassa, near Rh. Jimbau: *Mamit* S.33622 (SAR).

63. Paraphalaenopsis denevei: KALIMANTAN BARAT: only known from the type.

64. Paraphalaenopsis labukensis: SABAH: Mt. Kinabalu: 800-1000 m: *Beaman* 9066 (K) & 9350a (K); *Lohok* POC 87/0416 (SNP); *Lamb* AL 109/83 (TOC).

65. Paraphalaenopsis laycockii: KALIMANTAN TIMUR: locality unknown, cult. *Lamb* SAN 89614 (drawing in K).

66. Paraphalaenopsis serpentilingua: KALIMANTAN BARAT: only known from the type.

67. Phaius baconii: SABAH: Mt. Kinabalu: Pinosuk Plateau, 1400-1500 m: *Beaman* 9192 (K), 1400 m: *Beaman* 10712 (K); Penibukan, 1200 m: *Carr* SFN 26534 (SING), 1200 m: *J. & M.S. Clemens* 30848 (BM), 1500 m: *J. & M.S. Clemens* 30995 (BM), 1200-1500 m: *J. & M.S. Clemens* 31366 (BM, K), 1200 m: *J. & M.S. Clemens* 40253 (BM), 1500 m: *J. & M.S. Clemens* 51716 (K).

68. Phaius borneensis: KALIMANTAN TIMUR: Sungai Susuk, 10-20 m: *Kostermans* 5444 (BO) & 5539 (BO). SABAH: Mt. Kinabalu: Penataran River, 500-600 m: *Lamb* AL 551/86 (K); Mesilau River, 1500 m: *RSNB* 7030 (K).

69. Phaius reflexipetalus: SABAH: Mt. Kinabalu: Lohan River, 750 m: *Bacon* in *Lamb* AL 164/83 (K).

70. Phalaenopsis amabilis: SABAH: Banggi Island: *Fraser* 228 (K). Tenom District, Crocker Range: *Gibbs* 2704 (BM, K, US) & *Lamb* AL 120/83 (K). Mt.

Kinabalu: Lohan River, 700-800 m: *Beaman* 9077 (K) & 700 m: *Beaman* 9158 (K); Penibukan, 1200-1500 m: *J. & M.S. Clemens* 31626 (BM, BO); Tenompok, 1500 m: *J. & M.S. Clemens* 2867 (BM, BO, K); Poring, 600 m: *Lohok* POC 91/0332 (SNP).

71. Phalaenopsis cornucervi: KALIMANTAN BARAT: Pontianak: *Foerstermann* s.n. (W). KALIMANTAN SELATAN: Banjarmasin: *Delmaes* 21 (BO) & *Motley* 1011 (K). KALIMANTAN TIMUR: West Koetai, Benoewatoewa: *Endert* 1645 (L). Gesit Loewai, Mahakam: *van Gelder* 7 (BO). SABAH: Sandakan District, Ulu Dusun, Mile 32: *Lamb* SAN 81071 (SAN). SARAWAK: locality unknown: *Curtis* s.n. (W). Tatau: *Purseglove* P.5490 (K). Padawan, Bukit Pait, 61 km from Kuching: *Erwin & Paul* S.27435 (SAR). Lawas: *Clement* s.n. (BO). Batang River, Tinjar: *Fuchs* 21219 (Shell Research Herbarium). Bau, Lobang Angin: *Yii et al.* S.51272 (K, L, SAR).

72. Phalaenopsis fuscata: KALIMANTAN TIMUR: West Koetai, Long Ibok, 130 m: *Endert* 12565 (BO). SABAH: Sandakan Peninsula, Sungai Manjang, 20 m: *Lamb* s.n. (SAN). Tenom District, Crocker Range, 300 m: *Chiu & Lamb* s.n. (TOC). Rundum area, c. 900 m: *Lamb* AL 369/85 (K, TOC).

73. Phalaenopsis gigantea: SABAH: Tawau District, Merotai, *Ronnie Young*, many plants collected for cultivation (no herbarium collection). Tenom District, near Tomani: *Lamb* SAN 92261 (K), *Lamb* AL 96/83 (K) & *Lamb* AL 119/83 (TOC).

74. Phalaenopsis maculata: KALIMANTAN TIMUR: Koetai, near Long Sele, *Schlechter* 13480 (B, destroyed). SABAH: Ranau District, Sungai Merotai & Sungai Apas, sight records by *Lamb*. Mt. Trus Madi, 800 m: *Lamb* AL 1162/89 (K). Tawau District, Sungai Ulu Imbak: *Tiong* SAN 89086 (K, SAN). Danum Valley, 100-300 m: *Vermeulen & Lamb* 425 (K). SARAWAK: 1st Division, Bidi Cave: *J. & M.S. Clemens* 20710 (L, SAR) & 20728 (L), *Hewitt* s.n. (SAR). Bukit Rawan, Tebabang area, 630 m: *Awa & Paie* S.45280 (SAR). Bau: *Anderson* s.n. (SAR). Sungai Brang, 450 m: *Haviland* s.n. (SAR). Baram, *Sarawak Museum* 917 (SAR). 7th Division, Bukit Batu Tiban, Ulu Sungai Balleh, 980 m: *Yii et al.* S.52151 (AAU, K, L, MEL, SAR, SING). Mt. Mulu National Park, Sungai Tapin, 90 m: *Nielsen* 232 (AAU).

75. Phalaenopsis modesta: KALIMANTAN SELATAN: *Winkler* 2458 (AMES, L). KALIMANTAN TIMUR: West Koetai: *Endert* 4789 (BO). SABAH: Mt. Kinabalu: *Carr* 3746 (AMES); Dallas, 900 m: *J. & M.S. Clemens* 29302 (AMES, BM, K, L, NY); Lohan River, 700-900 m: *Beaman* 9260 (K), 600 m: *Clemens* 3372 (K) & *Lamb* s.n. (SAN); Sayap: *Lohok* POC 92/0295 (SNP). Tawau District (Elphinstone Province), Tawau: *Elmer* 20869 (K, L, M). Tenom District, 250 m: *Lamb* AL 260/84 (K). Interior Zone, Sungai Sapulut: *Lohok* in *Lamb* AL 207/84 (K, TOC). Beluran District, Sg. Monyed: *Lamb* SAN 84150 (SAN); Mt. Trus Madi, near Kaingaran, 700 m: *Lamb & Surat* in *Lamb* AL 1162/89 (K).

76. Phalaenopsis pantherina: SABAH: Labuan, *Burbidge* s.n. (W). Tenom District, Paling-Paling Hills, kampong Malaing, 600-800 m: *Surat & Lamb* in *Lamb* AL 439/85 (K). SARAWAK: Marudi and Miri, sight records by *Lamb*.

77. Phalaenopsis sumatrana: KALIMANTAN BARAT: Pontianak: *Coomans de Ruiter* s.n. (BO). SABAH: Keningau District: *Lamb* G6 (TOC). Tenom District, Mentailung River: *Lamb* C11 (K). SARAWAK: Lawas River: *Boxall* s.n. (W). 1st Division, Bukit Rawan, Tebakang area: *Awa & Paie* S.45280 (K, L, SAR).

78. Phalaenopsis violacea: KALIMANTAN: *Coomans de Ruiter* s.n. (BO). Sungai Semak: *Teijsmann* s.n. (BO). SARAWAK: *Boxall* s.n. (W). Lundu District, cult. Kuching: *Au Yong Nang Yip* (K, photo.).

79. Plocoglottis acuminata: BRUNEI: Belait District, Teraja: *Forman* 1071 (K, SAN). KALIMANTAN TIMUR: Tabang, Belayan River: *Forman* 510 (K). SABAH: Track to Long Miau, 900 m: *Lamb* AL 870/87 (K). Mt. Kinabalu: Mekedeu River, 400-500 m: *Beaman* 9647 (K). SARAWAK: Mt. Dulit, near Long Kapa: *Richards* 1035 (K), *Synge* 66 (K) & *Synge* 141 (K). Bidi Cave, near Kuching: *J. & M.S. Clemens* 20719 (K, SAR). Mt. Mulu National Park, common on alluvial flats to Camp 5 and at Park Headquarters: sight records by *Lamb*.

80. Porphyroglottis maxwelliae: KALIMANTAN BARAT: Pontianak: cult. *Bogor* (BO), *Coomans de Ruiter* 7028/23 (BO), *Haartrop* s.n. (BO), & *van Mueller* s.n. (BO). SABAH: Near Nabawan, 400-500 m: *Lamb* 121/83 (K).

81. Pristiglottis hasseltii: SABAH: Mt. Kinabalu: Kilembun Basin: *J. & M.S. Clemens* 32583 (BM); Tenompok, 1500 m: *Carr* 3736, SFN 28055 (K, SING). Crocker Range, Kimanis Road, 1200 m: *Lamb* AL 633/86 (K), & *Lohok & de Vogel* 8065 (L). SARAWAK: Mt. Matang, c. 600 m: *Haviland* 464 (SAR). Lambir National Park, ridge SW. of Bukit Lambir, 400 m: *Burtt* 11624 (E, SAR). Ulu Limbang, Sungai Pa Mario, near Mt. Batu Lawi, 1530 m: *Awa & Lee* S.44366 (SAR). Near Mt. Batu Lawi, Batu Buli, 1600 m: *Lamb* AL SAR 204/55 (K).

82. Pteroceras fragrans: SABAH: Mt. Kinabalu: Lohan River, 800 m: *Bailes & Cribb* 676 (K), *Clements* 3359 (K) & 3385 (K). Tambunan District: *Comber* 130 (K). Tambunan Valley: *Lamb* AL 398/85 (K). Tenom District, Kallang Falls, 300 m: *Lamb* AL 1483/92 (TOC). SARAWAK: Mt. Matang: *Ridley* s.n. (SING). Locality unknown, *native collector* BUR.SCI.1703 (AMES).

83. Renanthera bella: SABAH: Mt. Kinabalu: *Madani* SAN 89454 (K, L, SAN), *Bailes & Cribb* 665 (K) & *Grell* s.n. (K). Lahad Datu District, Bukit Silam: *Dransfield* 5849 (K).

84. Renanthera elongata: KALIMANTAN: Sungai Samak: *Teysmann* s.n. (BO). Nama Era, Putusiban, 100 m: *Afriastini* 934 a & b (B). KALIMANTAN SELATAN: Banjarmasin: *Motley* 867 (K). SABAH: Tenom District, Sapong, c. 250 m: *Kanis & Samuel* SAN 56143 (SAN). Paling-Paling Hills, 300 m: *Lamb* AL 74/83 (K). Kuala Penyu, Tg. Toulak/Mansud: *Amin* SAN 103019 (SAN); Hs. Kepayau: *Amin* SAN 115593 (SAN). SARAWAK: Bako National Park, Tanjong Sapi, 15 m: *Purseglove* P.5629 (K, L, SAR, SING). Kuching: *Haviland* (K), *Hewitt* 1157 (SAR) & *Seal* S.15334 (SAR). Lundu Road, Resam, path to Mt. Besi: *Paie* S.46097 (SAR). Bintulu, Sungai Labang: *Ashton* S.18138 (K, SAR).

85. Renanthera isosepala: SABAH: Labuk District, Kuala Labuk: *Dransfield* s.n. (K). Dent Peninsula, near Tungku, sea level: *Hepburn* s.n. (cult. TOC).

86. Renanthera matutina: SABAH: Tawau District, Tawau Hills National Park, Sungai Merotai: 1989 Sabah Parks Expedition, *Lim & Saharan* s.n. (SNP). Keningau District, near Nabawan, c. 550 m: *Lamb* AL 80/83 (K) & *Lamb* SAN 87452 (K, SAN); Benaud Road, 8-10 mile Sook, Nabawan: *Lamb* SAN 87452 (SAN). Tenom District: Kallang, c. 350 m: *Lamb* AL 1180/89 (K, TOC). Mt. Kinabalu: Poring, c. 600 m: *Lohok* POC 87/0254 (SNP).

87. Renantherella histrionica subsp. auyongii: SARAWAK: Damai Beach near Mt. Santubong: *Au Yong Nang Yip*, cult. *Lamb* AL SAR 265/86 (K).

88. Spathoglottis gracilis: SABAH: Mt. Kinabalu: Bambangan River, 1700 m: *RSNB* 1323 (K); Dapatan/Marai Parai Spurs, 1500 m: *Gibbs* 4076 (BM); Kinateki River, 1200 m: *Haviland* 1291 (K); Kinateki River/Marai Parai, 900-1500 m: *Collenette* A 2 (BM); Lohan River, 800-1000 m: *Beaman* 9070a (K), 800-1000 m: *Beaman* 9340 (K); Mahandei River Head, 1400 m: *Carr* SFN 26397 (SING); Mamut Copper Mine, 1400-1500 m: *Beaman* 10341 (K); Marai Parai, 1700 m: *J. & M.S. Clemens* 30809 (BM), 1400 m: *J. & M.S. Clemens* 32361 (BM); Marai Parai Spur, 1700 m: *Bailes & Cribb* 827 (K) & 1400 m: *Lamb* SAN 93371 (K); Penibukan, 1500 m: *J. & M.S. Clemens* 31010 (BM), 1200-1500 m: *J. & M.S. Clemens* 31514 (BM), 1500 m: *J. & M.S. Clemens* 51718 (BM) & 1400 m: *Lamb* AL 45/83 (K). Lahad Datu District, Mt. Nicola, c. 850 m: *Lamb* AL 1445/92 (K). SARAWAK: 4th Division, Lambir National Park: *George* S.40440 (SAR).

89. Spongiola lohokii: SABAH: Interior Zone, Kinabatangan District, 300-500 m: *Lamb* AL 316/85 (K).

90. Trichoglottis bipenicillata: KALIMANTAN BARAT: Kapuas: *Teysmann* 10902 (BO). SABAH: Mt. Kinabalu: Penataran ridge, 600 m: *Lamb* AL 185/84 (K). Tenom District, ridge above Kallang Falls, c. 600 m: *Lamb & Surat* in *Lamb* AL 952/88 (K) & *Lamb* AL 1255/90 (TOC).

91. Trichoglottis retusa: SABAH: Sandakan Peninsula, Ulu Dusun, 40 m: *Lamb* SAN 91564 (K). Sepilok Forest Reserve: *Saikeh Lantoh* SAN 87599 (K, SAN) & *Tham Chee Keong* SAN 91038 (SAN). Tawau District, Tanjung Batu: *Orolfo* 4673 (SING). Tenom District, Paling-Paling Hills: *Lamb* AL 559/86 (K, TOC). SARAWAK: Kuching: *Haviland* 1893 (SING).

92. Trichoglottis scapigera: SABAH: Keningau District, Keningau to Tenom Road: *Lamb* AL 55/83 (K). Tenom District, Crocker Range, 350 m: *Lamb* AL 1082/88 (TOC). SARAWAK: Mt. Matang: *Ridley* s.n. (K). Ulu Stirau, Labang, Bintulu: *Ashton* S.18084 (K).

93. Trichoglottis smithii: SABAH: Sook Plain: *Alphonso* s.n. (K). Tenom District, Tenom: *Collenette* 2267 (K). Paling-Paling Hills: *Lamb* SAN 87145 (SAN) & *Lamb* C9 (K). Mt. Kinabalu: Bukit Hempuen, 800-1000 m: *Cribb* 89/26 (K); Lohan River, 700 m: *Lamb* SAN 93366 (K), 700-900 m: *Beaman* 9254 (K) & 800-1000 m: *Beaman* 10006 (K). SARAWAK: Sungai Tru, Ulu Tinjar, c. 300 m: *Richards* 2637 (L, SING). Kuching: *Haviland* 3146 (K).

94. Trichoglottis winkleri var. minor: SABAH: Mt. Kinabalu: Poring: *Cribb* 89/60 (K); Bukit Hempuen, 600-700 m: *Lamb & Surat* in *Lamb* AL 1241/90 (K, TOC) & *Lohok* POC 87/0139 (SNP). Tenom District, Crocker Range, 300 m: *Lamb* AL 36/82 (K).

95. Tropidia saprophytica: SABAH: Mt. Kinabalu: Marai Parai, Haye-Haye/ Tahubang Rivers: *Lamb & Phillipps* SNP 3019 (K) & *Lamb* AL 747/87 (K). SARAWAK: 4th Division, Mt. Mulu National Park: SE. of Camp 1: *Nielsen* 336 (C). Sungai Tutuh opposite Long Tao, Ridge Camp: *Hansen* 437 (C, K). West Ridge by Camp 4: *Argent & Coppins* 1135 (AMES, E). West Ridge: *Parris* 78/39 (CGE).

96. Vanda dearei: KALIMANTAN BARAT: Bukit Krangau, Kapuas, Sekayan River: *Hort. Bogor* 930.I.11 (BO). Telawa, *Laboleum* s.n. (BO). KALIMANTAN TIMUR: W. Koetai, Lakoen: *Endert* 1892 (BO). SABAH: Kinabatangan District, Batu Puteh: *Lamb* AL 106/83 (K). Interior Zone, Sungai Sapulut, c. 300 m: *Lamb* AL 205/84 (K). Lamag District, Sg. Pin, c. 20 m: *Dewol & Harun* SAN 89914 (SAN). SARAWAK: locality unknown: *Hose* 1893 (SING). 1st Division, Bukit Sua, Kuching, Simangang Road: *Burtt & Martin* 4727 (SAR).

97. Vanda hastifera var. hastifera: KALIMANTAN BARAT: Pontianak: cult. *Smith* (L). SARAWAK: Kuching: *Hewitt* s.n. (K). SABAH: Sandakan District, Mile 58, Telupid: *Dewol & Alexius* SAN 79465 (SAN). Cult. Jesselton: *Woolley* s.n. (K, photograph). BORNEO: without precise locality: *Curtis* s.n. (K).

97a. Vanda hastifera var. gibbsiae: SABAH: Ranau District, Limbang: *Collenette* 533 (K). Tambunan District, Sinsuron Road, near Tambunan: *Lamb* AL 63/83 (K). Mt. Kinabalu: Minitinduk Gorge, 800 m: *Carr* SFN 26692 (K, SING); Kiau, 900 m: *Carr* SFN 26760 (SING); Dallas, 900 m: *Clemens* 26307 (BM, K); Mt. Kinabalu, locality unknown: *Haslam* s.n. (AMES, BM, K).

98. Vanda helvola: SABAH: Mt. Kinabalu: Tenompok, 1500 m: *J.&.M.S. Clemens* 29009 (BM, K); Bundu Tuhan, 1000 m: *Brentnall* 103 (K); Kundasang, 800 m: *Jukian & Lamb* in *Lamb* AL 33/82 (K).

99. Vanda lamellata: SABAH: locality unknown: *Hose* s.n. (SING). Kota Belud District, Mantanani Besar Island, 70 m: *Dewol & Kumin* SAN 89999 (K, L, SAN) & 40 m: *Lajangah* SAN 36103 (SAN). Kudat District: *Keith* 7147 (SING). Banggi Island: *Lamb* SAN 93458 (TOC), *Walton* s.n. (SING) & *Boden-Kloss* 19277 (SING). Tigabu Island: *Fraser* 249 (K).

100. Vanda scandens: SABAH: Pensiangan District, Witti Range: *Lamb* AL 73/ 83 (K). Nabawan, near secondary school, c. 500 m: *Lamb* s.n. (TOC). SARAWAK: 1st Division, Lobang Angin, Bau: *Yii et al.* S.41994 (SAR).

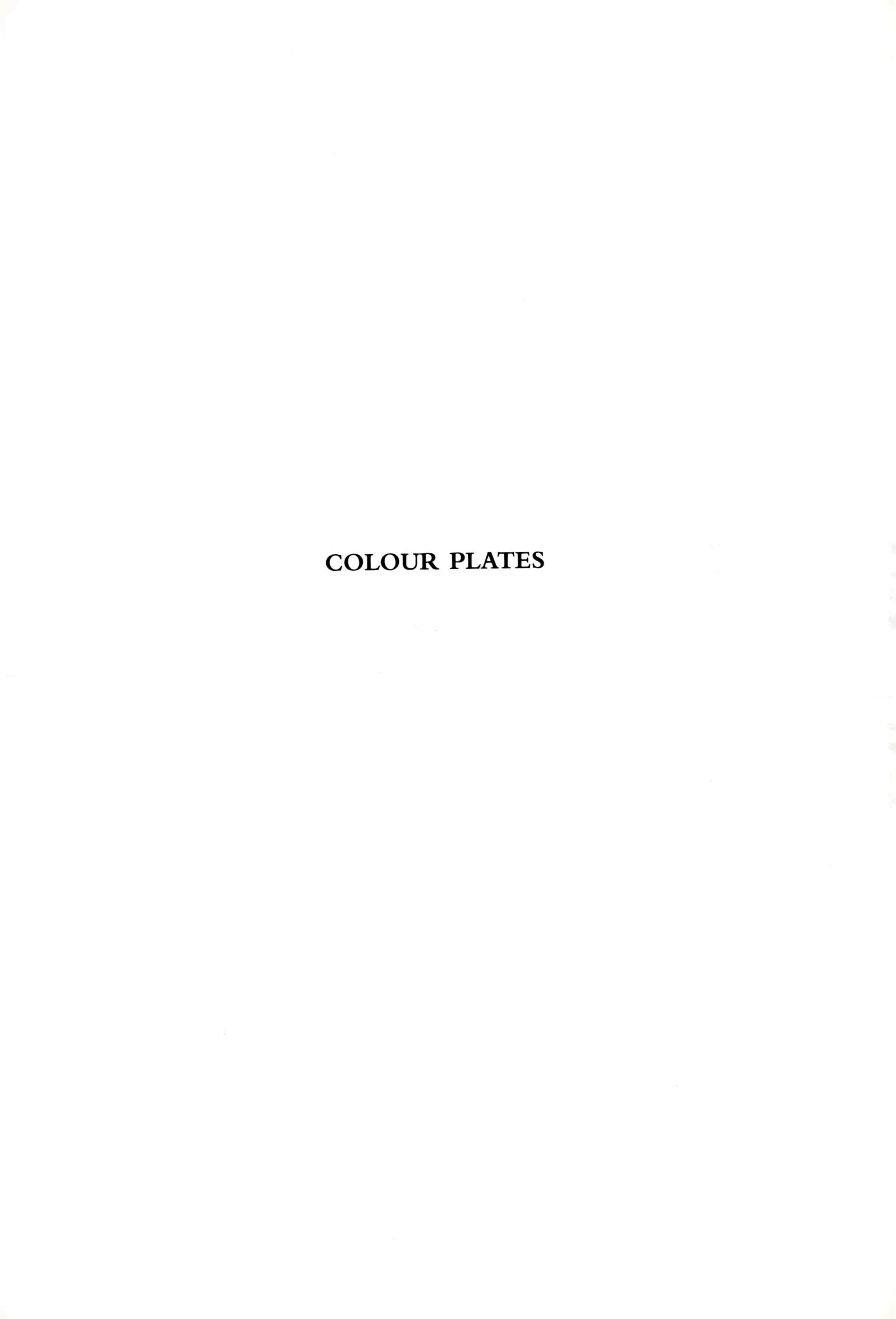

COLOUR PLATES

PLATE 1

A. *Acanthephippium lilacinum* J.J. Wood & C.L. Chan (1)

B. *Anoectochilus longicalcaratus* J.J. Sm. (2)

C. *Arachnis breviscapa* (J.J. Sm.) J.J. Sm. (3)

D. *Arachnis flosaeris* (L.) Rchb. f.(4)

E. *Arachnis hookeriana* (Rchb. f.) Rchb. f. (5)

F. *Arundina graminifolia* (D. Don) Hochr. (6)

A
B
C
D
E
F

PLATE 2

A. *Arundina graminifolia* (D. Don) Hochr. (6)

B. *Bulbophyllum mandibulare* Rchb. f. (7)

C. *Bulbophyllum microglossum* Ridl. (8)

D. *Bulbophyllum nabawanense* J.J. Wood & A. Lamb (9)

E. *Bulbophyllum pugilanthum* J.J. Wood (10)

F. *Calanthe crenulata* J.J. Sm. (11)

A
B
C
D
E
F

PLATE 3

A. *Calanthe sylvatica* (Thouars) Lindl. (12)

B. *Calanthe truncicola* Schltr. (13)

C. *Calanthe vestita* Lindl. (14)

D. *Ceratochilus jiewhoei* J.J. Wood & Shim (15)

E. *Chrysoglossum reticulatum* Carr (16)

F. *Chrysoglossum reticulatum* Carr (16)

A
B
C
D
E
F

PLATE 4

A. *Cleisostoma discolor* Lindl. (17)

B. *Collabium bicameratum* (J.J. Sm.) J.J. Wood (18)

C. *Collabium simplex* Rchb. f. (19)

D. *Collabium simplex* Rchb. f. (19)

E. *Corybas pictus* (Blume) Rchb. f. (20)

F. *Corybas pictus* (Blume) Rchb. f. (20)

A
B
C
D
E
F

PLATE 5

A. *Cymbidium elongatum* J.J. Wood, D.J. DuPuy & Shim (21)

B. *Cymbidium finlaysonianum* Lindl. (22)

C. *Cymbidium rectum* Ridl. (23)

D. *Cystorchis aphylla* Ridl. (24)

E. *Dendrobium maraiparense* J.J. Wood & C.L. Chan (25)

F. *Dendrobium microglaphys* Rchb. f. (26)

A
B
C
D
E
F

PLATE 6

A. *Dendrobium olivaceum* J.J. Sm. (27)

B. *Dendrobium piranha* C.L. Chan & P.J. Cribb (28)

C. *Dendrobium sandsii* J.J. Wood & C.L. Chan (29)

D. *Dendrobium sculptum* Rchb. f. (30)

E. *Dendrobium singkawangense* J.J. Sm. (31)

F. *Dendrobium spectatissimum* Rchb. f. (32)

A
B
C
D
E
F

PLATE 7

A. *Dimorphorchis lowii* (Lindl.) Rolfe (33)

B. *Dimorphorchis rossii* Fowlie (34)

C. *Dimorphorchis lowii* (Lindl.) Rolfe (lower flower) (33)

D. *Dimorphorchis lowii* (Lindl.) Rolfe (upper flower) (33)

A

B

C

D

PLATE 8

A. *Dimorphorchis rossii* Fowlie (lower flower) (34)

B. *Dimorphorchis rossii* Fowlie (upper flower) (34)

C. *Dimorphorchis rossii* var. *graciliscapa* A. Lamb & Shim (lower flower) (34a)

D. *Dimorphorchis rossii* var. *graciliscapa* A. Lamb & Shim (upper flower) (34a)

E. *Dimorphorchis rossii* var. *tenomensis* A. Lamb (lower flower) (34b)

F. *Dimorphorchis rossii* var. *tenomensis* A. Lamb (upper flower) (34b)

A
B
C
D
E
F

PLATE 9

A. *Dyakia hendersoniana* (Rchb. f.) Christenson (35)

B. *Epigeneium longirepens* (Ames & C. Schweinf.) Seidenf. (37)

C. *Epigeneium kinabaluense* (Ridl.) Summerh. (36)

D. *Epigeneium treacherianum* (Rchb. f. ex Hook. f.) Summerh. (38)

E. *Eria ignea* Rchb. f. (39)

F. *Eria ornata* (Blume) Lindl. (40)

PLATE 10

A. *Eulophia graminea* Lindl. (41)

B. *Eulophia spectabilis* (Dennst.) Suresh (42)

C. *Eulophia zollingeri* (Rchb. f.) J.J. Sm. (43)

D. *Gastrochilus patinatus* (Ridl.) Schltr. (44)

E. *Habenaria setifolia* Carr (45)

F. *Kingidium deliciosum* (Rchb. f.) Sweet (46)

A
B
C
D
E
F

PLATE 11

A. *Liparis latifolia* (Blume) Lindl. (47)

B. *Luisia curtisii* Seidenf. (48)

C. *Malaxis lowii* (E. Morren) Ames (49)

D. *Mischobulbum scapigerum* (Hook. f.) Schltr. (50)

E. *Nephelaphyllum aureum* J.J. Wood (51)

F. *Nephelaphyllum flabellatum* Ames & C. Schweinf. (52)

A
B
C
D
E
F

PLATE 12

A. *Ornithochilus difformis* (Wall. ex Lindl.) Schltr. var. difformis (53)

B. *Ornithochilus difformis* var. *kinabaluensis* J.J. Wood, A. Lamb & Shim (53a)

C. *Paphiopedilum bullenianum* (Rchb. f.) Pfitzer (54)

D. *Paphiopedilum hookerae* (Rchb.f.) Stein var. *volonteanum* (Sander ex Rolfe) Kerch. (55)

E. *Paphiopedilum javanicum* (Reinw. ex Lindl.) Pfitzer var. *virens* (Rchb. f.) Stein (56)

F. *Paphiopedilum lawrenceanum* (Rchb. f.) Pfitzer (57)

A
B
C
D
E
F

PLATE 13

A. *Paphiopedilum philippinense* (Rchb. f.) Stein (58)

B. *Paphiopedilum rothschildianum* (Rchb. f.) Stein (59)

C. *Paphiopedilum sanderianum* (Rchb. f.) Stein (60)

D. *Paphiopedilum supardii* Braem & Loeb (61)

E. *Papilionanthe hookeriana* (Rchb. f.) Schltr. (62)

F. *Paraphalaenopsis denevei* (J.J. Sm.) A.D. Hawkes (63)

A
B
C
D
E
F

PLATE 14

A. *Paraphalaenopsis labukensis* Shim, A. Lamb & C.L. Chan (64)

B. *Paraphalaenopsis serpentilingua* (J.J. Sm.) A.D. Hawkes (66)

C. *Paraphalaenopsis laycockii* (M.R. Hend.) A.D. Hawkes (65)

D. *Phaius baconii* J.J. Wood & Shim (67)

E. *Phaius borneensis* J.J. Sm. (68)

F. *Phaius reflexipetalus* J.J. Wood & Shim (69)

A
B
C
D
E
F

PLATE 15

A. *Phalaenopsis amabilis* (L.) Blume (70)

B. *Phalaenopsis cornucervi* (Breda) Blume & Rchb. f. (71)

C. *Phalaenopsis fuscata* Rchb. f. (72)

D. *Phalaenopsis gigantea* J.J. Sm. (73)

E. *Phalaenopsis maculata* Rchb. f. (74)

F. *Phalaenopsis modesta* J.J. Sm. (75)

A
B
C
D
E
F

PLATE 16

A. *Phalaenopsis modesta* J.J. Sm. (75)

B. *Phalaenopsis pantherina* Rchb. f. (76)

C. *Phalaenopsis sumatrana* Korth. & Rchb. f. (77)

D. *Phalaenopsis sumatrana* Korth. & Rchb. f. (77)

E. *Phalaenopsis sumatrana* Korth. & Rchb. f. (77)

F. *Phalaenopsis violacea* Witte (78)

A
B
C
D
E
F

PLATE 17

A. *Phalaenopsis violacea* Witte (78)

B. *Plocoglottis acuminata* Blume (79)

C. *Porphyroglottis maxwelliae* Ridl. (80)

D. *Porphyroglottis maxwelliae* Ridl. (80)

E. *Pristiglottis hasseltii* (Blume) Cretz. & J.J. Sm. (81)

F. *Pteroceras fragrans* (Ridl.) Garay (82)

A
B
C
D
E
F

PLATE 18

A. *Renanthera bella* J.J. Wood (83)

B. *Renanthera elongata* (Blume) Lindl. (84)

C. *Renanthera isosepala* Holttum (85)

D. *Renanthera matutina* (Blume) Lindl. (86)

E. *Renantherella histrionica* (Rchb. f.) Ridl. subsp. *auyongii* (Christenson) Senghas (87)

F. *Spathoglottis gracilis* Rolfe ex Hook. f. (88)

A
B
C
D
E
F

PLATE 19

A. *Spongiola lohokii* J.J. Wood & A. Lamb (89)

A

PLATE 20

A. *Spongiola lohokii* J.J. Wood & A. Lamb (89)

B. *Trichoglottis bipenicillata* J.J. Sm. (90)

C. *Trichoglottis retusa* Blume (91)

D. *Trichoglottis scapigera* Ridl. (92)

E. *Trichoglottis smithii* Carr (93)

F. *Trichoglottis winkleri* J.J. Sm. var. *minor* J.J. Sm. (94)

A
B
C
D
E
F

PLATE 21

A. *Tropidia saprophytica* J.J. Sm. (95)

B. *Vanda dearei* Rchb. f. (96)

C. *Vanda dearei* Rchb. f. (96)

D. *Vanda hastifera* Rchb. f. var. *hastifera* (97)

E. *Vanda hastifera* var. *gibbsiae* (Rolfe) Cribb (97a)

F. *Vanda helvola* Blume (98)

A
B
C
D
E
F

PLATE 22

A. *Vanda helvola* Blume (98)

B. *Vanda lamellata* Lindl. (99)

C. *Vanda scandens* Holttum (100)

A

B

C

GLOSSARY OF TERMS

ACRANTHOUS: applied to a sympodium with a main axis of annual portions of successive axes, each beginning with scale-leaves, and ending with an inflorescence.

ACULEATE: prickle-shaped.

ACUMEN: a tapering point.

ACUMINATE: having a gradually tapering point.

ACUTE: distinctly and sharply pointed, but not drawn out.

ADNATE: with one organ united to another.

ADVENTITIOUS: applied to roots which do not arise from the radicle or its subdivisions, but from a node on the stem, etc.

AMPLEXICAUL: clasping the stem.

ANASTOMOSE: when one vein unites with another, the connection forming a reticulation.

ANGULATE: more or less angular.

ANTHER: that part of the stamen in which the pollen is produced.

ANTHESIS: that period between the opening of the bud and the withering of the stigma or stamens.

ANTICOUS: the fore-part, i.e. that most remote or turned away from the axis.

ANTROSE: turned backwards, directed upwards.

APHYLLOUS: without leaves.

APICAL (inflorescence): borne at the top of the stem or pseudobulb.

APICULATE: furnished with a short and sharp, but not stiff, point.

APICULE: a short and sharp, but not stiff, point.

APPLANATE: flattened out or horizontally expanded.

APPRESSED: lying flat for the whole length of the organ.

ARCUATE: curved like a bow.

ARISTATE: awned.

AURICLE: a small lobe or ear.

AXILLARY (inflorescence): borne in the axil, i.e. the junction between petiole and stem.

BASAL (inflorescence): at the base of an organ or part such as the stem or pseudobulb).

BIFID: divided into 2 shallow segments, usually at the apex.

BIFURCATE: forked.

BRACT: a frequently leaf-like organ (often very reduced or absent) bearing a flower, inflorescence or partial inflorescence in its axil.

BULLATE: blistered or puckered.

BURSICLE: The pouch-like expansion of the stigma into which the caudicle of the pollinarium is inserted.

CADUCOUS: falling off early.

CAESPITOSE: tufted.

CALCEIFORM or **CALCEOLATE**: slipper-shaped.

CALLUS: a thickened area on the labellum.

CAMPANULATE or **CAMPANULIFORM**: bell-shaped.

CANALICULATE: channelled, with a longitudinal groove.

CAPILLARY (spur): slender, hair-like.

CAPITATE: like a pin-head, or knols.

CARINATE: keeled.

CATAPHYLL: the early leaf-forms of a plant or shoot, as cotyledons, bud-scales, rhizomescales, etc.

CAUDATE: tailed.

CAUDICLE: the lower stalk-like part of a pollinium, attaching the pollen-masses to the sticky disc or viscidium.

CAULESCENT: becoming stalked, where the stalk is clearly apparent.

CAULINE: borne on the stem.

CERACEOUS or **CEREOUS**: waxy.

CHAFFY: furnished with small membranous scales.

CHARTACEOUS: papery.

CILIATE: fringed with hairs.

CLAVATE: club-shaped, thickened towards the apex.

CLAW: the conspicuously narrowed and attenuate base of an organ.

CLINANDRIUM: the anther-bed, that part of the column in which the anther is concealed.

COCHLEATE: shell-shaped.

COLUMN: an advanced structure composed of a continuation of the flower-stalk, together with the upper part of the female reproductive organ (pistil) and the lower part of the male reproductive organ (stamen).

COMPLANATE: flattened, compressed.

CONDUPLICATE: folded face to face.

CONNATE: united, congenitally or subsequently.

CONNIVENT: coming into contact or converging.

CORDATE: heart-shaped.

CORIACEOUS: leathery.

COROLLA: the inner whorl of the flower.

CORYMBOSE (inflorescence): ± flat-topped.

CRENATE: scalloped, toothed with crenatures.

CRENULATE: crenate, but the toothing themselves small.

CRISPATE: curled.

CRISTATE: crested.

CUCULLATE: hooded or hood-shaped.

CUNEATE or **CUNEIFORM**: wedge-shaped.

CUSPIDATE: tipped with a sharp, rigid point.

CYMBIFORM: boat-shaped.

CYMOSE: a broad divaricately branched inflorescence, of determinate or centrifugal type.

DECURRENT: running down, as when leaves are prolonged beyond their insertion, and thus run down the stem.

DEFLEXED: bent outwards.

DENTATE: toothed.

DENTICULATE: minutely toothed.

DIAPHANOUS: permitting the light to shine through.

DISC: the face of any organ.

DISTICHOUS: leaves or flowers borne in spikelets alternating in 2 opposite ranks.

DIVARICATE: extremely divergent.

ECHINATE: prickly.

ELLIPTIC: ellipse-shaped, oblong with regularly rounded ends.

EMARGINATE: notched, usually at the apex.

ENSIFORM: sword-shaped.

ENTIRE: simple and with smooth margins.

EPICHILE: the terminal part of the lip when it is distant from the basal portion.

EPIPHYTE: a plant growing on another plant, but not parasitic.

EROSE: bitten or gnawed.

EXSERT: protrude beyond.

FALCATE: sickle-shaped.

FARINOSE: mealy.

FASCICLE: a small bunch or bundle.

FASCICULATE: clustered or bundled.

FIBRILLAE: small fibres.

FILIFORM: thread-like.

FIMBRIATE: fringed.

FLABELLATE: fan-shaped.

FLEXUOSE: bent alternately in opposite directions.

FORCIPATE: forked like pincers.

FRACTIFLEX: zigzag.

FURFURACEOUS: scurfy, having soft scales.

FUSIFORM: spindle-shaped.

GALEATE: helmet-shaped.

GEMINATE: paired.

GENICULATE: abruptly bent like a knee-joint.

GIBBOUS: pouched, more convex in one place than another.

GLABROUS: hairless.

GLAUCOUS: covered with a bluish-grey or seagreen bloom.

GLOBOSE: ± spherical.

GYROSE: curved backward and forward in turn.

HAMATE: hooked at the tip.

HASTATE: spear-or halbert-shaped, with the basal lobes turned outward.

HETERANTHOUS: an apical inflorescence produced on a separate shoot which does not develop to produce a pseudobulb and leaves.

HIPPOCREPIFORM: horseshoe-shaped.

HIRSUTE: hairy.

HISPID: bristly.

HYALINE: colourless or translucent.

HYPOCHILE: the basal portion of the lip.

HYSTERANTHOUS: used of an apical inflorescence produced after the pseudobulbs and leaves have developed.

IMBRICATE: overlapping.

INCUMBENT: lying on or against.

INFLORESCENCE: the disposition of the flowers on the floral axis or the flowers, bracts and floral axis in toto.

LABELLUM: a lip, used here for the enlarged, often highly modified abaxial petal of the orchid flower.

LACERATE: torn, or irregularly cleft.

LACINIATE: deeply slashed into narrow divisions.

LAMELLA: a membrane or septum.

LANCEOLATE: narrow, tapering to each end, lance-shaped.

LATERAL (inflorescence): borne on or near the side of pseudobulb or stem, usually in the axils of the bracts or leaves.

LAXLY: loose, distant.

LIGULATE or **LINGUIFORM**: tongue or strapshaped.

LIGULE: a tongue-like outgrowth.

LINEAR: at least 12 times longer than broad, with the sides ± parallel.

LIP: a modified abaxial petal, the labellum.

LITHOPHYTIC: growing upon stones and rocks.

LORATE: thong- or strap-shaped.

LUNATE: half-moon-shaped.

MAMMILLATE: teat-shaped.

MENTUM: a chin-like projection formed by the sepals and extended column-foot.

MESOCHILE: the intermediate portion of the lip when divided into three portions.

MONOPODIAL: growth which continues from a terminal bud from season to season.

MUCRO: a sharp terminal point.

MUCRONATE: possessing a mucro.

MULTICOSTATE: many-ribbed.

MURICATE: rough, with short and hard tubercular excrescences.

MYCORRHIZAL: of roots having a symbiotic relationship with certain fungi.

NAVICULAR: boat-shaped.

NERVOSE: prominently nerved.

NODE: that part of a stem which normally has a leaf or a whorl of leaves.

NON-RESUPINATE: of a flower that is not turned upside down.

OBCUNEATE: inversely wedge-shaped.

OBLANCEOLATE: tapering towards the base more than towards the apex.

OBLONC: much longer than broad, with nearly parallel sides.

OBOVATE: inversely ovate.

OBSCURE: dark or dingy in tint; uncertain in affinity or distinctiveness; hidden.

OBTUSE: blunt or rounded at the apex.

OVATE: egg-shaped, broader at the base.

PALEACEOUS: chaffy.

PANDURATE: fiddle-shaped.

PANICLE: a much-branched inflorescence.

PAPILLOSE: covered with soft superficial glands or protuberances, i.e. papillae.

PAPYRACEOUS: papery.

PECTINATE: combed.

PEDICEL: the stalk of a single flower in an inflorescence.

PEDUNCLE: the stalk bearing an inflorescence or solitary flower.

PULLUCID: wholly or partially transparent.

PELTATE: disc-shaped, the stalk arising from the under-surface.

PENDENT: hanging.

PERIANTH: the outer, sterile whorls of a flower, often differentiated into calyx and corolla.

PILOSE: softly hairy.

PLICATE: folded in many pleats.

PLURICOSTATE: many ribbed.

PLURISULCATE: many grooved or furrowed.

POLLINARIUM: the male reproductive system of an orchid.

POLLINIA: pollen-masses.

PORATE: set with pores.

PORRECT: directed outward and forward.

PRAEMORSE: bitten off at the apex.

PROTERANTHOUS: of an apical inflorescence produced before the pseudobulbs and leaves on the same shoot.

PROTOCORM: the embryo before primary differentiation is complete.

PSEUDOBULB: a swollen aerial stem.

PUBERULENT: slightly hairy.

PUBESCENT: softly hairy or downy.

PYRIFORM: pear-shaped.

QUADRATE: four-cornered, square.

RACEME: a single, elongate, indeterminate inflorescence with pedicellate flowers.

RECLINATE: turned or bent downward.

RECURVED: curved backward or downward.

REFLEXED: abruptly bent or turned downward or backward.

RENIFORM: kidney-shaped.

REPENT: prostrate and rooting.

RESUPINATE: upside-down, or apparently so.

RETICULATE: netted.

RETRORSE: directed backward or downward.

RETUSE: shallowly notched at a rounded apex.

REVOLUTE: rolled back from the margin.

RHACHIS: the axis of an inflorescence or compound leaf.

RHIZOME: underground stem bearing scale leaves and adventitious roots.

ROSTELLUM: the often beak-like sterile third stigma lying between the functional stigmas and stamen.

ROSTRATE: beaked.

ROSULATE: in rosettes.

ROTUND: rounded in outline.

RUGOSE: wrinkled.

RUGULOSE: somewhat wrinkled.

RUPICOLOUS: growing on or amongst rocks.

SACCATE: with a conspicuous hollow swelling.

SAGITTATE: arrow-head-shaped.

SAPROPHYTE: a plant which obtains its food materials by absorption of complex organic chemicals from the soil; often without chlorophyll.

SCANDENT: climbing.

SCAPE: a leafless peduncle arising directly from a rosette of basal leaves.

SCARIOUS: thin, dry and membranous.

SCORPIOID: a coiled, determinate inflorescence.

SECUND: directed to one side only.

SEPTATE: divided by partitions.

SEPTUM: a partition.

SERRATE: with sharp, ± regular teeth, like a saw.

SESSILE: not stalked.

SETACEOUS: thread- or bristle-like.

SETOSE: bristly.

SINUATE: with a deep wavy margin.

SPATHACEOUS: bearing a spathe.

SPATHE: a large bract sheathing an inflorescence.

SPATHULATE: oblong and attenuated at the base, like a spatula.

SPICATE: like a spike, disposed in a spike.

SPUR: a long, usually nectar-containing, tubular projection of a perianth-segment, commonly the lip.

STAMINODE: a sterile stamen, often modified in shape and size.

STELIDIA: column teeth.

STELLATE: star-shaped.

STIGMA: the receptive part(s) of the gynoecium, i.e. female sex organs, on which the pollen germinates.

STIPE (pl. STIPITES): the stalk-like support of an organ, e.g. pollinium.

STRIGOSE: beset with sharp-pointed appressed straight and stiff hairs or bristles; hispid.

SUBCYLINDRIC: half cylindric.

SUBEQUITANT (leaves): half folded sharply inwards from the midrib.

SUBSPREADING: half or somewhat having a gradually outward direction.

SUBQUADRATE: half square, somewhat square.

SUBSIMILAR: somewhat similar.

SUBULATE: awl-shaped.

SUBTEND: to extend under, or be opposite to.

SULCATE: grooved or furrowed.

SUPERPOSED: one above the other.

SYMPODIAL: growth in which each new shoot is determinate and terminates in a potential inflorescence or solitary flower.

SYNANTHOUS: when pseudobulb, leaves and apical inflorescence are produced together.

SYNSEPALUM: two sepals united together, e.g. in *Paphiopedilum* .

TERETE: circular in transverse section, cylindric and usually tapering.

TESSELLATE: chequered.

TRIDENTATE: three-toothed.

TRIFID: three-cleft.

TRIQUETROUS: three-edged.

TRULLATE: trowel-shaped.

TRUNCATE: ending abruptly, as though broken off.

TUBER: a thickened and short subterranean branch, beset with buds or 'eyes'.

UMBEL: a usually flat-topped inflorescence in which the pedicels arise from the same point on the peduncle.

UNCIFORM: hook-shaped.

UNCINATE: hooked.

UNDULATE: waved.

UNGUICULATE: contracted at the base into a claw.

VALVATE: margins of sepals or petals not overlapping.

VELAMEN: a parchment-like sheath or layer of spiral-coated air-cells on the root which may act as protective insulation.

VENOSE: having veins.

VENTRICOSE: swollen or inflated on one side.

VERRUCOSE: warty.

VISCIDIUM: a sticky disc or plate joined to the pollinium, enabling it to adhere to an insect's body during cross-fertilisation.

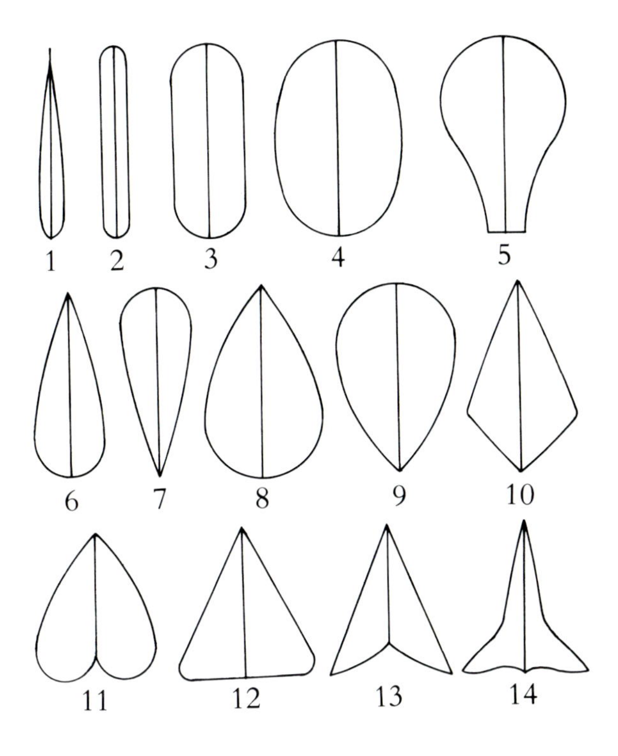

Leaf shapes
1 - *Subulate* 2 - *Linear* 3 - *Oblong* 4 - *Elliptic*
5 - *Spathulate* 6 - *Lanceolate* 7 - *Oblanceolate*
8 - *Ovate* 9 - *Obovate* 10 - *Trullate* 11 - *Cordate*
12 - *Triangular* 13 - *Sagittate* 14 - *Hastate*

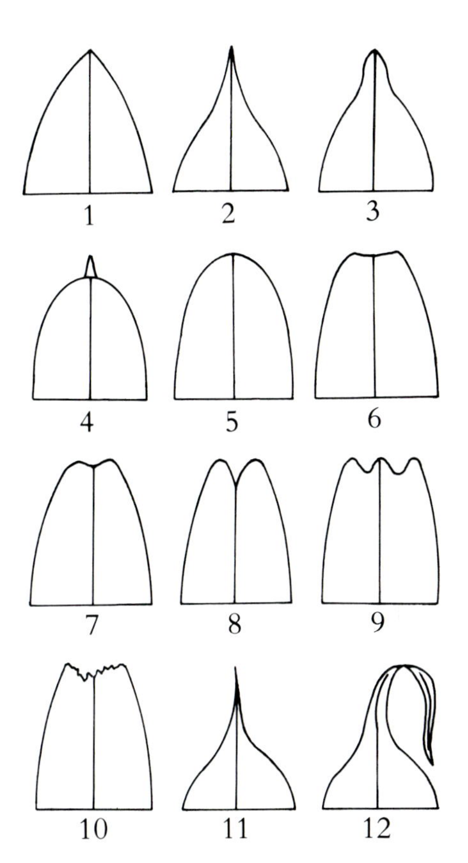

Leaf apices
1 - *Acute*
2 - *Acuminate*
3 - *Apiculate*
4 - *Mucronate*
5 - *Obtuse*
6 - *Truncate*
7 - *Retuse*
8 - *Emarginate*
9 - *Tridentate*
10 - *Praemorse*
11 - *Setose*
12 - *Caudate*

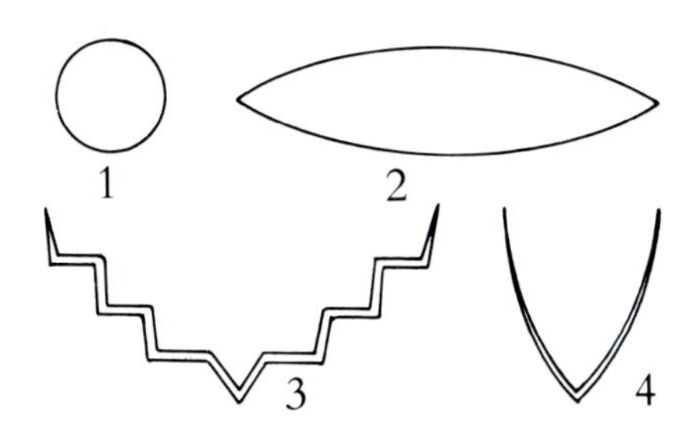

Leaf cross-sections
1 - *Terete*
2 - *Bilaterally compressed*
3 - *Plicate*
4 - *Conduplicate*

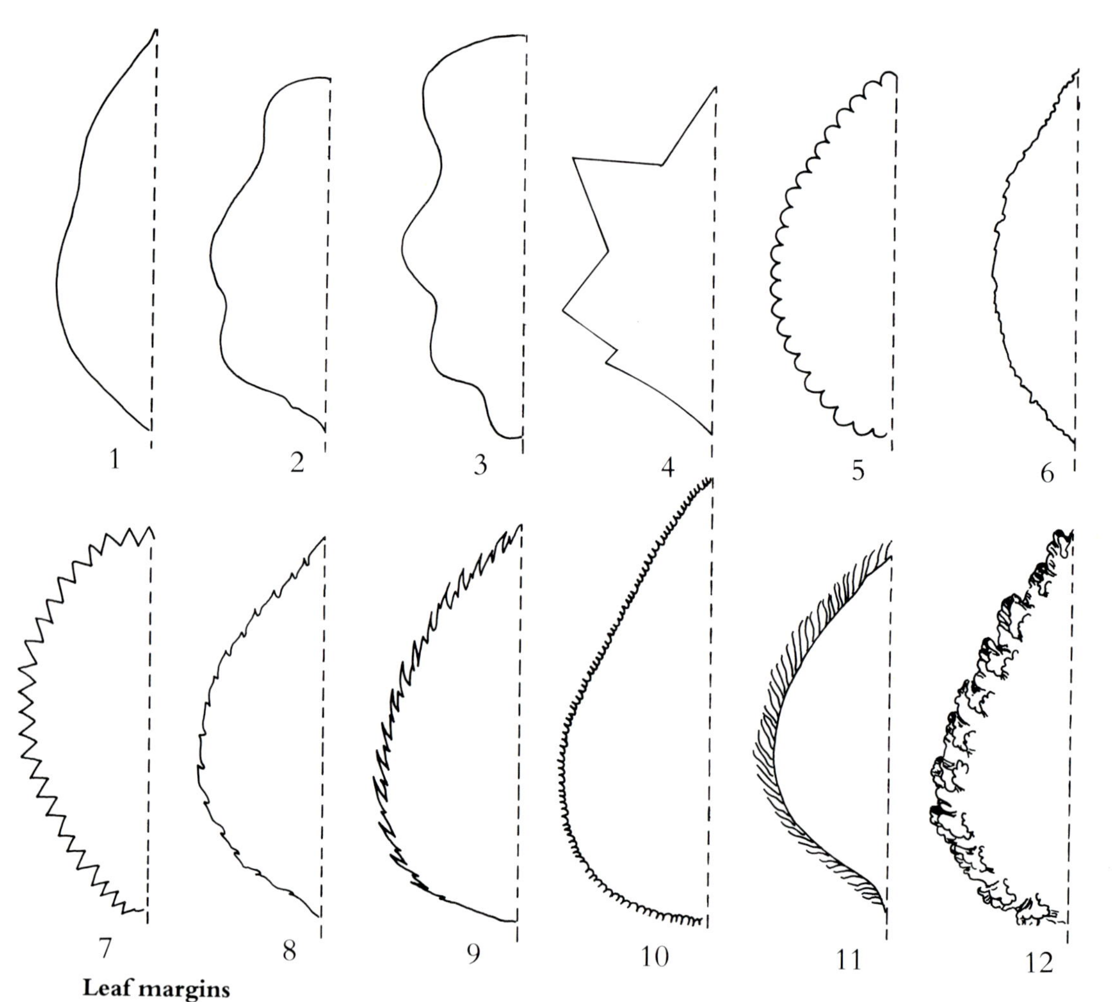

Leaf margins

1 - *Entire* 2 - *Undulate* 3 - *Sinuate* 4 - *Angulate* 5 - *Crenate* 6 - *Erose* 7 - *Dentate*
8 - *Serrate* 9 - *Doubly serrate* 10 - *Fimbriate* 11 - *Ciliate* 12 - *Crispate*

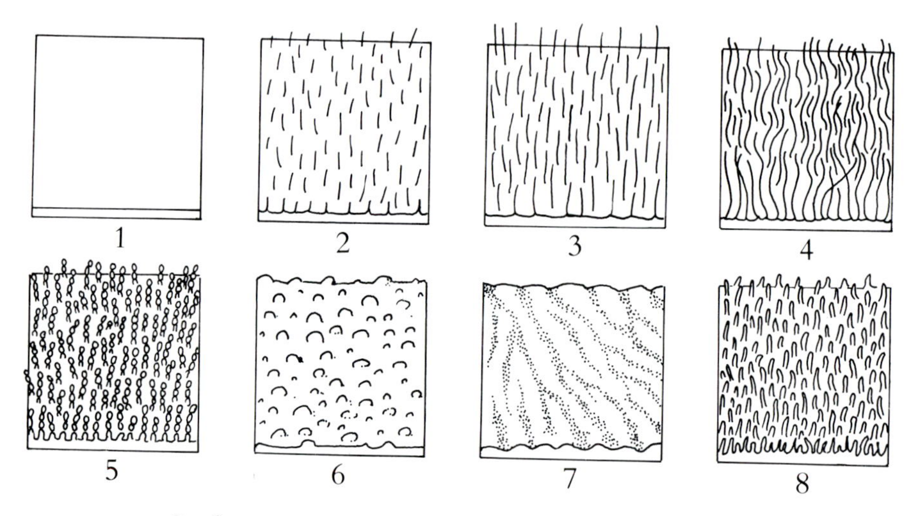

Surface textures (all much magnified)

1 - *Glabrous* 2 - *Pilose* 3 - *Hirsute* 4 - *Woolly*
5 - *Farinose* 6 - *Verrucose* 7 - *Rugulose* 8 - *Papillose*

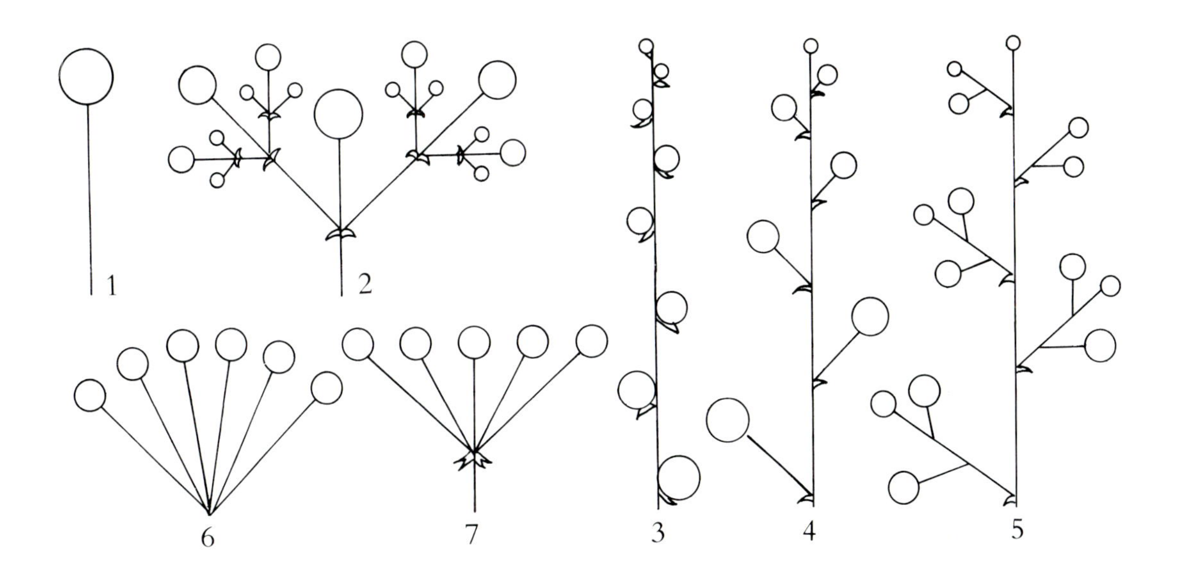

Inflorescence types
1 - *Single-flowered* 2 - *Cymose* 3 - *Spicate* 4 - *Racemose*
5 - *Paniculate* 6 - *Fasciculate* 7 - *Umbellate*

Inflorescence position
1 - **Basal**, *Spathoglottis* 2 - **Axillary**, *Vanda* 3 - **Apical**, *Paphiopedilum*

INDEX TO ORCHID SCIENTIFIC NAMES

(Compiled by K.M. Wong)

(numbers in bold type indicate pages with detailed treatment; numbers within brackets indicate pages with illustrations)

A

B

C

D

E

F

G

H

K

L

M

N

O

P

R

S

T